AF588037

Synthesis Lectures on Computer Vision

This series publishes on topics pertaining to computer vision and pattern recognition. The scope follows the purview of premier computer science conferences, and includes the science of scene reconstruction, event detection, video tracking, object recognition, 3D pose estimation, learning, indexing, motion estimation, and image restoration. As a scientific discipline, computer vision is concerned with the theory behind artificial systems that extract information from images. The image data can take many forms, such as video sequences, views from multiple cameras, or multi-dimensional data from a medical scanner. As a technological discipline, computer vision seeks to apply its theories and models for the construction of computer vision systems, such as those in self-driving cars/navigation systems, medical image analysis, and industrial robots.

Sijia Liu · Yang Liu · Nathalie Baracaldo
Editors

Machine Unlearning for Governance of Foundation Models

Springer

Editors
Sijia Liu
Department of Computer Science
and Engineering
Michigan State University
Sunnyvale, CA, USA

Yang Liu
Department of Computer Science
and Engineering
University of California, Santa Cruz
Scotts Valley, CA, USA

Nathalie Baracaldo
Almaden Research Center
IBM Research—Almaden
San Jose, CA, USA

ISSN 2153-1056 ISSN 2153-1064 (electronic)
Synthesis Lectures on Computer Vision
ISBN 978-3-032-17281-5 ISBN 978-3-032-17282-2 (eBook)
https://doi.org/10.1007/978-3-032-17282-2

This Springer imprint is published by the registered company Springer Nature Switzerland AG
The registered company address is: Gewerbestrasse 11, 6330 Cham, Switzerland

Preface

The rapid rise of foundation models such as large language models (LLMs) and diffusion models (DMs) has transformed the landscape of artificial intelligence. Yet, their unprecedented capabilities come with equally unprecedented risks: the retention of sensitive data, copyrighted content, harmful instructions, and other undesirable knowledge within their vast parameter spaces. *Machine unlearning (MU)* has thus emerged as a critical research frontier, providing the tools to selectively, efficiently, and reliably erase such information from trained models without sacrificing their broader utility.

Back in June 2024, the authors (Sijia Liu, Yang Liu, and Nathalie Baracaldo) delivered a widely attended CVPR'24 tutorial on "Machine Unlearning in Computer Vision: Foundations and Applications", where we introduced *unlearning* as an *enabler* for trustworthy computer vision. We envisioned unlearning as a central intervention and controllable operation for building AI systems that remain safe, adaptable, and compliant over time, especially when *removing* outdated, private, or harmful information from pre-trained models becomes essential. In the months following this tutorial, research in LLM unlearning quickly accelerated, fueled by the urgent need to ensure responsibility and safety of LLMs. The rapid growth of MU in generative models has produced a wide spectrum of algorithms, evaluation frameworks, and application scenarios, yet also revealed the need for a unified and in-depth introduction to the field across modalities.

This book was conceived to fill that gap. It adopts a *model-data-optimization* tri-design perspective, complemented by *assessments and applications*, to offer a comprehensive and structured treatment of machine unlearning that blends algorithmic rigor, practical guidance, and cross-disciplinary insights. From the academic perspective, unlearning poses fundamental technical challenges in optimization, data curation, model adaptation, and evaluation, calling for novel theoretical frameworks and principled benchmarks. From the industrial perspective, it has rapidly become a practical necessity for deploying AI systems securely and responsibly, enabling privacy compliance, intellectual property protection, fairness, and AI safety in highstakes domains. The book reflects this dual academic-industrial perspective.

With contributions from both the primary authors and invited external collaborators, the book is organized into two main parts as follows:

- Part I develops optimization principles behind unlearning objectives, utility preservation, and robustness.
- Part II takes a data-centric view, exploring coreset selection, data integrity, and label manipulation.
- Part III focuses on model-level interventions, from sparsity and attribution methods to auxiliary model construction for large-scale systems.
- Part IV addresses the evaluation challenges of faithfulness, instability, and robustness.
- Part V connects unlearning to its applications in making AI systems trustworthy and governable.

California, USA
August 2025

Sijia Liu
Yang Liu
Nathalie Baracaldo

Acknowledgements Making this book possible required time and effort. First of all, Sijia Liu, Yang Liu, and Nathalie Baracaldo want to thank all the book chapter contributors. Without their amazing research and chapter contributions, this book would not have been possible.

Sijia Liu wishes to express his deepest gratitude to all his Ph.D. students in the OPTML Lab @ Michigan State University who contributed to this book through their research and insights. He also thanks the Computer Science and Engineering Department at Michigan State University for its strong academic environment and institutional support. Sijia gratefully acknowledges the funding agencies and organizations that have supported his lab's research on machine unlearning, including National Science Foundation, Army Research Office, Open Philanthropy, Center for AI Safety, IBM, Cisco, Amazon, DSO National Laboratories, and Lawrence Livermore National Laboratory. Their generous support has enabled many of the advances discussed in this book. Finally, Sijia extends his heartfelt gratitude to his wife Fangrong for her unwavering support and encouragement, which provided essential strength throughout the book writing.

Yang would like to thank his Ph.D. students at UC Santa Cruz, whose creativity and dedication are a constant source of inspiration. Their original research has contributed to several chapters of this book. He is also grateful for the support and mentorship provided by the Computer Science and Engineering Department at UC Santa Cruz. Special thanks goes to Sijia for introducing him to the topic of unlearning and for the many conversations and collaborations that followed. He is indebted as well to all of his collaborators on unlearning, whose insights and efforts have significantly shaped this work. Finally, Yang wishes to express his deepest gratitude to his family—Tess and KK—for their love and support. This book would not have been possible without you.

Nathalie would like to thank her family for their continuous support in all her projects. Matthias and Santiago, you guys rock! Nathalie also wants to thank David Cox for his continuous support and enthusiasm for cutting-edge research efforts. Magic happens when people come together to make new ideas grow! Special thanks to Sijia for being a great collaborator in this and other research endeavors.

The authors would also like to thank Springer Nature for supporting this book project, with special gratitude to Susanne Filler for inviting us to contribute this book and to Prasanna Kumar Narayanasamy for his kind assistance and guidance throughout the preparation process.

Competing Interests The authors have no competing interests to declare that are relevant to the content of this manuscript.

Contents

Contributors

Farhan Ahmed IBM Research, San Jose, CA, USA

Hadi Amiri University of Massachusetts Lowell, Lowell, MA, USA

Nathalie Baracaldo IBM Research, San Jose, CA, USA

Shiyu Chang University of California, Santa Barbara, CA, USA

Yiwei Chen Michigan State University, East Lansing, MI, USA

Jiali Cheng University of Massachusetts Lowell, Lowell, MA, USA

Zonglin Di University of California, Santa Cruz, CA, USA

Minxin Du The Hong Kong Polytechnic University, Hong Kong, China

Chongyu Fan Michigan State University, East Lansing, MI, USA

Anisa Halimi IBM Research Europe, Dublin, Ireland

Jiabao Ji University of California, Santa Barbara, CA, USA

Jinghan Jia Michigan State University, East Lansing, East Lansing, USA

Swanand Ravindra Kadhe IBM Research, San Jose, CA, USA

Chris Yuhao Liu University of California, Santa Cruz, USA

Sijia Liu Michigan State University, East Lansing, USA

Yang Liu University of California, Santa Cruz, USA

Yujian Liu University of California, Santa Barbara, CA, USA

Inkit Padhi IBM Research, San Jose, CA, USA

Soumyadeep Pal Michigan State University, East Lansing, MI, USA

Ambrish Rawat IBM Research Europe, Dublin, Ireland

Changsheng Wang Michigan State University, East Lansing, MI, USA

Yaxuan Wang University of California, Santa Cruz, Santa Cruz, CA, USA

Dennis Wei IBM Research, San Jose, CA, USA

Xiaoyu Xu The Hong Kong Polytechnic University, Hong Kong, China

Xiang Yue Carnegie Mellon University, Pittsburgh, PA, USA

Yihua Zhang Michigan State University, East Lansing, MI, USA

An Introduction to Machine Unlearning 1

Sijia Liu, Yang Liu and Nathalie Baracaldo

Abstract

As foundation models expand in scale and societal influence, the ability to faithfully remove undesired data, knowledge, or capabilities has become critical. This challenge, known as *machine unlearning (MU)*, involves the irreversible removal of sensitive data, copyrighted content, and harmful behaviors from trained models. In this chapter, we present MU as both a scientific frontier and a societal necessity. We distinguish it from shallow model editing and emphasize the central triad of challenges: achieving unlearning effectiveness/robustness (erasing the target) and preserving non-target utility to avoid collateral damage. To frame this book, we adopt a tri-design perspective, spanning model, data, and optimization, which together provides the methodological and practical foundations of MU. We also situate unlearning in applied contexts, underscoring its role in AI safety, trustworthy deployment, and compliance-driven governance. MU is not peripheral, but foundational for safe, reliable, and governable AI.

S. Liu (✉)
Michigan State University, East Lansing, USA
e-mail: liusiji5@msu.edu

Y. Liu
University of California, Santa Cruz, USA
e-mail: yangliu@ucsc.edu

N. Baracaldo
IBM Research, San Jose, USA
e-mail: baracald@us.ibm.com

S. Liu et al. (eds.), *Machine Unlearning for Governance of Foundation Models*, Synthesis Lectures on Computer Vision, https://doi.org/10.1007/978-3-032-17282-2_1

1.1 Defining Machine Unlearning

Machine unlearning (MU) refers to the set of algorithmic techniques designed to remove the influence of specific data, concepts, or behaviors (i.e., the "unlearning target") from a trained machine learning (ML) model, without requiring full retraining from scratch [1–3]. In its most common form, MU aims to produce an updated model whose outputs and internal representations are indistinguishable from those of a model trained without the "forget set" (i.e., the data to be erased), and/or to prevent the model from exhibiting the undesired behaviors (e.g., predictions or generations) linked to that unlearning target.

To make MU more intuitive, imagine it as *performing surgery on an AI model*. Just as a skilled surgeon removes a tumor while carefully preserving the surrounding healthy tissue, unlearning excises harmful, biased, or private knowledge from a trained model without damaging its valuable capabilities. Another way to think about it is like *patching software*: when engineers discover a bug, they release a targeted fix rather than rewriting the entire program from scratch. Similarly, when AI models exhibit unwanted or unsafe behaviors, unlearning techniques act as precise patches, erasing the harmful influence while keeping the rest of the system intact.

Importantly, MU is not merely a counterpoint to learning; it is increasingly viewed as a critical component of the trusted ML *model life cycle*, ensuring that models remain safe, compliant, and adaptable post-deployment. Before delving into detailed applications in Sect. 1.1.1, we highlight representative use cases where MU plays a central role, spanning instance-level, concept-level, and behavior-level forgetting, while noting that these levels often interact and reinforce one another.

- *Objectionable private information*: Removing private or sensitive data accidentally ingested during model training to prevent privacy violations.
- *Regulatory compliance*: Supporting legal mandates such as General Data Protection Regulation (GDPR) [4], where users can request the removal of their data from deployed models at any time.
- *Data licensing and ownership*: Enforcing evolving data licenses or intellectual property restrictions, e.g., removing copyrighted text, images, or code when licenses expire or become more restrictive.
- *Domain- or geography-specific constraints*: Erasing data or behaviors disallowed in specific jurisdictions or applications without affecting the model's global capabilities.
- *Poisoned, toxic, or biased data*: Removing malicious backdoors, model bias, toxic language generation, or harm production capabilities, e.g., malware production from a code-generation model.

As these examples illustrate, machine unlearning serves as a principled mechanism for erasing harmful data influence and adapting model behaviors after deployment. Rather than relying solely on retraining from scratch, a process that is often prohibitively expensive and environmentally unsustainable, MU provides a scalable and targeted alternative. It enables practitioners to continuously align foundation models with evolving legal, ethical, and safety requirements, while maintaining their utility for approved applications. This perspective naturally motivates the methodological foundations and optimization techniques discussed in the following sections, setting the stage for a deeper exploration of unlearning as a cornerstone of trustworthy and governable AI systems.

1.1.1 Why Unlearning Matters for Foundation Models?

Foundation models, such as large language models (LLMs) and diffusion models (DMs), have transformed AI by enabling zero-shot reasoning, cross-domain generation, and rapid adaptation to new tasks. However, their remarkable capabilities also carry a critical risk: the inadvertent memorization, storage, and reproduction of sensitive, harmful, or otherwise undesired content from training data. With their vast and often entangled knowledge stores, including copyrighted materials, personal information, and hazardous instructions, these models require effective unlearning strategies. MU offers a practical and principled solution to surgically remove unwanted knowledge while preserving the model's legitimate abilities.

In what follows, we demonstrate that MU is a core operational capability for foundation models. It serves as *(i)* the backbone of responsible AI deployment enabling models to meet legal obligations by complying with privacy regulations and intellectual property rights, *(ii)* to maintain security resilience by removing backdoors and poisoned data, and *(iii)* to uphold ethical alignment by preventing the generation of harmful or unsafe content. Beyond these safeguards, MU also supports sustainable AI development, making it possible to deliver cost-effective, targeted model control without the expense of full retraining.

First, one of the most pressing motivations for MU is the need to *protect privacy and comply with data protection regulations*. For example, LLMs trained on large-scale Internet data can reproduce personal identification information (such as names, phone numbers, addresses, or chat logs) if such details were present in their training corpus. Similarly, DMs (diffusion models) trained on images from online platforms can inadvertently regenerate identifiable photographs or other sensitive visual content. Privacy laws such as the General Data Protection Regulation (GDPR) [4] and the California Consumer Privacy Act (CCPA) [5] explicitly mandate the removal of personal data upon request. MU enables compliance by not only deleting raw data but also eliminating its learned influence in the model's parameters, preventing both direct reproduction and indirect inference.

Intellectual property (IP) concerns further underscore the need for MU. Many foundation models are trained on datasets that contain copyrighted works or material protected by restrictive licenses. Changes in licensing agreements or legal disputes may require the removal of such content from a model's capabilities. For instance, an LLM might be required to "forget" passages from copyrighted books or confidential corporate documents [6], while a DM might need to unlearn an artist's distinctive style once usage rights are withdrawn [7].

Another known example is the escalating conflict between content creators and AI companies, leading to high-profile lawsuits such as The New York Times versus OpenAI [8]. Without MU, enforcing these content removals would demand retraining the entire model from scratch, a process that is both financially and operationally prohibitive at the scale of modern foundation models.

Security is another key driver for MU adoption. For example, open-weight models are vulnerable to data poisoning [9], where malicious actors intentionally inject harmful or manipulative content into training data. In LLMs, poisoned samples could implant backdoor triggers that cause the model to generate malicious code or biased text when prompted with specific keywords. In DMs, poisoned images could cause the model to insert hidden patterns or manipulate visual styles in undesired ways. MU acts as a post-hoc defense, enabling the targeted removal of these malicious patterns without the cost or disruption of complete retraining, thereby restoring trust in the model's outputs.

MU is also essential for ensuring safety and ethical alignment. LLMs have been shown to produce harmful instructions, such as guides for illicit activities or unsafe technical procedures [10], while DMs can generate biased, sensitive, or violent imagery, even when such outputs are unintended [11]. By applying MU, such biased or unsafe capabilities can be surgically removed while preserving the model's broader generative strengths. This level of fine-grained control is critical for deployed systems that interact with the public, where lapses in safety or ethics can lead to severe reputational damage, legal consequences, and societal harm.

As an illustrative case, Fig. 1.1 shows how MU can be applied to a DM to mitigate the risk of generating nudity in its outputs. In this example, the input prompt "attractive male, character design, painting by Gaston Bussière" causes the original DM to generate an image containing sensitive content (nudity). This represents a clear safety concern for deployment, particularly in applications accessible to a broad audience. By applying the MU algorithm SalUn [12], the model is selectively modified to remove its tendency to generate such unsafe outputs, while preserving its ability to produce relevant, high-quality images. After unlearning, the same input prompt yields a normal, fully clothed depiction, aligning with safety requirements. This case illustrates MU's role as a surgical intervention for generative models, targeting specific undesired behaviors without compromising the broader creative capabilities of the system.

Finally, *MU plays a central role in the lifecycle management of foundation models.* These models operate in dynamic environments where data distributions, societal norms, and legal requirements evolve over time. What is considered acceptable knowledge at deployment may become outdated, biased, or unsafe months later. MU provides a sustainable, controllable mechanism for continuous refinement, enabling models to adapt to new regulations, cultural shifts, and emerging threats without the expense and environmental impact of retraining.

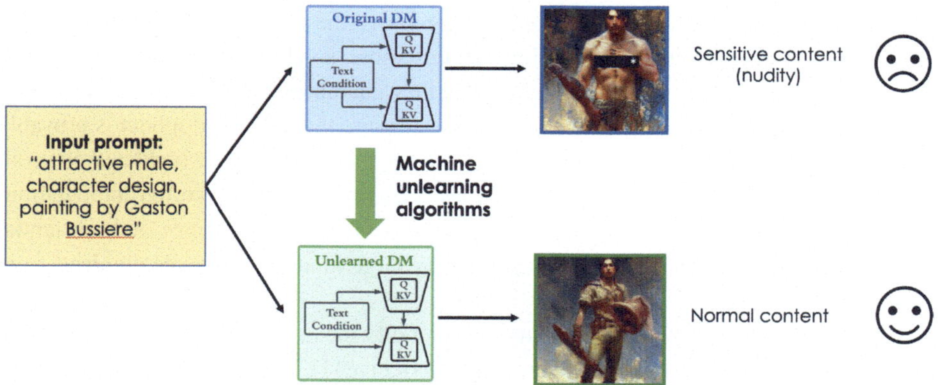

Fig. 1.1 Example of applying the MU algorithm SalUn [12] to a diffusion model (MU) to prevent the generation of nudity, transforming sensitive outputs into safe, normal content

1.1.2 Formalizing Machine Unlearning

Machine unlearning (MU) can be formalized as the task of modifying a trained model so that it behaves as if specific data or behaviors–collectively referred to as the *unlearning target*–had never contributed to its training, or as the deliberate degradation of the model's capabilities associated with that unlearning target. Let $\mathcal{D} = \mathcal{D}_\mathrm{r} \cup \mathcal{D}_\mathrm{f}$ denote the original training set, where $\mathcal{D}_\mathrm{r}$ is the *retain set* and $\mathcal{D}_\mathrm{f}$ is the *forget set*. A pre-trained model $f_{\boldsymbol{\theta}}$, parameterized by $\boldsymbol{\theta}_0$, is transformed into a new model $f_{\boldsymbol{\theta}}$ (parameterized by $\boldsymbol{\theta}$) using $\mathcal{D}$, with the goal of satisfying two necessary conditions:

- *First*, for any input $\mathbf{x}$ drawn from the forget distribution $\mathcal{D}_\mathrm{f}$, the predictions of $f_{\boldsymbol{\theta}}$ should approximate those of an ideal model $f_{\boldsymbol{\theta}^*}$ trained from scratch on the retain set D_r only: $f_{\boldsymbol{\theta}}(\mathbf{x}) \approx f_{\boldsymbol{\theta}^*}(\mathbf{x})$ for $\forall \mathbf{x} \in \mathcal{D}_\mathrm{f}$. In the context of unlearning for removal at the *capability level*, this condition can alternatively be approximated by enforcing a behavioral divergence $f_{\boldsymbol{\theta}}(\mathbf{x}) \neq f_{\boldsymbol{\theta}_0}(\mathbf{x})$ for $\forall \mathbf{x} \in \mathcal{D}_\mathrm{f}$.
- *Second*, for any input $\mathbf{x}$ from the retain distribution $\mathcal{D}_\mathrm{r}$, $f_{\boldsymbol{\theta}}$ should preserve the performance and behavior of the original model: $f_{\boldsymbol{\theta}}(\mathbf{x}) \approx f_{\boldsymbol{\theta}_0}(\mathbf{x})$ for $\forall \mathbf{x} \in \mathcal{D}_\mathrm{r}$.

These conditions can be quantified in multiple ways, through output distribution divergence (e.g., Kullback–Leibler (KL) divergence), internal feature similarity, or preference-based alignment scores, depending on the specific unlearning methods.

Exact Versus Approximate Unlearning

To satisfy the above conditions, MU methods can be broadly divided into *exact unlearning* and *approximate unlearning* approaches.

Exact unlearning methods aim to *certifiably* remove the influence of the forget set by retraining the model, or sub-models within an ensemble, solely on the retain set [13–15]. These methods offer two clear advantages. First, they guarantee that no catastrophic forgetting of unrelated information occurs. Second, they ensure the unlearned model is provably equivalent to one trained from scratch on $\mathcal{D}_r$, making them suitable for regulated environments where compliance must be auditable. However, exact unlearning is often computationally prohibitive. It requires full access to training datasets, original pipelines, and substantial compute resources, making it impractical for very large models or models obtained from third-party providers.

In contrast, *approximate unlearning* methods modify the original model directly without retraining, offering a computationally efficient alternative [1]. They require no access to the original training data and scale effectively to foundation models where even partial removal of harmful or copyrighted content is valuable. This efficiency comes at a cost: approximate methods lack formal guarantees, may leave residual traces of the forget set, and risk unintentionally degrading unrelated capabilities if the optimization is not carefully regularized. Nevertheless, because of their scalability and practicality, approximate unlearning dominates real-world applications involving LLMs and diffusion models, where rapid risk mitigation often outweighs the need for provable guarantees.

In this book, most of the unlearning methodologies discussed are categorized as approximate unlearning, due to their scalability, flexibility, and proven effectiveness in applications.

Unified Optimization Objective Formulation

A central measure of MU quality is *unlearning effectiveness*: the extent to which a model's responses, internal states, and behaviors no longer expose any trace of the unlearning target, even under adversarial probing. Equally essential is *utility preservation*, which ensures that the model continues to perform its intended tasks and retain its broader capabilities outside the forgetting scope. Without preserving utility, unlearning risks producing a model that is safe but unusable. Achieving balance among them is inherently challenging, which makes MU most appropriately framed as a bi-objective optimization problem.

$$\underset{\boldsymbol{\theta}}{\text{minimize}}\ \ell_f(\boldsymbol{\theta}) + \gamma \ell_r(\boldsymbol{\theta}), \tag{1.1}$$

where ℓ_f promotes removal of the target influence, ℓ_r enforces utility preservation, $\lambda > 0$ controls the balance between these competing objectives, and the optimization is initialized from the original model $\boldsymbol{\theta} = \boldsymbol{\theta}_0$.

Problem (1.1) may appear to be a straightforward model editing task, finding parameter updates that maximize forgetting without causing unnecessary collateral damage to retained capabilities. In reality, achieving *deep forgetting*–where the model's internal representations and generative pathways are permanently altered so that the forgotten content cannot be reconstructed–is far more challenging, especially for foundation models. Knowledge in LLMs and DMs is highly entangled across parameters and layers, with individual facts, styles, or behaviors dispersed across many weights and feature subspaces. Conventional model editing approaches may appear to suppress undesired content, but they often result in only superficial forgetting. See Sect. 1.2.5 for more details. In practice, such strategies can fail when faced with paraphrased prompts or indirect queries. For instance, unlearned models via finetuning can be resurrected with small benign inputs, effectively overturning the intended forgetting [16]. Similarly, conventional model editing techniques can make localized behavioral changes but rarely guarantee irreversible knowledge removal [17], as adversaries can exploit alternative recall pathways.

Therefore, *robustness*, the ability to sustain forgetting even under indirect or reformulated prompts, paraphrased queries, or alternate input modalities, is important for MU. Achieving deep forgetting requires restructuring the model's parameter space so thoroughly that the unlearning target is erased from all reachable representational states. At the same time, this restructuring must safeguard the model's unrelated skills and capabilities, ensuring that forgetting is both irreversible and utility-preserving.

In brief, MU is not merely an afterthought or optional enhancement; it is a high-stakes capability that demands rigorous balancing of effectiveness, robustness, and utility preservation. In the following sections, we build on this foundation to examine the major methodological families for achieving MU, analyzing their algorithmic principles, and real-world applicability.

1.2 Overview of Machine Unlearning Method Families

MU methods span a diverse range of algorithmic strategies, each designed to achieve the dual objectives of effective forgetting and utility preservation under different constraints. These methods can be broadly grouped into several families, distinguished by their underlying principles and the types of interventions they apply, ranging from direct parameter optimization and influence-based weight adjustment to randomized overwriting and input-level control. This section provides an overview of these method families, highlighting their core ideas and representative techniques.

1.2.1 From Gradient Ascent to General Optimization Divergence-Driven Methods

A foundational approach to MU is to apply *gradient ascent (GA)* on the forget set, effectively taking training steps to reverse the learning signal tied to the data or behaviors that need erasing [18]. While this idea elegantly captures the principle of pushing the model's behavior away from the original (referred to as "*optimization divergence*"), it comes with steep practical challenges. In particular, fine-tuning the model in this way often leads to utility degradation or even model collapse [19], as it becomes difficult to judge when to stop the unlearning update and overshooting can result in excessive divergence from the model's useful capabilities.

This limitation motivates a more controlled strategy: framing unlearning as a bi-objective optimization problem (1.1), where one simultaneously promotes forgetting and preserves overall performance. Within this paradigm, *gradient difference (GradDiff)* [20, 21] represents a method that balances the gradients of the forget and retain objectives of (1.1), providing finer control over the forgetting process and helping preserve utility.

Beyond GradDiff, the divergence-driven principle has inspired more nuanced approaches like *negative preference optimization (NPO)* [22]. Derived from direct preference optimization (DPO) [23], NPO treats content from the forget set as negative examples, shifting the model's behavior to deliberately diverge from its original responses on these inputs, while maintaining preferences elsewhere.

Further recent developments build on this theme. The *simple NPO method (SimNPO)* [24] rethinks NPO by cutting reliance on a reference model (i.e., the original pre-unlearning model), reducing what is called reference model bias. SimNPO simplifies optimization, normalizes gradient behavior across diverse forget data, and exhibits greater robustness against relearning attacks [16] compared to its predecessor NPO. Other methods further enrich this family. For instance, a *normalized gradient difference approach (NGDiff)* [25] introduces adaptive learning rates to maintain stable trade-offs between forgetting and retention, delivering steady performance across language model benchmarks.

These divergence-driven methods highlight a key progression: from brute-force gradient manipulations toward principled, stability-aware optimization frameworks, enabling effective forgetting while safeguarding the model's broader capabilities post unlearning.

1.2.2 From Influence Function–Based Unlearning to Second-Order Unlearning

The influence function has its origins in classic statistical diagnostics, originally introduced to trace the effect that individual training points have on learned model predictions [26], which demonstrated how the influence of perturbations to training data on model prediction could be projected into changes in model parameters, enabling a theoretical understanding of data influence without needing model training. Adapted to the setting of machine unlearning, this influence function approach, referred to as *influence unlearning* [27], allows us to approximate the impact of removing a subset of training data (i.e., the forget set) through a single, influence-based parameter update. This provides a computationally efficient alternative to model updating-based unlearning approaches, such as GA and its variants.

Yet, the classical influence-function method relies heavily on local linear assumptions. In high-capacity models such as LLMs and DMs, parameter interactions are highly non-linear and interdependent, so a single one-shot correction often fails to fully eliminate the influence of the forget set. However, the influence-function formulation also implicitly reveals that a *second-order* update, incorporating curvature information via the Hessian, is necessary to more faithfully align the unlearned model's behavior with that of an exactly unlearned model, one that has been retrained from scratch solely on the retain set.

This insight directly motivates the development of *second-order unlearning (SOUL)* [28]. SOUL bridges influence-function-based strategies with second-order optimization, leveraging techniques inspired by the Sophia optimizer [29] to perform iterative curvature-aware updates rather than relying on the single-shot correction used in the classical influence-function method. By repeatedly applying second-order optimization steps, SOUL more accurately tracks the parameter adjustments that would result from retraining solely on the retain set, thereby achieving a closer alignment with the behavior of an exactly unlearned model. This iterative refinement yields a more stable convergence toward the desired forgetting behavior and produces markedly improved trade-offs between unlearning efficacy and utility preservation.

Building on the need for more precise parameter updates, the influence function approach need not be limited to attributing the influence of specific forget-set examples and erasing that influence directly. They can also serve as a principled mechanism for identifying *which* parameters are most responsible for encoding the target content and which must be preserved to maintain retained behaviors. This perspective is operationalized in the *weight attribution-guided unlearning framework (WAGLE)* [30]. Framed within a bi-level optimization structure [31], WAGLE computes closed-form attribution scores for each weight, quantifying its impact on both forgetting and retention objectives. These scores are then used to construct a targeted weight selection mask, ensuring that unlearning updates are concentrated on the parameters most critical for removing the unwanted knowledge while avoiding unnecessary changes to parameters essential for other capabilities. This weight attribution strategy serves two purposes: (i) it minimizes collateral damage to model utility by avoiding broad, indiscriminate updates, and (ii) it provides a modular design that can be integrated with various unlearning algorithms.

The above lineage, from influence functions to second-order optimization, and finally to weight attribution, demonstrates a progressive shift toward more precise, efficient, and interpretable unlearning mechanisms. Each step brings us closer to deep forgetting, by targeting the complex, intertwined representations stored in generative models without compromising their valuable capabilities.

1.2.3 From Random Label to Random Feature Approaches

Another direct way to make a model "unlearn" an association is to corrupt the supervision signal for the forget set. In *random label-based unlearning*, samples in the forget set are reassigned to random targets and the model is fine-tuned so that the original mapping is overwritten. This idea is especially natural for discriminative models where labels are explicit. For instance, *saliency unlearning (SalUn)* [12] marries random re-labeling with weight saliency: it first identifies the subset of parameters most responsible for the forget association and then fine-tunes only those salient weights on the randomly re-labeled forget data. SalUn also extends to conditional diffusion models by pairing the forget concept (prompt) with mismatched images so the generator learns to decouple the concept from its visual realization.

For generative models, explicit labels are not always available, so a related technique is *random feature-based unlearning*. Instead of corrupting output labels, the internal feature representations of the forget set are redirected toward unrelated, randomly chosen target values. This approach aims to break the internal representational pathways that produce the undesired content. A notable method in this category is *representation misdirection unlearning (RMU)* [10], where a specific intermediate layer is chosen and the hidden activations for forget examples are shifted toward random or semantically irrelevant vectors. At the same time, the retain examples are anchored by preserving their original representations.

This "misdirection" disrupts the internal computation pathways that lead to unwanted outputs, making it harder for the model to regenerate the forgotten content even under indirect or adversarial prompting. RMU has been used to remove capabilities such as generating unsafe biological or cyber content in LLMs [10]. The key design choice is selecting the right layer and strength for the feature corruption: too early in the network and the effect may get washed out by later layers, too late and it may harm output fluency or quality.

In practice, both random label and random feature approaches are attractive because they are simple, model-agnostic, and computationally efficient. However, they come with trade-offs. If the corruption is too weak, the model may not fully forget; if too strong or applied indiscriminately, it can cause severe utility loss. Two common safeguards are (1) localizing updates to high-attribution or salient parameters like SalUn [12], and (2) enforcing constraints, either at the representation level or in the output distribution, to preserve non-target skills. With these safeguards in place, random label and random feature approaches could provide an effective, scalable path to unlearning for both discriminative and generative foundation models.

1.2.4 Input Probing and Prompting-Based Approaches

Input-level unlearning replaces weight updates with prompt-side interventions. Instead of changing parameters, these methods learn or craft inputs that steer a frozen model away from the forget target. This paradigm is attractive when model weights are inaccessible (e.g., API-only, enterprise deployments) and when compute or downtime constraints make fine-tuning impractical. Existing studies [1, 32] explicitly contrast this track with model-based unlearning and note two practical upsides: it works under black-box access and is parameter-efficient because the learnable objects are prompts or lightweight guards rather than full model weights; guardrail methods (prompting + filtering) can even achieve unlearning comparable to fine-tuning on several benchmarks [32].

One branch of this family focuses on *in-context unlearning (ICUL)* [33], which injects modified versions of the forget examples, such as mismatched labels or contradictory instructions, directly into the model's inference-time context. By manipulating only the inputs, ICUL can replicate or surpass the forgetting achieved by gradient-based unlearning, all while avoiding weight updates. Another approach, *soft-prompt unlearning (SPUL)* [34], learns a small number of trainable prompt tokens that are prepended to the input, guiding the model to ignore or suppress forgotten content. Because these prompts are lightweight and the model remains frozen, SPUL achieves a favorable trade-off between forgetting and utility preservation, while remaining computationally efficient. Other methods, such as *embedding-corrupted prompts (ECO)* [32], detect forget-related queries through a classifier and automatically inject corrupted embedding sequences that push the model into an "unlearned" output space. This technique scales effectively from small to very large LLMs, offering effective forgetting with minimal side effects.

The main strengths of these input-level methods are their speed, reversibility, and modularity. They can be deployed without downtime, toggled on or off per use case, and updated independently of the base model. However, because the underlying parameters remain unchanged, these methods typically lack the guarantees of deep forgetting. The knowledge associated with the forget target is still present in the model and can resurface under paraphrased prompts, prompt injection attacks, or other adversarial manipulations. For this reason, while input probing and prompting are invaluable in black-box and rapid-response contexts, they are best viewed as complements to parameter-level unlearning rather than complete substitutes, especially when robustness against motivated adversaries is a requirement.

The input probing and prompting based method relates to LLM Guardrails. Guardrailing, which accesses prompts before using them as inputs to the model, has been widely applied to modern LLMs to prevent adversaries with harmful incentives [35–43]. Notable guardrail-based unlearning approaches include in-context unlearning [33] and prompt-based guardrailing [44], both of which require no additional fine-tuning to achieve unlearning to some extent. [33] leverage modern LLMs' ability in in-context learning by prepending a small number of positive and negative samples in the prompt to steer the model's response based on those samples. [44] guard the unlearning target via prompt injection, which inserts fixed instructions in the prompt to the LLM. Both methods can only be applied to instruction-tuned models and rely on an LLM's ability to follow instructions. Prepending such instructions also leads to significant performance degradation on regular tasks, as shown in [44].

1.2.5 From Model Editing to Deep Forgetting

Model editing and machine unlearning share the similar goal of modifying a model's behavior, but their motivations, guarantees, and operational principles differ in several ways. Model editing, whether through fine-tuning, task vectors [45], or other targeted interventions, focuses on altering model outputs for specific queries or domains without necessarily removing underlying knowledge. In contrast, unlearning aims for deep forgetting: ensuring that the model behaves as if the targeted data or capability had never been part of its training, ideally making recovery of that knowledge impossible even under adversarial probing.

A key example of the gap between editing and unlearning emerges in weight-based task vector approaches [45]. Here, model updates corresponding to a specific task are isolated and then added or subtracted from the base model to create a global behavioral shift. While this technique can efficiently "subtract" unwanted capabilities, its global nature often makes it too coarse for unlearning in generative models, where knowledge is highly entangled across parameters and layers [6]. Applying a task vector subtraction may overshoot, harming unrelated abilities, or undershoot, leaving partial traces of the forgotten content intact. This suggests that effective unlearning requires *locality*, focusing parameter changes on the subspaces most responsible for the undesired behavior, rather than broad, indiscriminate weight shifts.

Local editing methods address this need for precision by restricting changes to weights or attention heads most relevant to the forget scope. While this can help preserve utility, these edits can be shallow: they suppress the surface form of knowledge without restructuring deeper representational pathways. This leaves the model vulnerable to indirect recovery through paraphrasing, reasoning chains, or prompt injection. Consequently, while model editing provides useful inspiration, especially for rapid deployment in controlled contexts, effective unlearning for foundation models requires going beyond localized surface changes to reconfigure the parameter space in a way that more faithfully eliminates the reachable representations of the targeted content.

In this book, we will elaborate on the optimization-data-model tri-design principles that are central to achieving effective unlearning.

1.3 Significance of Unlearning: A User-Inspired and Industrial Perspective

As demonstrated in Sect. 1.1.1, machine unlearning is rapidly emerging as a cornerstone technology for AI safety and trustworthy AI, bridging cutting-edge research with pressing industrial needs. In high-stakes domains, ranging from generative media to enterprise decision-support systems, unlearning provides the capability to surgically remove unsafe, biased, or unauthorized knowledge from large-scale foundation models without degrading their general performance. This is not merely a research curiosity; it is a critical operational capability that enables AI systems to comply with evolving legal requirements, maintain public trust, and safeguard against adversarial misuse.

From an AI safety perspective, unlearning addresses the challenge of post-deployment risk mitigation. Foundation models deployed in open, dynamic environments can unintentionally acquire or reveal capabilities that pose security, ethical, or societal risks. For example, in text generation, unlearning can be used to remove harmful instructions for weapon design; in image generation, it can filter out unsafe or non-consensual content generation capabilities; and in code generation, it can eliminate backdoors or insecure coding patterns. Unlike static, one-time alignment, unlearning offers a dynamic safety mechanism, allowing deployed models to be continuously refined in response to newly discovered vulnerabilities or shifts in acceptable use policies.

From a trustworthy AI and compliance perspective, unlearning enables enterprises to operationalize privacy rights and intellectual property protections. Rather than retraining massive models, an approach that is often prohibitively costly and environmentally unsustainable, industrial teams can apply targeted unlearning to meet legal obligations quickly and at lower cost. This selective capability removal also facilitates regulatory audits by offering measurable evidence of forgetting, strengthening corporate accountability in AI governance.

In industry, unlearning further supports resilience and lifecycle management for foundation models. As product requirements evolve, companies can repurpose models while ensuring that outdated or undesired behaviors are erased, avoiding reputational risks and reducing maintenance overhead. For example, a conversational AI service could adapt to new content guidelines by unlearning outdated language usage, or a vision model could remove culturally insensitive image associations without compromising its general classification or generative utility.

Therefore, unlearning transforms foundation models from static assets into continuously governable systems, a shift that is essential for safely scaling AI in real-world applications. Its dual role in mitigating safety risks and enabling responsible, regulation-compliant innovation makes it a linchpin of future industrial AI deployment strategies.

1.4 Why this Book? A Model-Data-Optimization Tri-design Perspective

Machine unlearning is a multidimensional design challenge that touches the very core of how modern AI systems are built, optimized, and deployed. This book takes a *tri-design perspective*, viewing unlearning through the intertwined lenses of *model-based*, *data-based*, and *optimization-based* approaches, while also grounding the discussion in *application-driven* motivations for AI safety and trustworthiness.

From the *model perspective*, we investigate how knowledge is modularized, stored, and intertwined within the architecture of foundation models. Unlearning in these systems is rarely about a single parameter, the influence of knowledge concepts or model behaviors is often distributed across multiple layers, attention heads, and latent subspaces. This makes modular unlearning a critical design principle: rather than applying blunt, global parameter changes that risk harming unrelated capabilities, we aim to identify and selectively adjust only the model components most responsible for the targeted knowledge. Achieving this requires deep architectural awareness, including mapping the pathways through which the forget set's signals propagate and pinpointing "influence modules" whose modification produces localized forgetting effects. Techniques such as parameter attribution, structural probing, and representation analysis help reveal these influence modules, enabling more surgical and scope-bounded interventions. By isolating the affected submodules and preserving others, modular unlearning not only improves utility retention but also reduces the risk of catastrophic side effects like emergent capability degradation. Furthermore, this model-based approach supports reusability and composability, allowing models to be "patched" or "rolled back" for different unlearning needs without full retraining.

From the *data perspective*, we focus on the role of the forget dataset, its quality, and the strategies used to manipulate or augment it to achieve reliable unlearning. Forget data is not just a list of items to erase, it encodes the very signals that shaped the model's internal representations. Data-based unlearning research therefore examines both precise target specification (ensuring the forget set truly reflects the undesired capability or knowledge) and boundary definition (ensuring unrelated abilities remain intact). One increasingly important tool here is *data corruption*, used both for training and evaluation. In training, corruption can weaken the model's ability to reconstruct forget content by replacing it with randomized or misleading signals. In evaluation, controlled corruption of forget data offers a way to test unlearning robustness: if a model can recover the original content from partially corrupted inputs, it suggests that forgetting is incomplete and that deeper representational removal is needed. Beyond corruption, the data perspective also includes *coreset selection*, identifying the smallest, most influential subset of forget data that achieves the desired forgetting effect, as well as data attribution, which determines which training samples contributed most to the undesired capability. Ultimately, the data-based perspective treats unlearning as a data–model interaction problem, where success depends on both the specificity and the representational leverage of the forget data.

From the *optimization perspective*, unlearning is fundamentally an objective balancing problem, removing targeted knowledge or behaviors while preserving the model's utility on the retain set. This tension creates several unique challenges compared to standard training or fine-tuning. On one side, strong forgetting pressure risks catastrophic over-forgetting, where the model's outputs degrade broadly beyond the intended scope; on the other, overly cautious updates may leave residual traces of the forget content. Designing optimization procedures that navigate this trade-off is therefore central to robust and efficient unlearning. Another challenge is robustness: even if a model appears to have forgotten, it may still be vulnerable to extraction attacks (e.g., targeted prompting or fine-tuning) that recover the erased information. Robust unlearning requires adversarially aware optimization, incorporating worst-case evaluation objectives or min–max training schemes that simulate malicious recovery attempts during the unlearning process. This reframes unlearning as an adversarial defense problem, where the optimization target is not only effectiveness but also resilience under hostile conditions. In essence, the optimization perspective treats unlearning as a multi-objective, potentially adversarial optimization problem that must simultaneously (1) remove targeted capabilities, (2) safeguard unrelated ones, (3) resist recovery attempts, and

(4) remain computationally feasible for large-scale deployment. Mastering this balance is the cornerstone of scalable, trustworthy, and regulation-ready unlearning systems.

The rest of the book reflects this layered approach.

- Part I on optimization foundations address the unlearning-utility trade-off, robustness, and efficiency in algorithm design.
- Part II focuses on data-centric chapters to examine unlearning through coresets, data integrity, and label-smoothing-based strategies.
- Part III includes model-centric chapters to explore sparsity, saliency, attribution, and auxiliary prediction as avenues for unlearning.
- Part IV includes evaluation-focused chapters to highlight both the progresses and limitations of current benchmarks, questioning the stability of retraining and the faithfulness of representational deletion.
- Part V covers application-driven chapters to connect these advances to AI safety, fairness, multimodal tasks, and industrial deployment scenarios.

By uniting the optimization, data, model, evaluation, and application perspectives, this book provides both a methodological framework and a practical toolkit for designing unlearning techniques.

References

1. Liu, S., Yao, Y., Jia, J., Casper, S., Baracaldo, N., Hase, P., Yao, Y., Liu, C.Y., Xu, X., Li, H., et al.: Rethinking machine unlearning for large language models. Nat. Mach. Intell., pp. 1–14 (2025)
2. Cao, Y., Yang, J.: Towards making systems forget with machine unlearning. In: 2015 IEEE Symposium on Security and Privacy, pp. 463–480. IEEE (2015)
3. Bourtoule, L., Chandrasekaran, V., Choquette-Choo, C.A., Jia, H., Travers, A., Zhang, B., Lie, D., Papernot, N.: Machine unlearning. In: 2021 IEEE Symposium on Security and Privacy (SP), pp. 141–159. IEEE (2021)
4. Hoofnagle, C.J., van der Sloot, B., Borgesius, F.Z.: The European Union general data protection regulation: what it is and what it means. Inf. Commun. Technol. Law **28**(1), 65–98 (2019)
5. Illman, E., Temple, P.: California consumer privacy act. Bus. Lawyer **75**(1), 1637–1646 (2019)
6. Shi, W., Lee, J., Huang, Y., Malladi, S., Zhao, J., Holtzman, A., Liu, D., Zettlemoyer, L., Smith, N.A., Zhang, C.: MUSE: machine unlearning six-way evaluation for language models. In: The Thirteenth International Conference on Learning Representations (2025). https://openreview.net/forum?id=TArmA033BU
7. Zhang, Y., Fan, C., Zhang, Y., Yao, Y., Jia, J., Liu, J., Zhang, G., Liu, G., Kompella, R.R., Liu, X., Liu, S.: Unlearncanvas: stylized image dataset for enhanced machine unlearning evaluation in diffusion models. In: The Thirty-Eight Conference on Neural Information Processing Systems Datasets and Benchmarks Track (2024). https://openreview.net/forum?id=t9aThFL1lE
8. Mullin, B., Grant, N.: New York Times sues OpenAI and Microsoft over A.I. use of copyrighted work. The New York Times (2023). https://www.nytimes.com/2023/12/27/business/media/new-york-times-open-ai-microsoft-lawsuit.html. Accessed 2025-06-26

9. Fan, J., Yan, Q., Li, M., Qu, G., Xiao, Y.: A survey on data poisoning attacks and defenses. In: 2022 7th IEEE International Conference on Data Science in Cyberspace (DSC), pp. 48–55. IEEE (2022)
10. Li, N., Pan, A., Gopal, A., Yue, S., Berrios, D., Gatti, A., Li, J.D., Dombrowski, A.K., Goel, S., Mukobi, G., Helm-Burger, N., Lababidi, R., Justen, L., Liu, A.B., Chen, M., Barrass, I., Zhang, O., Zhu, X., Tamirisa, R., Bharathi, B., Herbert-Voss, A., Breuer, C.B., Zou, A., Mazeika, M., Wang, Z., Oswal, P., Lin, W., Hunt, A.A., Tienken-Harder, J., Shih, K.Y., Talley, K., Guan, J., Steneker, I., Campbell, D., Jokubaitis, B., Basart, S., Fitz, S., Kumaraguru, P., Karmakar, K.K., Tupakula, U., Varadharajan, V., Shoshitaishvili, Y., Ba, J., Esvelt, K.M., Wang, A., Hendrycks, D.: The WMDP benchmark: measuring and reducing malicious use with unlearning. In: Proceedings of the 41st International Conference on Machine Learning, Proceedings of Machine Learning Research, vol. 235, pp. 28525–28550. PMLR (2024)
11. Zhang, Y., Jia, J., Chen, X., Chen, A., Zhang, Y., Liu, J., Ding, K., Liu, S.: To generate or not? Safety-driven unlearned diffusion models are still easy to generate unsafe images... for now. In: European Conference on Computer Vision, pp. 385–403. Springer (2024)
12. Fan, C., Liu, J., Zhang, Y., Wei, D., Wong, E., Liu, S.: SalUn: empowering machine unlearning via gradient-based weight saliency in both image classification and generation. In: International Conference on Learning Representations (2024)
13. Thudi, A., Deza, G., Chandrasekaran, V., Papernot, N.: Unrolling SGD: understanding factors influencing machine unlearning. arXiv preprint arXiv:2109.13398 (2021)
14. Kadhe, S.R., Halimi, A., Rawat, A., Baracaldo, N.: FairSISA: ensemble post-processing to improve fairness of unlearning in LLMs. arXiv preprint arXiv:2312.07420 (2023)
15. Guo, C., Goldstein, T., Hannun, A., Van Der Maaten, L.: Certified data removal from machine learning models. arXiv preprint arXiv:1911.03030 (2019)
16. Hu, S., Fu, Y., Wu, S., Smith, V.: Unlearning or obfuscating? Jogging the memory of unlearned LLMs via benign relearning. In: The Thirteenth International Conference on Learning Representations (2025). https://openreview.net/forum?id=fMNRYBvcQN
17. Hase, P., Bansal, M., Kim, B., Ghandeharioun, A.: Does localization inform editing? Surprising differences in causality-based localization versus knowledge editing in language models. Adv. Neural Inf. Process. Syst. **36**, 17643–17668 (2023)
18. Thudi, A., Deza, G., Chandrasekaran, V., Papernot, N.: Unrolling SGD: understanding factors influencing machine unlearning. In: 2022 IEEE 7th European Symposium on Security and Privacy (EuroS&P), pp. 303–319. IEEE (2022)
19. Zhang, R., Lin, L., Bai, Y., Mei, S.: Negative preference optimization: from catastrophic collapse to effective unlearning. In: First Conference on Language Modeling (2024). https://openreview.net/forum?id=MXLBXjQkmb
20. Liu, B., Liu, Q., Stone, P.: Continual learning and private unlearning. In: Conference on Lifelong Learning Agents, pp. 243–254. PMLR (2022)
21. Maini, P., Feng, Z., Schwarzschild, A., Lipton, Z.C., Kolter, J.Z.: TOFU: a task of fictitious unlearning for LLMs. In: First Conference on Language Modeling (2024). https://openreview.net/forum?id=B41hNBoWLo
22. Zhang, Y., Jia, J., Chen, X., Chen, A., Zhang, Y., Liu, J., Ding, K., Liu, S.: To generate or not? Safety-driven unlearned diffusion models are still easy to generate unsafe images... for now. In: European Conference on Computer Vision (ECCV) (2024)
23. Rafailov, R., Sharma, A., Mitchell, E., Manning, C.D., Ermon, S., Finn, C.: Direct preference optimization: your language model is secretly a reward model. In: Thirty-Seventh Conference on Neural Information Processing Systems (2023). https://openreview.net/forum?id=HPuSIXJaa9
24. Fan, C., Liu, J., Lin, L., Jia, J., Zhang, R., Mei, S., Liu, S.: Simplicity prevails: rethinking negative preference optimization for LLM unlearning. arXiv preprint arXiv:2410.07163 (2024)

25. Bu, Z., Jin, X., Vinzamuri, B., Ramakrishna, A., Chang, K.W., Cevher, V., Hong, M.: Unlearning as multi-task optimization: a normalized gradient difference approach with an adaptive learning rate. arXiv preprint arXiv:2410.22086 (2024)
26. Koh, P.W., Liang, P.: Understanding black-box predictions via influence functions. In: International Conference on Machine Learning, pp. 1885–1894. PMLR (2017)
27. Jia, J., Liu, J., Ram, P., Yao, Y., Liu, G., Liu, Y., Sharma, P., Liu, S.: Model sparsity can simplify machine unlearning. In: Thirty-Seventh Conference on Neural Information Processing Systems (2023)
28. Jia, J., Zhang, Y., Zhang, Y., Liu, J., Runwal, B., Diffenderfer, J., Kailkhura, B., Liu, S.: SOUL: unlocking the power of second-order optimization for LLM unlearning. In: Al-Onaizan, Y., Bansal, M., Chen, Y.N. (eds.) Proceedings of the 2024 Conference on Empirical Methods in Natural Language Processing, pp. 4276–4292. Association for Computational Linguistics, Miami, Florida, USA (2024). https://doi.org/10.18653/v1/2024.emnlp-main.245. https://aclanthology.org/2024.emnlp-main.245/
29. Liu, H., Li, Z., Hall, D.L.W., Liang, P., Ma, T.: Sophia: a scalable stochastic second-order optimizer for language model pre-training. In: The Twelfth International Conference on Learning Representations (2024). https://openreview.net/forum?id=3xHDeA8Noi
30. Jia, J., Liu, J., Zhang, Y., Ram, P., Baracaldo, N., Liu, S.: WAGLE: strategic weight attribution for effective and modular unlearning in large language models. In: The Thirty-Eighth Annual Conference on Neural Information Processing Systems (2024). https://openreview.net/forum?id=VzOgnDJMgh
31. Zhang, Y., Khanduri, P., Tsaknakis, I., Yao, Y., Hong, M., Liu, S.: An introduction to bilevel optimization: foundations and applications in signal processing and machine learning. IEEE Signal Process. Mag. **41**(1), 38–59 (2024)
32. Liu, C.Y., Wang, Y., Flanigan, J., Liu, Y.: Large language model unlearning via embedding-corrupted prompts. In: The Thirty-Eighth Annual Conference on Neural Information Processing Systems (2024). https://openreview.net/forum?id=e5icsXBD8Q
33. Pawelczyk, M., Neel, S., Lakkaraju, H.: In-context unlearning: language models as few-shot unlearners. In: Proceedings of the 41st International Conference on Machine Learning (2024)
34. Bhaila, K., Van, M.H., Wu, X.: Soft prompting for unlearning in large language models. In: Chiruzzo, L., Ritter, A., Wang, L. (eds.) Proceedings of the 2025 Conference of the Nations of the Americas Chapter of the Association for Computational Linguistics: Human Language Technologies (Volume 1: Long Papers), pp. 4046–4056. Association for Computational Linguistics, Albuquerque, New Mexico (2025). https://doi.org/10.18653/v1/2025.naacl-long.204. https://aclanthology.org/2025.naacl-long.204/
35. Rebedea, T., Dinu, R., Sreedhar, M., Parisien, C., Cohen, J.: Nemo guardrails: a toolkit for controllable and safe LLM applications with programmable rails. arXiv preprint arXiv:2310.10501 (2023)
36. Inan, H., Upasani, K., Chi, J., Rungta, R., Iyer, K., Mao, Y., Tontchev, M., Hu, Q., Fuller, B., Testuggine, D., et al.: Llama guard: LLM-based input-output safeguard for human-AI conversations. arXiv preprint arXiv:2312.06674 (2023)
37. Yuan, Z., Xiong, Z., Zeng, Y., Yu, N., Jia, R., Song, D., Li, B.: RigorLLM: resilient guardrails for large language models against undesired content. arXiv preprint arXiv:2403.13031 (2024)
38. Markov, T., Zhang, C., Agarwal, S., Nekoul, F.E., Lee, T., Adler, S., Jiang, A., Weng, L.: A holistic approach to undesired content detection in the real world. In: Proceedings of the AAAI Conference on Artificial Intelligence, pp. 15009–15018 (2023)
39. Lees, A., Tran, V.Q., Tay, Y., Sorensen, J., Gupta, J., Metzler, D., Vasserman, L.: A new generation of perspective API: efficient multilingual character-level transformers. In: Proceedings of the 28th ACM SIGKDD Conference on Knowledge Discovery and Data Mining, pp. 3197–3207 (2022)

40. Dong, Y., Mu, R., Jin, G., Qi, Y., Hu, J., Zhao, X., Meng, J., Ruan, W., Huang, X.: Building guardrails for large language models. arXiv preprint arXiv:2402.01822 (2024)
41. Wang, Y., Singh, L.: Adding guardrails to advanced chatbots. arXiv preprint arXiv:2306.07500 (2023)
42. Goyal, S., Hira, M., Mishra, S., Goyal, S., Goel, A., Dadu, N., Kirushikesh, D., Mehta, S., Madaan, N.: LLMGuard: guarding against unsafe LLM behavior. In: Proceedings of the AAAI Conference on Artificial Intelligence, vol. 38, pp. 23790–23792 (2024)
43. Chu, Z., Wang, Y., Li, L., Wang, Z., Qin, Z., Ren, K.: A causal explainable guardrails for large language models. arXiv preprint arXiv:2405.04160 (2024)
44. Thaker, P., Maurya, Y., Smith, V.: Guardrail baselines for unlearning in LLMs. arXiv preprint arXiv:2403.03329 (2024)
45. Ilharco, G., Ribeiro, M.T., Wortsman, M., Schmidt, L., Hajishirzi, H., Farhadi, A.: Editing models with task arithmetic. In: The Eleventh International Conference on Learning Representations (2023). https://openreview.net/forum?id=6t0Kwf8-jrj

Part I

The Optimization Lens on Machine Unlearning

2 Optimization Foundations for Machine Unlearning

Sijia Liu

Abstract

This chapter develops the optimization foundations of machine unlearning (MU). It begins with the design of forget objectives, covering formulations such as gradient difference (GradDiff), negative preference optimization (NPO), its simple variant SimNPO, and representation misdirection unlearning (RMU). These span output-level suppression and representation-level disruption, each with distinct trade-offs in stability, boundedness, and structural guarantees. Building on these objectives, the chapter introduces unlearning-aware optimizers, linking influence-function methods to second-order optimization. Classical influence-based unlearning is extended into an iterative second-order framework that accelerates convergence and stabilizes the forgetting—retaining trade-off. Finally, the chapter examines bi-level optimization (BLO) as a principled means of balancing forgetting with utility. By disentangling lower-level forget objectives from upper-level retain objectives, BLO achieves state-of-the-art performance, enabling effective unlearning without undermining general utility. Collectively, these advances establish a cohesive foundation for optimization-driven unlearning, transforming MU from an ad hoc practice into a systematic methodology for safe, reliable, and governable AI.

2.1 Optimization Problem for Machine Unlearning

Unlearning tasks can take various forms, but they almost always revolve around a specific subset of data or behaviors that must be removed, what we call the forget set ($\mathcal{D}_\mathrm{f}$). To ensure the model does not collapse or lose unrelated capabilities in the process, unlearning is often

S. Liu (✉)
Michigan State University, East Lansing, MI, USA
e-mail: liusiji5@msu.edu

S. Liu et al. (eds.), *Machine Unlearning for Governance of Foundation Models*, Synthesis Lectures on Computer Vision, https://doi.org/10.1007/978-3-032-17282-2_2

paired with a retain set ($\mathcal{D}_\mathrm{r}$), a collection of non-forgotten data or behaviors that anchor the model's preserved utility. Together, these two sets define the dual pressures at the heart of machine unlearning: removing targeted knowledge while maintaining overall competence.

From this perspective, machine unlearning can be formalized as a regularized optimization problem, expressed in (1.1): $\underset{\boldsymbol{\theta}}{\text{minimize}}\ \ell_\mathrm{f}(\boldsymbol{\theta}) + \gamma \ell_\mathrm{r}(\boldsymbol{\theta})$, where the objective is to simultaneously balance the goals of forgetting and retaining. On one side, the optimization must aggressively suppress the influence of the forget set, ensuring that sensitive, copyrighted, or harmful information no longer shapes the model's representations or outputs—even when subjected to adversarial probing or indirect queries. On the other side, it must safeguard the retain set, preventing unnecessary degradation of fluency, reasoning skills, and task performance in domains unrelated to the unlearning target.

In the regularized optimization formulation, the retain loss $\vdash_\mathrm{r}$ typically mirrors the standard training loss computed over the retain set, ensuring that core capabilities are preserved. By contrast, the design of the forget loss ℓ_f is far more subtle and challenging, as it dictates how the model is encouraged to discard knowledge associated with the forget set. Different formulations of ℓ_f lead to fundamentally different optimization strategies. In fact, as discussed in Sect. 1.2, these formulations underpin the major families of unlearning methods. Therefore, a central pillar of optimization-driven unlearning lies in the design of the forget objective ℓ_f, since it fundamentally dictates how the model diverges from the forget distribution while maintaining utility. We will elaborate on this critical aspect in the next section.

2.2 Forget Objective Design

In this section, we examine several representative families of forget objectives, organized by whether they operate at the *output* level (e.g., prediction suppression, or divergence from reference responses) or at the *representation* level (e.g., feature randomization).

Gradient Difference (GradDiff) [1, 2]. The forget loss is defined as the negative cross-entropy (CE) objective on the forget set:

$$\ell_\mathrm{f} = -\ell_\mathrm{CE}(\boldsymbol{\theta}; \mathcal{D}_\mathrm{f}), \tag{2.1}$$

where ℓ_CE denotes the sequence-level cross-entropy loss. Minimizing ℓ_f thus corresponds to gradient ascent (GA) on the forget set $\mathcal{D}_\mathrm{f}$, actively discouraging the model from fitting the forgotten data [3, 4]. If the retain objective remains the standard CE loss $\ell_\mathrm{r} = \ell_\mathrm{CE}(\boldsymbol{\theta}; \mathcal{D}_\mathrm{r})$, then combining it with the forget objective yields a single regularized optimization that simultaneously applies gradient ascent on the forget set and gradient descent on the retain set [1–3].

Negative Preference Optimization (NPO) [5]. NPO extends the direct preference optimization (DPO) framework [6] to unlearning by treating forget data as negative examples.

Unlike GradDiff, whose GA-based forget loss is unbounded from below, NPO introduces a bounded forget loss and an adaptive gradient weighting, providing more stable and controlled forgetting dynamics. Formally, the NPO forget loss evaluated at a forget data (x, y) is defined as

$$\ell_{\mathrm{f}}(y|x;\boldsymbol{\theta}) = -\frac{2}{\beta}\log\sigma\left(-\beta\log\frac{\pi_{\boldsymbol{\theta}}(y|x)}{\pi_{\mathrm{ref}}(y|x)}\right), \tag{2.2}$$

where $\pi_{\boldsymbol{\theta}}(y|x)$ is the prediction probability of the model $\boldsymbol{\theta}$ given the input-response pair (x, y), π_{ref} is the reference model (the checkpoint prior to unlearning), $\sigma(\cdot)$ is the sigmoid function, and $\beta > 0$ controls temperature. The corresponding gradient takes the form

$$\nabla_{\boldsymbol{\theta}}\ell_{\mathrm{f}}(y|x;\boldsymbol{\theta}) = w_{\boldsymbol{\theta}}(x, y)\cdot\nabla_{\boldsymbol{\theta}}\log\pi_{\boldsymbol{\theta}}(y|x), \quad w_{\boldsymbol{\theta}}(x, y) = \frac{2\pi_{\boldsymbol{\theta}}(y|x)^{\beta}}{\pi_{\boldsymbol{\theta}}(y|x)^{\beta} + \pi_{\mathrm{ref}}(y|x)^{\beta}}, \tag{2.3}$$

where $\nabla_{\boldsymbol{\theta}}$ denotes the gradient with respect to $\boldsymbol{\theta}$, and $w_{\boldsymbol{\theta}}(x, y)$ is known as a gradient weight smoothing scheme.

The above reveals the two mechanisms behind NPO. By construction, the forget loss is bounded below by zero, unlike the GA-based forget loss $\ell_{\mathrm{f}} = -\log\pi_{\boldsymbol{\theta}}(y|x)$ which has no lower bound. Minimizing it toward zero corresponds to pushing $\pi_{\boldsymbol{\theta}}(y|x)$ well below $\pi_{\mathrm{ref}}(y|x)$, ensuring the model assigns much lower probability to generating forget responses than the reference model. At the same time, the adaptive weight $w_{\boldsymbol{\theta}}(x, y)$ moderates the update size. Since in most forgetting scenarios $\pi_{\boldsymbol{\theta}}(y|x) < \pi_{\mathrm{ref}}(y|x)$, this weight remains below one, leading to more gradual divergence compared to GA, which implicitly uses weight one.

Thus, NPO reframes forgetting as a preference-learning task, transforming the unstable gradient ascent formulation into a bounded and adaptively smoothed optimization. This combination prevents collapse, regulates divergence speed, and preserves utility more effectively, making it one of the most widely adopted forget objectives in recent large-scale unlearning studies.

Simple Negative Preference Optimization (SimNPO) [7]. Existing work [7] has pointed out that the incorporation of the reference model π_{ref} in NPO (2.2) may introduce an implicit bias in the unlearning objective. Specifically, the optimization tends to emphasize enlarging the distance from the reference model rather than directly accounting for the inherent difficulty of unlearning individual samples. At first glance, forcing $\pi_{\boldsymbol{\theta}}(y|x) \ll \pi_{\mathrm{ref}}(y|x)$ appears desirable, given that π_{ref} corresponds to the initial model prior to unlearning. However, this reliance on π_{ref} can overshadow sample-specific challenges, resulting in an uneven allocation of unlearning effort across the forget set. These sample-specific unlearning difficulties are particularly pronounced when the forget data vary in sequence length or exhibit different levels of memorization by the reference model [7].

Moreover, the presence of the reference model also complicates the effectiveness of NPO's gradient weight smoothing $w_{\boldsymbol{\theta}}(x, y)$ in (2.3). In particular, the adaptive weight $w_{\boldsymbol{\theta}}(x, y) = \frac{2\pi_{\boldsymbol{\theta}}(y|x)^{\beta}}{\pi_{\boldsymbol{\theta}}(y|x)^{\beta}+\pi_{\mathrm{ref}}(y|x)^{\beta}}$ in (2.3) reduces to $w_{\boldsymbol{\theta}}(x, y) \approx 1$ at the early optimization stage, since $\boldsymbol{\theta}$ is initialized from π_{ref}. As a result, the intended smoothing effect fails to reflect data-dependent variability, leaving NPO initially indistinguishable from standard gradient ascent. SimNPO circumvents these issues by removing the dependence on the reference model, simplifying the formulation while more faithfully capturing the intrinsic difficulty of unlearning each sample.

Taking the inspiration of reference model-free simple preference optimization (SimPO) [8], one can mitigate the reference model bias in NPO by replacing its reward formulation $\beta \log(\pi_{\boldsymbol{\theta}}(y|x)/\pi_{\text{ref}}(y|x))$ in (2.2) with the SimPO-based reward formulation $(\beta/|y|)\log(\pi_{\boldsymbol{\theta}}(y|x))$. This modification transforms (2.2) into the *SimNPO loss*:

$$\ell_{\text{f}}(\boldsymbol{\theta}) = \mathbb{E}_{(x,y)\in\mathcal{D}_{\text{f}}}\left[-\frac{2}{\beta}\log\sigma\left(-\frac{\beta}{|y|}\log\pi_{\boldsymbol{\theta}}(y|x) - \gamma\right)\right] \tag{2.4}$$

where $\gamma \geq 0$ is the reward margin parameter, inherited from SimPO, which defines the margin of preference for a desired response over a dispreferred one. However, one can set $\gamma = 0$ as a larger γ requires greater compensation to suppress token prediction, which may accelerate the utility drop during unlearning.

Representation Misdirection Unlearning (RMU) [9]. The RMU-based forget objective operates by disrupting the intermediate representations that the model forms for forget data samples. Specifically, for each input $x \in \mathcal{D}_{\text{f}}$ designated for forgetting, RMU encourages the internal feature extractor $M_{\boldsymbol{\theta}}(\cdot)$ to map x not to meaningful semantic features, but instead to random vectors $\mathbf{v}$ drawn from a uniform distribution. This yields:

$$\ell_{\text{f}}(\boldsymbol{\theta}; \mathcal{D}_{\text{f}}) = \mathbb{E}_{x\in\mathcal{D}_{\text{f}}}[\|M_{\boldsymbol{\theta}}(x) - c \cdot \mathbf{v}\|_2^2], \tag{2.5}$$

where the hyperparameter c plays a role in controlling the magnitude of the randomized activations, thereby determining how aggressively the model is pushed to erase structured information from unsafe data. Similarly, the choice of which intermediate layer to randomize is equally important, since different layers capture distinct levels of abstraction, from low-level features to high-level semantics. Together, these design choices determine the delicate balance between effective forgetting and the preservation of general model utility.

Compared with output-level approaches such as GradDiff or NPO, RMU operates directly in the representation space, forcing the model to overwrite unsafe features with randomized noise. This intervention ensures that the model discards meaningful internal encodings of the forget set, thereby erasing their semantic footprint from the learned representation space. By disrupting the model's capacity to form coherent representations of unsafe inputs, RMU offers a more structural guarantee of forgetting than approaches that only suppress output probabilities.

2.3 Unlearning-Aware Optimizer Design

A standard approach to minimizing the regularized unlearning objective in (1.1) is to apply first-order gradient-based methods, such as stochastic gradient descent (SGD) or adaptive momentum methods (e.g., Adam). While effective in conventional training, these optimizers are not inherently tailored to the unique requirements of unlearning. Therefore, one can also design optimizers specialized for unlearning. In this section, we will demonstrate the connection between influence function in unlearning and the second-order optimizer, Sophia (Second-order Clipped Stochastic Optimization) [10]. Influence function naturally captures the sensitivity of model parameters to individual training samples, which aligns closely with the goal of precisely removing the effect of forget data. Leveraging this perspective, second-order optimizers, through approximations of the Hessian, can provide a principled mechanism to accelerate and stabilize unlearning.

2.3.1 Preliminaries on Influence Unlearning

Influence unlearning is a one-shot technique that leverages the influence function framework [11–13] to estimate and remove the contribution of the forget set $\mathcal{D}_\mathrm{f}$ from a pre-trained model $\boldsymbol{\theta}_0$. Unlike *iterative* optimization methods, which repeatedly update model parameters via gradient steps, influence unlearning applies a single corrective update to the parameter vector $\boldsymbol{\theta}_0$. This update is derived by approximating how the parameters would have shifted if the forget data had been excluded during the original training process, thereby directly counteracting their influence in weight space. While theoretically elegant and computationally appealing, the practical use of influence unlearning has so far been restricted to small-scale vision models and relatively simple architectures [13–15]. This is because influence unlearning relies on several strong ap- proximations in its derivation and computation, as elaborated on below.

Let $\boldsymbol{\theta}^*$ denote a retrained model from scratch on the retain set $\mathcal{D}_\mathrm{r}$, *i.e.*, the solution to the optimization problem $\min_{\boldsymbol{\theta}} \sum_{i=1}^{N} \ell(y_i|x_i; \boldsymbol{\theta})$ with random initialization, where ℓ is the cross-entropy loss for training, $(x_i, y_i) \in \mathcal{D}_\mathrm{r}$ is training data point, and N is the total number of training data points in $\mathcal{D}_\mathrm{r}$. The *objective of influence unlearning* is to derive the weight modification from the pre-trained model $\boldsymbol{\theta}_0$ to the retrained model $\boldsymbol{\theta}^*$, *i.e.*, $\boldsymbol{\theta}^* - \boldsymbol{\theta}_0$. To this end, a *weighted* training problem is introduced:

$$\boldsymbol{\theta}(\mathbf{w}) := \arg\min_{\boldsymbol{\theta}} \ell(\boldsymbol{\theta}, \mathbf{w}), \;\; \ell(\boldsymbol{\theta}, \mathbf{w}) = \sum_{i=1}^{N} [w_i \ell(y_i|x_i; \boldsymbol{\theta})] \tag{2.6}$$

where w_i represents the introduced data influence weight. If the data point (x_i, y_i) is removed from the training set, then w_i takes a value of 0. By the definition of (2.6), the pretrained and retrained models $\boldsymbol{\theta}_0$ and $\boldsymbol{\theta}^*$ can be expressed as

$$\boldsymbol{\theta}_0 = \boldsymbol{\theta}(\mathbf{1}), \quad \boldsymbol{\theta}^* = \boldsymbol{\theta}(\mathbf{w}_{\text{MU}}), \tag{2.7}$$

where $\boldsymbol{\theta}(\mathbf{1})$ entails training over the entire training set with weights $\mathbf{w} = \mathbf{1}$. Here $\mathbf{1}$ denotes the all-one vector. Similarly, given the unlearning-specific weighting scheme, $\mathbf{w}_{\text{MU}} = \mathbf{1}_{\mathcal{D}_\text{r}}$, $\boldsymbol{\theta}(\mathbf{w}_{\text{MU}})$ corresponds to the retrained model post unlearning. Here $\mathbf{1}_{\mathcal{D}_\text{r}}$ denotes an element-wise indicator function that takes the value 1 if the data point belongs to the retain set $\mathcal{D}_\text{r}$ and 0 otherwise. Based on (2.7), influence unlearning then aims to derive:

$$\Delta(\mathbf{w}_{\text{MU}}) = \boldsymbol{\theta}(\mathbf{w}_{\text{MU}}) - \boldsymbol{\theta}(\mathbf{1}). \tag{2.8}$$

The derivation of (2.8) is highly non-trivial as the retrained model $\boldsymbol{\theta}^*$ cannot be directly obtained and is implicitly defined through the optimization problem $\min_{\boldsymbol{\theta}} \ell(\boldsymbol{\theta}, \mathbf{w}_{\text{MU}})$. To proceed, the influence function approach [11–13] simplifies (2.8) by applying a first-order Taylor expansion to $\boldsymbol{\theta}(\mathbf{w}_{\text{MU}})$ at $\mathbf{w} = \mathbf{1}$:

$$\Delta(\mathbf{w}_{\text{MU}}) = \boldsymbol{\theta}(\mathbf{w}_{\text{MU}}) - \boldsymbol{\theta}(\mathbf{1}) \approx \frac{d\boldsymbol{\theta}(\mathbf{w})}{d\mathbf{w}} |_{\mathbf{w}=\mathbf{1}} (\mathbf{w}_{\text{MU}} - \mathbf{1}), \tag{2.9}$$

where $\frac{d\boldsymbol{\theta}(\mathbf{w})}{d\mathbf{w}}$ denotes the full derivative of $\boldsymbol{\theta}(\mathbf{w})$ with respect to (w.r.t.) $\mathbf{w}$, and is known as *implicit gradient* [16]. Utilizing the implicit function theorem [17], the closed form of the influence unlearning formula (2.9) can be given by [13, Proposition 1]:

$$\boldsymbol{\theta}(\mathbf{w}_{\text{MU}}) \approx \boldsymbol{\theta}_0 + \mathbf{H}^{-1} \nabla_{\boldsymbol{\theta}} \ell(\boldsymbol{\theta}, \mathbf{1} - \mathbf{w}_{\text{MU}}) |_{\boldsymbol{\theta}=\boldsymbol{\theta}_0}, \tag{2.10}$$

where $\ell(\boldsymbol{\theta}, \mathbf{w})$ represents the $\mathbf{w}$-weighted training loss (2.6), $\mathbf{H}^{-1}$ stands for the inverse of the second-order derivative (*i.e.*, Hessian matrix) $\nabla_{\boldsymbol{\theta},\boldsymbol{\theta}} \ell(\boldsymbol{\theta}, \mathbf{1}/N)$ evaluated at $\boldsymbol{\theta}_0$, $\nabla_{\boldsymbol{\theta}} \ell$ denotes the gradient of ℓ, and $\mathbf{1} - \mathbf{w}_{\text{MU}}$ yields $\mathbf{1} - \mathbf{1}_{\mathcal{D}_\text{r}}$, which captures the data weight on the forget set $\mathcal{D}_\text{f}$. To compute (2.10), one must determine the inverse-Hessian gradient product. However, exact computation is often computationally prohibitive. To address this challenge, numerical approximations such as the WoodFisher approximation [18] are often employed to estimate the inverse-Hessian gradient product.

2.3.2 From Influence Unlearning to Second-Order Optimization

An *intriguing observation* from (2.10) is that influence unlearning conforms to the generic form of second-order (SO) optimization [19]. As in Newton's method, one uses a SO approximation of a loss function ℓ to locate its minima. This yields a descent algorithm based on a Newton step [20]:

$$\boldsymbol{\theta}_{t+1} = \boldsymbol{\theta}_t \underbrace{-\eta_t \mathbf{H}_t^{-1} \mathbf{g}_t}_{\text{Newton step}}, \tag{2.11}$$

where t represents the iteration index of Newton's method, $\boldsymbol{\theta}_{t+1}$ denotes the currently updated optimization variables, $\eta_t > 0$ is the learning rate, and $\mathbf{H}_t$ and $\mathbf{g}_t$ represent the Hessian matrix and the gradient of the loss ℓ, respectively, evaluated at $\boldsymbol{\theta}_t$.

The structural similarity between the influence unlearning formulation in (2.10) and the Newton step in second-order optimization (2.11) naturally motivates the question: can we integrate second-order optimization principles into influence unlearning, thereby extending it from a one-shot weight adjustment into an iterative and more effective unlearning procedure?

If we can transition from the static, one-shot nature of influence unlearning to a dynamic, iterative optimization process, we anticipate that the diminished accuracy resulting from the approximations used in influence unlearning (2.10) will be mitigated through the iterative engagement of the learning process. However, we still face the computational challenge posed by the Hessian inversion in (2.11). Therefore, *we need to select a practically feasible second-order optimization method for unlearning.*

Sophia (Second-order Clipped Stochastic Optimization) [10], a simple scalable second-order optimizer, is well-suited since it utilizes a simple diagonal matrix estimate of the Hessian and has shown its effectiveness in large-model training. Sophia modifies the vanilla Newton's method to

$$\boldsymbol{\theta}_{t+1} = \boldsymbol{\theta}_t - \eta_t \text{clip}(\mathbf{m}_t / \max\{\gamma \mathbf{h}_t, \epsilon\}, 1), \tag{2.12}$$

where $\mathbf{m}_t \leftarrow \beta_1 \mathbf{m}_{t-1} + (1 - \beta_1)\mathbf{g}_t$ is the exponential moving average (EMA) of the FO (first-order) gradient with parameter $\beta_1 > 0$, $\mathbf{h}_t$ denotes the EMA of the Hessian diagonal estimates obtained from the diagonal of the Gauss-Newton matrix [10], and the clipping operation $\text{clip}(\boldsymbol{\theta}, a)$ limits the magnitude of each element in vector $\boldsymbol{\theta}$ to a maximum of a, thereby preventing excessively large updates that could destabilize the optimization process. In (2.12), both the clipping operation $\text{clip}(\cdot, \cdot)$ and the division operation $\cdot/\cdot$ are all performed element-wise, and $\gamma > 0$ and $\epsilon > 0$ are additional parameters in the clipping operation. In (2.12), if the clipping operation is absent with $\gamma = 1$ and $\epsilon \to 0$, then the Sophia update (2.12) simplifies to the Newton update (2.11) utilizing the diagonal Hessian estimate for $\mathbf{H}$.

Next, we can link influence unlearning (2.10) with Sophia, leading to second-order optimizer for unlearning, known as SOUL [21]. Recall from (2.10) and (2.6) that the change in data weights $(\mathbf{1} - \mathbf{w}_{\text{MU}})$ encodes the influence of the forget set $\mathcal{D}_\text{f}$ in model training. Therefore, we can interpret the term $\mathbf{H}^{-1}\nabla_{\boldsymbol{\theta}}\ell(\boldsymbol{\theta}_0, \mathbf{1} - \mathbf{w}_{\text{MU}})$ in (2.10) as a second-order optimization step over the *forget set.*

2.3.3 Iterative Gains with Second-Order Unlearning

We next empirically demonstrate the advantages of SOUL (second-order optimization-based unlearning) over first-order methods such as GA and GradDiff [1, 2], by examining the dynamics of unlearning and retaining convergence across training epochs. Figure 2.1 reports the evolution of forget accuracy (lower is better, indicating more effective unlearning) and

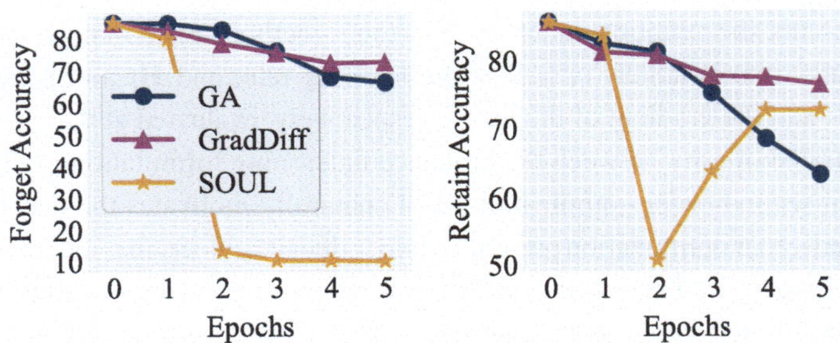

Fig. 2.1 Unlearning performance versus optimization epochs using different optimizers in TOFU unlearning [3]. Left: forget accuracy versus epochs; right: retain accuracy versus epochs

retain accuracy (higher is better, indicating stronger utility preservation) during the TOFU unlearning task [3].

The results reveal a clear gap: both GA and GradDiff converge more slowly in terms of unlearning compared to SOUL. While GradDiff is relatively effective at retaining accuracy, its forgetting performance lags behind. GA, on the other hand, drives stronger forgetting but causes a severe collapse in retention by the final epoch. In contrast, SOUL rapidly achieves lower forget accuracy and simultaneously stabilizes retain accuracy. This balance is attributed to its second-order optimization design: by explicitly accounting for the curvature of the loss landscape with respect to the forget set as in (2.10), SOUL accelerates unlearning convergence. Moreover, the adaptive step size inherent in the second-order update enables the model to recover and rewind retention performance, avoiding the utility degradation seen in GA.

Furthermore, examining SOUL at its first epoch reveals another important insight in Fig. 2.1: The initial update is similar to the classical influence unlearning step, but the outcome demonstrates limited unlearning effectiveness. This limitation stems from the inherent weakness of one-time influence unlearning, which lacks the iterative optimization power necessary to refine and stabilize the forgetting process. In contrast, SOUL extends influence unlearning into a fully iterative, optimization-driven framework, thereby enhancing both convergence and efficacy.

2.4 Balancing Utility with Bi-Level Optimization

In the most commonly-used formulation of machine unlearning, the balance between unlearning effectiveness and utility preservation is typically achieved by regularizing the forget loss ℓ_{f} with the retain loss ℓ_{r}, weighted by a tradeoff parameter γ in (1.1). By tuning γ, practitioners attempt to enforce forgetting while mitigating the degradation of model utility. However, this scalar regularization often provides only a coarse control mechanism, failing to adapt to the varying levels of difficulty across different forget and retain samples. This motivates the need for more principled approaches, such as bi-level optimization (BLO), that explicitly disentangle and coordinate the objectives of forgetting and retention.

BLO (bi-level optimization) is a class of optimization problems involving two nested levels (*upper-* and *lower-level*), where the objective and variables of the *upper-level* problem depend on the optimizer of the *lower-level* one. The canonical formulation of BLO is given by:

$$\underbrace{\underset{\boldsymbol{\theta}}{\text{minimize}} \;\; f(\boldsymbol{\theta}, \boldsymbol{\phi}^*(\boldsymbol{\theta}))\, \boldsymbol{\theta},}_{\text{Upper-level optimization over}} \quad \text{subject to} \;\; \underbrace{\boldsymbol{\phi}^*(\boldsymbol{\theta}) \in \underset{\boldsymbol{\phi}}{\arg\min}\, g(\boldsymbol{\theta}, \boldsymbol{\phi})\, \boldsymbol{\phi},}_{\text{Lower-level optimization over}} \tag{2.13}$$

where $\boldsymbol{\theta} \in \mathbb{R}^m$ denotes the upper-level variable, $\boldsymbol{\phi} \in \mathbb{R}^n$ is the lower-level variable, and $\boldsymbol{\phi}^*(\boldsymbol{\theta})$ is a corresponding optimal solution of the lower-level problem. It is evident that the upper- and lower-level problems are inherently coupled, since the optimality of $\boldsymbol{\phi}^*(\boldsymbol{\theta})$ depends on $\boldsymbol{\theta}$.

In the context of machine unlearning, the bilevel structure can be naturally interpreted through the lens of forget—retain tradeoffs. The lower-level objective corresponds to the forget objective, whose role is to drive the model toward erasing the influence of the designated forget set and thereby produce an unlearned model. Built on top of this, the upper-level objective operates over the solution space defined by the lower level and is tasked with ensuring that the resulting model still preserves utility, *i.e.*, performance on the retain set and alignment with the desired distributional behavior. This coupling highlights the central challenge of unlearning: the unlearned solution must simultaneously satisfy a strong erasure constraint while maintaining generalization, which motivates the adoption of BLO as a principled framework.

Based on the above, we can frame the machine unlearning problem (1.1) as the BLO problem

$$\underset{\boldsymbol{\theta} \in \Phi^*}{\text{minimize}} \;\; \ell_{\mathrm{r}}(\boldsymbol{\theta}; \mathcal{D}_{\mathrm{r}}), \quad \text{subject to} \;\; \Phi^* = \underset{\boldsymbol{\theta}}{\arg\min}\, \ell_{\mathrm{f}}(\boldsymbol{\theta}; \mathcal{D}_{\mathrm{f}}), \tag{2.14}$$

where Φ^* denotes the set of optimal solutions to the lower-level problem, while the upper-level objective seeks to identify the most suitable unlearned model that minimizes the retain loss. We note that (2.14) represents a specific form of BLO, commonly referred to as a *simple bi-level problem*. In this formulation, the lower-level optimization variable coincides exactly with the upper-level variable, thereby eliminating the need to maintain two distinct parameter sets across the hierarchical structure.

Table 2.1 LLM unlearning performance of various methods on the WMDP benchmark

Method	Bio. Acc. ↓	Cyber Acc. ↓	MMLU ↑
Original	63.7	44.0	58.1
RMU [9]	31.2	28.2	57.1
NPO [5]	42.5	28.3	40.0
SimNPO [7]	41.6	32.2	47.1
BLUR-NPO	27.6	26.5	48.4
BLUR-RMU	26.9	26.6	57.0

Unlearning efficacy is measured by Bio. Acc. and Cyber Acc. (lower is better), while utility preservation is measured by MMLU (higher is better). The original model is given by the LLM Zephyr 7b beta model

To solve (2.14), a more tractable approach is to reformulate the lower-level problem as a stationary condition that characterizes its optimal solutions. By enforcing this condition as a constraint, we obtain the following constrained optimization problem [22]:

$$\underset{\boldsymbol{\theta}\in\Phi^*}{\text{minimize}} \;\; \ell_{\mathrm{r}}(\boldsymbol{\theta}; \mathcal{D}_{\mathrm{r}}), \quad \text{subject to } \nabla_{\boldsymbol{\theta}}\ell_{\mathrm{f}}(\boldsymbol{\theta}; \mathcal{D}_{\mathrm{f}}) = 0. \tag{2.15}$$

To ensure convergence toward a stationary point of the forget objective ℓ_{f}, namely satisfying $\nabla_{\boldsymbol{\theta}}\ell_{\mathrm{f}}(\boldsymbol{\theta}; \mathcal{D}_{\mathrm{f}}) = 0$, the update direction in the upper-level problem must remain aligned with the descent direction of ℓ_{f}. To achieve this, the BLO-based unlearning method, BLUR [22], proposes decomposing the retain gradient into its constructive and destructive components relative to the forget gradient. The destructive component, responsible for hindering progress toward minimizing ℓ_{f}, is selectively removed, while the non-destructive component is preserved to maintain utility. This leads to an upper-level update direction that can be expressed as a weighted combination of $\nabla_{\boldsymbol{\theta}}\ell_{\mathrm{f}}$ and $\nabla_{\boldsymbol{\theta}}\ell_{\mathrm{r}}$, where only the conflict-free portion of the retain gradient contributes. Such a bi-level construction allows the unlearning optimizer to faithfully reduce the forget loss while simultaneously controlling the retention loss, thereby providing a principled tradeoff between unlearning effectiveness and utility preservation.

To demonstrate the effectiveness of BLO-based unlearning, BLUR [22], Table 2.1 presents the performance of various unlearning methods on the WMDP benchmark [9], evaluating both unlearning efficacy (via Bio. and Cyber. accuracy, where lower is better) and utility preservation (via MMLU, where higher is better). As shown, baseline methods such as RMU and NPO achieve moderate unlearning but either sacrifice retention (e.g., NPO with MMLU = 40.0) or fail to sufficiently reduce forget accuracy (e.g., RMU with Bio. Acc. = 31.2). SimNPO offers a partial tradeoff improvement but still underperforms on Cyber tasks. In contrast, the BLUR-based variants (BLUR–NPO and BLUR–RMU, which adopt NPO- and RMU-type forget objectives, respectively) deliver the strongest unlearning performance across both tasks. Specifically, Bio. Acc. drops to 27.6/26.9 and Cyber Acc. to 26.5/26.6, marking substantial improvements over all baseline methods. Notably, BLUR–

RMU preserves downstream utility exceptionally well (MMLU = 57.0), nearly matching RMU (57.1) while achieving markedly stronger forgetting. These results underscore the effectiveness of the proposed bi-level optimization strategy, which simultaneously enhances unlearning effectiveness and utility retention.

References

1. Liu, B., Liu, Q., Stone, P.: Continual learning and private unlearning. In: Conference on Lifelong Learning Agents, pp. 243–254. PMLR (2022)
2. Yao, Y., Xu, X., Liu, Y.: Large Language Model Unlearning. arXiv preprint arXiv:2310.10683 (2023)
3. Maini, P., Feng, Z., Schwarzschild, A., Lipton, Z.C., Kolter, J.Z.: TOFU: a task of fictitious unlearning for LLMs. In: First Conference on Language Modeling (2024). https://openreview.net/forum?id=B41hNBoWLo
4. Thudi, A., Deza, G., Chandrasekaran, V., Papernot, N.: Unrolling sgd: understanding factors influencing machine unlearning. In: Proceedings of the 2022 IEEE 7th European Symposium on Security and Privacy (EuroS&P), pp. 303–319. IEEE (2022)
5. Zhang, R., Lin, L., Bai, Y., Mei, S.: Negative preference optimization: from catastrophic collapse to effective unlearning. In: First Conference on Language Modeling (2024). https://openreview.net/forum?id=MXLBXjQkmb
6. Rafailov, R., Sharma, A., Mitchell, E., Manning, C.D., Ermon, S., Finn, C.: Direct preference optimization: your language model is secretly a reward model. In: Thirty-Seventh Conference on Neural Information Processing Systems (2023). https://openreview.net/forum?id=HPuSIXJaa9
7. Fan, C., Liu, J., Lin, L., Jia, J., Zhang, R., Mei, S., Liu, S.: Simplicity prevails: rethinking negative preference optimization for llm unlearning. arXiv preprint arXiv:2410.07163 (2024)
8. Meng, Y., Xia, M., Chen, D.: Simpo: simple preference optimization with a reference-free reward. Adv. Neural. Inf. Process. Syst. **37**, 124198–124235 (2024)
9. Li, N., Pan, A., Gopal, A., Yue, S., Berrios, D., Gatti, A., Li, J.D., Dombrowski, A.K., Goel, S., Mukobi, G., Helm-Burger, N., Lababidi, R., Justen, L., Liu, A.B., Chen, M., Barrass, I., Zhang, O., Zhu, X., Tamirisa, R., Bharathi, B., Herbert-Voss, A., Breuer, C.B., Zou, A., Mazeika, M., Wang, Z., Oswal, P., Lin, W., Hunt, A.A., Tienken-Harder, J., Shih, K.Y., Talley, K., Guan, J., Steneker, I., Campbell, D., Jokubaitis, B., Basart, S., Fitz, S., Kumaraguru, P., Karmakar, K.K., Tupakula, U., Varadharajan, V., Shoshitaishvili, Y., Ba, J., Esvelt, K.M., Wang, A., Hendrycks, D.: The WMDP benchmark: measuring and reducing malicious use with unlearning. In: Proceedings of the 41st International Conference on Machine Learning, Proceedings of Machine Learning Research, vol. 235, pp. 28525–28550. PMLR (2024)
10. Liu, H., Li, Z., Hall, D.L.W., Liang, P., Ma, T.: Sophia: a scalable stochastic second-order optimizer for language model pre-training. In: The Twelfth International Conference on Learning Representations (2024). https://openreview.net/forum?id=3xHDeA8Noi
11. Koh, P.W., Liang, P.: Understanding black-box predictions via influence functions. In: International conference on machine learning, pp. 1885–1894. PMLR (2017)
12. Grosse, R., Bae, J., Anil, C., Elhage, N., Tamkin, A., Tajdini, A., Steiner, B., Li, D., Durmus, E., Perez, E., et al.: Studying large language model generalization with influence functions. arXiv preprint arXiv:2308.03296 (2023)
13. Jia, J., Liu, J., Ram, P., Yao, Y., Liu, G., Liu, Y., Sharma, P., Liu, S.: Model sparsity can simplify machine unlearning. In: Thirty-Seventh Conference on Neural Information Processing Systems (2023)

14. Izzo, Z., Smart, M.A., Chaudhuri, K., Zou, J.: Approximate data deletion from machine learning models. In: International Conference on Artificial Intelligence and Statistics, pp. 2008–2016. PMLR (2021)
15. Warnecke, A., Pirch, L., Wressnegger, C., Rieck, K.: Machine unlearning of features and labels. arXiv preprint arXiv:2108.11577 (2021)
16. Zhang, Y., Khanduri, P., Tsaknakis, I., Yao, Y., Hong, M., Liu, S.: An introduction to bilevel optimization: foundations and applications in signal processing and machine learning. IEEE Sig. Process. Mag. **41**(1), 38–59 (2024)
17. Krantz, S.G., Parks, H.R.: The Implicit Function Theorem: History, Theory, and Applications. Springer, Berlin (2002)
18. Singh, S.P., Alistarh, D.: Woodfisher: efficient second-order approximation for neural network compression. Adv. Neural. Inf. Process. Syst. **33**, 18098–18109 (2020)
19. Boyd, S.P., Vandenberghe, L.: Convex Optimization. Cambridge University Press, Cambridge (2004)
20. Bazaraa, M.S., Sherali, H.D., Shetty, C.M.: Nonlinear Programming: Theory and Algorithms. Wiley, Amsterdam (2013)
21. Jia, J., Zhang, Y., Zhang, Y., Liu, J., Runwal, B., Diffenderfer, J., Kailkhura, B., Liu, S.: SOUL: unlocking the power of second-order optimization for LLM unlearning. In: Al-Onaizan, Y., Bansal, M., Chen, Y.N. (eds.) Proceedings of the 2024 Conference on Empirical Methods in Natural Language Processing, pp. 4276–4292. Association for Computational Linguistics, Miami, Florida, USA (2024). https://doi.org/10.18653/v1/2024.emnlp-main.245
22. Reisizadeh, H., Jia, J., Bu, Z., Vinzamuri, B., Ramakrishna, A., Chang, K.W., Cevher, V., Liu, S., Hong, M.: Blur: a bi-level optimization approach for llm unlearning. arXiv preprint arXiv:2506.08164 (2025)

Robust Optimization in Machine Unlearning

3

Chongyu Fan

Abstract

This chapter examines the robustness challenges in machine unlearning, focusing on two major vulnerabilities: *relearning attacks*, where forgotten knowledge rapidly resurfaces under targeted fine-tuning, and *irrelevant fine-tuning*, where even unrelated training tasks can inadvertently undo unlearning effects. It first introduces unlearning robustness against worst-case relearning attacks as a *min–max optimization* problem, leveraging techniques such as sharpness-aware minimization and smoothness optimization to flatten the forget loss landscape and improve resistance to adversarial parameter updates. This chapter then addresses the broader challenge of arbitrary downstream fine-tuning by extending *invariant risk minimization* to the unlearning setting, enforcing environment-agnostic forgetting so that unlearning effects persist across diverse fine-tuning tasks. These techniques achieve substantially higher robustness against both direct relearning attacks and arbitrary downstream fine-tuning, ensuring that forgotten knowledge remains reliably erased under a wide range of post-unlearning scenarios.

3.1 Robustness Issues of Machine Unlearning

Despite rapid progress in machine unlearning, recent studies have uncovered fundamental robustness vulnerabilities in existing methods for generative models, including both large language models (LLMs) [1] and diffusion models (DMs) [2]. In this chapter, we focus primarily on LLM unlearning as the main use case, while noting that many of the challenges and insights generalize across modalities.

C. Fan (✉)
Michigan State University, East Lansing, MI, USA
e-mail: fanchon2@msu.edu

S. Liu et al. (eds.), *Machine Unlearning for Governance of Foundation Models*, Synthesis Lectures on Computer Vision, https://doi.org/10.1007/978-3-032-17282-2_3

Evidence from multiple studies highlights the robustness issues of machine unlearning. For example, it is shown in [3] that widely adopted evaluation protocols often fail to detect residual knowledge lingering in supposedly unlearned models. Complementing this finding, adversarial analyses in [4] demonstrated that fine-tuning on as few as ten unrelated examples can recover most of the previously erased information. Reinforcing these concerns, relearning attacks [5, 6] revealed that even a small amount of unlearning-related data can quickly restore harmful knowledge through lightweight fine-tuning. These vulnerabilities reflect a broader fragility: small weight modifications, such as parameter-efficient fine-tuning, can undo prior alignment or unlearning efforts [7–9]. Additional evidence includes erased concepts reemerging through neuron repurposing [10] and quantization-based attacks recovering forgotten knowledge [11]. Therefore, existing unlearning methods remain far from robust and highlight the need for more principled approaches.

3.1.1 Vulnerability to Relearning Attacks

Relearning attacks [5, 6] aim to reverse the effects of unlearning by fine-tuning the unlearned model on a small subset of the original forget set, thereby recovering the removed knowledge with minimal effort. Formally, a relearning attack can be modeled as

$$\min_{\boldsymbol{\delta}} \ell_{\text{relearn}}(\boldsymbol{\theta}_{\text{u}} + \boldsymbol{\delta} | \mathcal{D}_{\text{relearn}}), \tag{3.1}$$

where $\boldsymbol{\theta}_{\text{u}}$ denotes the unlearned model, while $\boldsymbol{\delta}$ represents the optimization variable corresponding to the model update introduced during the relearning process. The relearn set $\mathcal{D}_{\text{relearn}}$ may consist of data samples drawn from the same distribution as the forget set $\mathcal{D}_{\text{f}}$, such as the retain set or a much smaller subset of the original forget set. The relearning objective ℓ_{relearn} is designed to counteract the original forgetting objective, for example, by using the negative forget loss or the standard fine-tuning loss computed on $\mathcal{D}_{\text{relearn}}$.

As a motivating example, Fig. 3.1 illustrates the performance of the negative preference optimization (NPO)-based unlearning approach [12] in mitigating the malicious use of the LLM Zephyr-7B-beta on the WMDP (Weapons of Mass Destruction Proxy)-Bio dataset [13]. In this setting, lower model accuracy on the WMDP evaluation set indicates stronger unlearning. Accordingly, the unlearning effectiveness (UE) is defined as *1—Accuracy on the WMDP evaluation set*, so that higher UE values directly correspond to more effective unlearning. As shown in Fig. 3.1a, the NPO-unlearned model (termed 'Unlearn') achieves a much higher UE compared to the original model prior to unlearning (referred to as 'Origin'). And it effectively mitigates hazardous knowledge, as evidenced by the generation example in Fig. 3.1b. However, when a relearning attack is introduced by fine-tuning the unlearned model for a single epoch using only a few forget samples—specifically, 20, 40, or 60 samples (referred to as 'Relearn20', 'Relearn40', and 'Relearn60', respectively)—the unlearned model can be reverted, resuming the generation of harmful responses similar to 'Origin'.

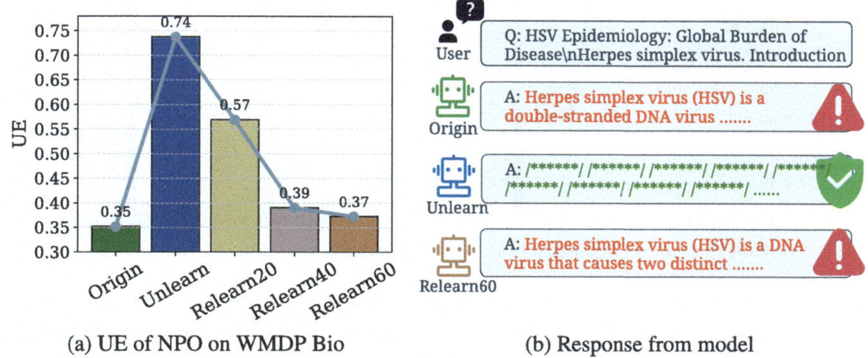

Fig. 3.1 Unlearning example on the WMDP Bio dataset before and after relearning attacks: **a** unlearning effectiveness (UE) of Zephyr-7B-beta ('Origin'), the NPO-unlearned model w/o relearning ('Unlearn'), and the relearned model from the unlearned one ('RelearnN'), where N represents the number of forget data samples used for relearning. **b** Response example of different models in (**a**)

3.1.2 Vulnerability to Irrelevant Fine-Tuning

While relearning attacks (3.1) primarily examine the worst-case robustness of unlearning (using a relearn set drawn from the same distribution as the original forget set), recent studies [3–5, 14, 15] reveal a broader and more concerning vulnerability. Specifically, knowledge erased through unlearning can be rapidly recovered through additional model fine-tuning, even when the fine-tuning data are entirely *unrelated* to the original forget set. This phenomenon, known as the *vulnerability of unlearned models to irrelevant fine-tuning*, highlights a fundamental weakness in existing unlearning methods: the erased knowledge remains latent and can resurface under seemingly innocuous model updates.

As a concrete illustration, Fig. 3.2 compares the unlearning effectiveness and downstream fine-tuning performance of models produced by the unlearning methods NPO [12] and RMU [13] across varying numbers of fine-tuning epochs on the downstream datasets GSM8K and AGNews. In this setup, the LLM Zephyr-7b-beta is first unlearned on the WMDP dataset [13] to mitigate harmful content generation, with performance measured as *1 minus accuracy on the WMDP-Bio evaluation set*, referred to as *forget quality*, so that higher values indicate stronger unlearning. For ease of comparison, the performance of a model fine-tuned directly from the original pre-trained checkpoint (i.e., 'Original') is also reported. The results reveal two consistent trends. First, although both NPO and RMU exhibit high forget quality prior to fine-tuning (epoch 0), their unlearning effectiveness steadily degrades as fine-tuning proceeds, even though the downstream tasks (GSM8K and AGNews) are unrelated to the original unlearning objective (WMDP). Second, with increasing epochs, the accuracy of all models, regardless of their unlearning method, progressively converges toward the performance of fine-tuning the 'Original' model. Notably, while RMU initially

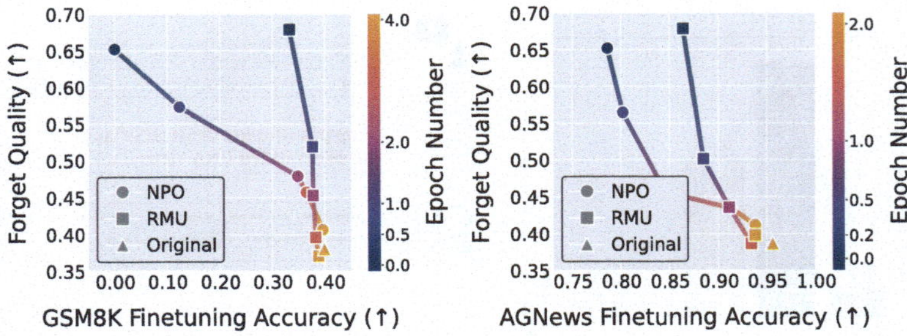

Fig. 3.2 Fine-tuning undermines existing unlearning methods. Performance evaluation of NPO and RMU applied to the LLM Zephyr-7b-beta for removing harmful knowledge generation on the WMDP dataset [13]. Unlearning effectiveness is measured by accuracy on the WMDP-Bio evaluation set, where lower accuracy indicates stronger forgetting. Accordingly, 'forget quality' is defined as 1 − evaluation accuracy, with higher values representing more effective unlearning. *(Left: GSM8K fine-tuning)* Forget quality and fine-tuning accuracy are shown for NPO- and RMU-unlearned models, together with the original (non-unlearned) model, under downstream fine-tuning on the GSM8K dataset. Fine-tuning epochs are indicated by color, ranging from 0 (no fine-tuning) to the number required to match the fully fine-tuned performance of the original model. Dots of the same position and color across different trajectories correspond to the same fine-tuning epoch. *(Right: AGNews fine-tuning)* Similar trajectories are presented for downstream fine-tuning on the AGNews dataset

achieves higher forget quality than NPO, it is less robust to downstream fine-tuning, as evidenced by its faster rate of *un*unlearning across epochs.

3.2 Min–Max (Robust) Optimization for Machine Unlearning

The previous examples highlight the robustness issues of current unlearning approaches and motivate a re-examination of the underlying optimization principles. In particular, there is a need to design robust optimization frameworks that explicitly account for adversarial relearning scenarios and improve the resilience of unlearned models.

3.2.1 Robust Unlearning via Sharpness-Aware Minimization (SAM)

Inspired by the relearning attack formulation in (3.1), improving unlearning robustness can be naturally viewed as an *adversary-defense game*. This setting parallels adversarial training [16], but with a crucial difference: instead of crafting adversarial *inputs*, the adversary here perturbs the *model weights* to counteract the forget objective.

Specifically, if the relearning objective is defined as the opposite of the forget objective $\ell_{\mathrm{relearn}} = -\ell_{\mathrm{f}}$, then incorporating the relearning adversary into the unlearning process yields the following min–max robust optimization problem:

$$\min_{\boldsymbol{\theta}} \underbrace{\max_{\|\boldsymbol{\delta}\|_p \leq \rho} \ell_{\mathrm{f}}(\boldsymbol{\theta}+\boldsymbol{\delta}|\mathcal{D}_{\mathrm{f}})}_{:=\ell_{\mathrm{f}}^{\mathrm{SAM}}(\boldsymbol{\theta})} +\lambda \ell_{\mathrm{r}}(\boldsymbol{\theta}|\mathcal{D}_{\mathrm{r}}), \tag{3.2}$$

where $\|\cdot\|_p$ denotes the ℓ_p norm ($p \geq 1$), with $p = 2$ as the default setting. And similar to adversarial training [16], the ability of the adversary (i.e., the 'follower') to disrupt the unlearned model (i.e., the 'leader') is constrained by $\|\boldsymbol{\delta}\|_p \leq \rho$, where $\rho > 0$ is a small constant.

Interestingly, the min-max optimization problem (3.2) aligns closely with the principles of sharpness-aware minimization (SAM) [17], with the SAM loss $\ell_{\mathrm{f}}^{\mathrm{SAM}}(\boldsymbol{\theta})$ applied to forget objective. Conventionally, SAM aims to enhance model generalization by explicitly considering the sensitivity of the loss landscape to weight perturbations, thereby encouraging *smoothness* optimization. Yet, SAM also resonates with the robust unlearning in (3.2). As shown by (3.2), SAM promotes the *flatness* of the forget loss landscape since it seeks a minimum that maintains a uniformly low loss across the neighborhood of the model. Therefore, SAM facilitates smoothness optimization in machine unlearning.

Based on the SAM algorithm [17], the *inner maximization* in (3.2) can be solved in *closed form* using linear approximation:

$$\begin{aligned} \boldsymbol{\delta}^*(\boldsymbol{\theta}) &:= \arg\max_{\|\boldsymbol{\delta}\|_2 \leq \rho} \ell_{\mathrm{f}}(\boldsymbol{\theta}+\boldsymbol{\delta}) \overset{(a)}{\approx} \arg\max_{\|\boldsymbol{\delta}\|_2 \leq \rho} \ell_{\mathrm{f}}(\boldsymbol{\theta}) + \boldsymbol{\delta}^\top \nabla_{\boldsymbol{\theta}} \ell_{\mathrm{f}}(\boldsymbol{\theta}) \\ &= \arg\max_{\|\boldsymbol{\delta}\|_2 \leq \rho} \boldsymbol{\delta}^\top \nabla_{\boldsymbol{\theta}} \ell_{\mathrm{f}}(\boldsymbol{\theta}) \overset{(b)}{=} \rho \frac{\nabla_{\boldsymbol{\theta}} \ell_{\mathrm{f}}(\boldsymbol{\theta})}{\|\nabla_{\boldsymbol{\theta}} \ell_{\mathrm{f}}(\boldsymbol{\theta})\|_2}, \end{aligned} \tag{3.3}$$

where for simplicity, $\mathcal{D}_{\mathrm{f}}$ is omitted in the notation of the forget loss, $^\top$ denotes the transpose, and $\nabla_{\boldsymbol{\theta}}$ represents the first-order derivative with respect to $\boldsymbol{\theta}$. In (3.3), approximation (a) follows from the first-order Taylor expansion of $\ell_{\mathrm{f}}(\boldsymbol{\theta}+\boldsymbol{\delta})$ with respect to $\boldsymbol{\delta}$ around $\mathbf{0}$. Equality (b) arises from the fact that the maximum cosine similarity is achieved when $\boldsymbol{\delta}$ is aligned with the gradient direction $\nabla_{\boldsymbol{\theta}} \ell_{\mathrm{f}}(\boldsymbol{\theta})$ and attains the largest permissible magnitude ρ.

By substituting the weight perturbation $\boldsymbol{\delta}^*(\boldsymbol{\theta})$ (3.3) into the SAM-based forget loss, the min–max optimization problem (3.2) is transformed into the following minimization problem:

$$\min_{\boldsymbol{\theta}} \ell_{\mathrm{f}}^{\mathrm{SAM}}(\boldsymbol{\theta}) = \min_{\boldsymbol{\theta}} \ell_{\mathrm{f}} \left(\boldsymbol{\theta} + \rho \frac{\nabla_{\boldsymbol{\theta}} \ell_{\mathrm{f}}(\boldsymbol{\theta})}{\|\nabla_{\boldsymbol{\theta}} \ell_{\mathrm{f}}(\boldsymbol{\theta})\|_2} \right). \tag{3.4}$$

To solve (3.4), note that the gradient of $\ell_{\mathrm{f}}^{\mathrm{SAM}}$ implicitly depends on the second-order derivative of $\ell_{\mathrm{f}}(\boldsymbol{\theta})$, i.e., the Hessian of ℓ_{f}. This establishes a connection between (3.4) and the curvature of the forget loss landscape with respect to $\boldsymbol{\theta}$. This insight can be further clarified by approximating ℓ_{f} in (3.4) using its first-order Taylor expansion at $\rho = 0$ [18]:

$$\ell_{\mathrm{f}}^{\mathrm{SAM}}(\boldsymbol{\theta}) = \ell_{\mathrm{f}} \left(\boldsymbol{\theta} + \rho \frac{\nabla_{\boldsymbol{\theta}} \ell_{\mathrm{f}}(\boldsymbol{\theta})}{\|\nabla_{\boldsymbol{\theta}} \ell_{\mathrm{f}}(\boldsymbol{\theta})\|_2} \right) \approx \ell_{\mathrm{f}}(\boldsymbol{\theta}) + \rho \|\nabla_{\boldsymbol{\theta}} \ell_{\mathrm{f}}(\boldsymbol{\theta})\|_2. \tag{3.5}$$

The minimization of the above objective function with a first-order optimizer then involves the Hessian of ℓ_f, which arises through the derivative of $\|\nabla_{\boldsymbol{\theta}} \ell_f(\boldsymbol{\theta})\|_2$:

$$\frac{d\|\nabla_{\boldsymbol{\theta}} \ell_f(\boldsymbol{\theta})\|_2}{d\boldsymbol{\theta}} = \frac{d(\|\nabla_{\boldsymbol{\theta}} \ell_f(\boldsymbol{\theta})\|_2^2)^{1/2}}{d\boldsymbol{\theta}} = \frac{1}{2}(\|\nabla_{\boldsymbol{\theta}} \ell_f(\boldsymbol{\theta})\|_2^2)^{-1/2}(2\mathbf{H}\nabla_{\boldsymbol{\theta}} \ell_f(\boldsymbol{\theta})) = \mathbf{H}\mathbf{v}, \quad (3.6)$$

where $\mathbf{H} = \nabla_{\boldsymbol{\theta},\boldsymbol{\theta}} \ell_f(\boldsymbol{\theta})$ is the Hessian matrix of the forget loss ℓ_f with respect to $\boldsymbol{\theta}$, and $\mathbf{v} = \frac{\nabla_{\boldsymbol{\theta}} \ell_f(\boldsymbol{\theta})}{\|\nabla_{\boldsymbol{\theta}} \ell_f(\boldsymbol{\theta})\|_2}$ denotes the normalized gradient direction. It is assumed here that $\nabla_{\boldsymbol{\theta}} \ell_f(\boldsymbol{\theta})$ is nonzero.

It is noteworthy that the quantity $\mathbf{Hv}$ in (3.6) is also employed in the *curvature regularization* [19] to enhance the adversarial robustness of discriminative models against input-level attacks. In that context, however, the Hessian $\mathbf{H}$ and gradient $\mathbf{v}$ are defined with respect to the model input, rather than the model parameters as in (3.6). Using a finite-difference approximation, $\mathbf{Hv}$ can be expressed as

$$\mathbf{H}\mathbf{v} \approx \frac{\nabla_{\boldsymbol{\theta}} \ell_f(\boldsymbol{\theta} + \mu \mathbf{v}) - \nabla_{\boldsymbol{\theta}} \ell_f(\boldsymbol{\theta})}{\mu}, \quad (3.7)$$

where $\mu > 0$ represents the discretization step, controlling the scale at which gradient variations are constrained to remain small. Based on (3.6) and (3.7), solving the problem (3.4) drives convergence toward a stationary point, which consequently reduces the curvature, i.e., $\|\mathbf{Hv}\|_2 \to 0$. This suggests that *reducing curvature*, and thereby *increasing the smoothness* of the forget loss surface, is beneficial to the resilience of LLM unlearning against the worst-case relearning attacks.

3.2.2 Broader Smoothness Optimization to Improve Unlearning Robustness

As analyzed above, the SAM-like optimization in (3.2) and (3.4) highlights the role of smoothness optimization in achieving robust unlearning. Building on this insight, the analysis can be extended to a broader range of smoothness optimization techniques, including randomized smoothing (RS), gradient penalty (GP), curvature regularization (CR), and weight averaging (WA).

First, RS transforms a non-smooth objective function into a smooth one by convolving it with a (smooth) Gaussian distribution function [20]. The underlying rationale is that the convolution of two functions produces a new function that is at least as smooth as the smoothest of the original functions. Let $\boldsymbol{\delta}$ represent a random perturbation vector sampled from the Gaussian distribution $\mathcal{N}(0, \sigma^2)$, where the mean is 0 and the variance is σ^2 for each independent and identically distributed (i.i.d.) variable component. Recall that SAM targets the worst-case (maximum) perturbation $\boldsymbol{\delta}$ in (3.3). In contrast, RS introduces a random perturbation, smoothing the optimization objective by averaging over random perturbations. This modifies the forget loss $\ell_f^{\mathrm{SAM}}(\boldsymbol{\theta})$ in (3.2) to:

$$\ell_{\mathrm{f}}^{\mathrm{RS}}(\boldsymbol{\theta}) = \mathbb{E}_{\boldsymbol{\delta} \sim \mathcal{N}(0,\sigma^2)}[\ell_{\mathrm{f}}(\boldsymbol{\theta} + \boldsymbol{\delta})]. \tag{3.8}$$

It is worth noting that in the context of adversarial robustness against input-level adversarial attacks, RS has been widely employed to smooth the model's *input*, offering (certified) robustness against such attacks [21].

Second, GP naturally originates from SAM, as demonstrated in (3.5). When incorporated as a regularization term in SAM's objective, this variant is referred to as penalty SAM [18]:

$$\ell_{\mathrm{f}}^{\mathrm{GP}}(\boldsymbol{\theta}) = \ell_{\mathrm{f}}(\boldsymbol{\theta}) + \rho \|\nabla_{\boldsymbol{\theta}} \ell_{\mathrm{f}}(\boldsymbol{\theta})\|_2. \tag{3.9}$$

In the context of adversarial robustness, applying a gradient norm penalty has also been shown to be beneficial for defending against adversarial attacks [22]. However, in this scenario, the gradient is computed with respect to the model's *input* rather than its weights.

Third, CR also naturally emerges as a variant of SAM, given by (3.6) and (3.7). Unlike SAM, which implicitly reduces curvature through its optimization process, CR explicitly penalizes the curvature in the forget loss. This direct penalization on (3.7) leads to the CR-based variant of SAM:

$$\ell_{\mathrm{f}}^{\mathrm{CR}}(\boldsymbol{\theta}) = \ell_{\mathrm{f}}(\boldsymbol{\theta}) + \gamma \|\nabla_{\boldsymbol{\theta}} \ell_{\mathrm{f}}(\boldsymbol{\theta} + \mu \mathbf{v}) - \nabla_{\boldsymbol{\theta}} \ell_{\mathrm{f}}(\boldsymbol{\theta})\|_2, \tag{3.10}$$

where $\gamma > 0$ is a regularization parameter, and recall that $\mathbf{v} = \frac{\nabla_{\boldsymbol{\theta}} \ell_{\mathrm{f}}(\boldsymbol{\theta})}{\|\nabla_{\boldsymbol{\theta}} \ell_{\mathrm{f}}(\boldsymbol{\theta})\|_2}$. Similar to RS and GP, curvature regularization, when applied to the loss surface with respect to *inputs*, is also a known technique for enhancing adversarial robustness [19].

Fourth, WA is a technique designed to enforce weight smoothness by averaging multiple model checkpoints collected along the training trajectory [23]. This is given by

$$\boldsymbol{\theta}_{\mathrm{WA},t} = \frac{\boldsymbol{\theta}_{\mathrm{WA},t} \cdot n + \boldsymbol{\theta}_t}{n+1}, \quad \boldsymbol{\theta}_t = \boldsymbol{\theta}_{t-1} + \Delta \boldsymbol{\theta}_t, \tag{3.11}$$

where t represents the training epoch index, and $\boldsymbol{\theta}_{\mathrm{WA},t}$ denotes the model parameters after applying WA at epoch t. The parameter n specifies the number of past checkpoints to be averaged. Additionally, $\boldsymbol{\theta}_t$ refers to the optimization variable for solving the SAM-based unlearning problem (3.2) at epoch t, while $\Delta \boldsymbol{\theta}_t$ represents the corresponding descent step used to update $\boldsymbol{\theta}$. As shown in [24], WA also enhances adversarial robustness against adversarial examples in discriminative models.

The role of the previously discussed smoothness optimization techniques (SAM, RS, GP, CR, and WA) in enhancing unlearning robustness can be examined from the perspective of the *loss landscape*.

The loss landscape characterizes the geometric surface of a loss function with respect to variations in model parameters [25]. For visualization, the loss sensitivity can be represented by a parametric model defined as $f(x, y) = \ell(\boldsymbol{\theta} + x \cdot \mathbf{r}_1 + y \cdot \mathbf{r}_2)$, where ℓ denotes the prediction loss, $\mathbf{r}_1$ and $\mathbf{r}_2$ are Gaussian random directions, and x and y are scalar coefficients controlling perturbation strength. The three-dimensional loss landscape is then obtained by

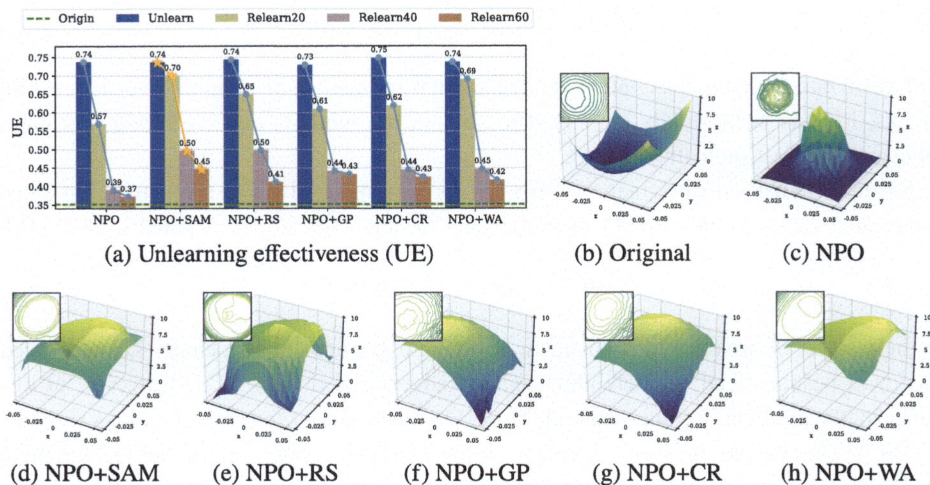

(a) Unlearning effectiveness (UE) (b) Original (c) NPO

(d) NPO+SAM (e) NPO+RS (f) NPO+GP (g) NPO+CR (h) NPO+WA

Fig. 3.3 Improved unlearning robustness by smoothness optimization-integrated NPO (including NPO+SAM, RS, GP, CR, or WA) compared to vanilla NPO on WMDP following the setup in Fig. 3.1. **a** Unlearning effectiveness of different models ('Unlearn' and 'RelearnN' that undergoes relearning with N examples) obtained from various NPO variants. **b** and **c** The prediction loss landscape of the original model and NPO-unlearned model on the forget set, where higher values around $x = y = 0$ indicate more effective unlearning. The 3D loss landscape is defined as $z = \ell(\boldsymbol{\theta} + x \cdot \mathbf{r}_1 + y \cdot \mathbf{r}_2)$, with $\boldsymbol{\theta}$ representing the unlearned model. **d–h** Similar loss landscape visualizations to (**b**), but with the unlearned model obtained using smooth variants of NPO

plotting the variation of $f(x, y)$ with respect to x and y. Smoothness is reflected in a relatively flat surface in the vicinity of the current model parameters.

Following the experimental setup in Figs. 3.1 and 3.3a presents the unlearning effectiveness (UE) of different models ('Unlearn' and 'RelearnN') using various unlearning methods. These include NPO and its smoothness-augmented variants, denoted as NPO + X, where X corresponds to techniques such as SAM, RS, GP, CR, or WA. When subjected to relearning attacks ('RelearnN'), the smooth variants consistently achieve higher UE compared to vanilla NPO. Among them, NPO + SAM provides the strongest robustness; for instance, under Relearn20, it attains a UE of 0.70, compared to 0.57 for the original NPO. Importantly, in the absence of relearning attacks ('Unlearn'), the incorporation of smoothness optimization does not compromise performance in the non-adversarial setting, as shown by the stable UE value of approximately 0.74. Figure 3.3b and c illustrate the prediction loss landscape of the original and NPO-unlearned models on the forget set $\mathcal{D}_\mathrm{f}$. The z-axis denotes the prediction loss, with higher values indicating more effective unlearning (i.e., poorer predictive performance on $\mathcal{D}_\mathrm{f}$). At the point $x = y = 0$, the NPO-unlearned model attains a higher prediction loss than the original model, signifying successful unlearning. However, the surrounding loss landscape remains sharp, revealing that the unlearning process offers limited robustness. In contrast, Fig. 3.3d–h display the landscapes of NPO models enhanced with SAM, RS, GP, CR, and WA. These variants produce significantly smoother landscapes compared to Fig. 3.3c, indicating improved stability of the unlearned solutions.

3.3 Invariance Optimization for Robust Unlearning

The robust optimization framework in the previous section strengthens unlearning against *worst-case* attacks by explicitly countering weight perturbations that maximize the forget loss. While effective against adversarial relearning, this approach still assumes knowledge of the attacker's objective and focuses primarily on local robustness around the unlearned model parameters. In practice, however, *irrelevant fine-tuning*—even on datasets entirely unrelated to the forget set—can rapidly restore forgotten knowledge, as shown in Fig. 3.2. This vulnerability reveals a key limitation: Robustness alone is insufficient because it protects only against anticipated perturbations, whereas post-unlearning fine-tuning introduces unstructured and potentially unknown parameter updates.

To address the above broader challenge, we seek a model whose *unlearning effect remains invariant* to additional fine-tuning, regardless of the fine-tuning dataset or objective. In other words, beyond minimizing the worst-case forget loss, we aim to learn representations whose unlearning behavior persists under diverse downstream training environments.

This naturally motivates integrating *invariant risk minimization (IRM)* [26, 27] into machine unlearning. Originally developed for domain generalization, IRM enforces that a single predictor remains optimal across multiple training environments. By interpreting each fine-tuning task as a distinct environment, IRM offers a principled way to achieve: removing harmful knowledge via the forget objective and ensuring the unlearning effect persists across arbitrary post-unlearning fine-tuning tasks via invariance regularization. The following sections formalize this connection and introduce the method *Invariant LLM Unlearning (ILU)* [9], which integrates IRM principles into unlearning optimization to defend against both adversarial relearning and unstructured fine-tuning.

3.3.1 Preliminaries Invariant Risk Minimization (IRM)

IRM is designed to identify an *invariant* (i.e., training environment-agnostic) model, aiming to learn universal representations that yield optimal predictions across diverse environments [26–28]. By interpreting fine-tuning tasks as additional unlearning environments, integrating IRM with unlearning is expected to improve the invariance of the unlearned model and thereby enhance its robustness against fine-tuning. This motivates framing LLM unlearning within the IRM paradigm.

Following the setup in [26], let $\mathcal{D}_i$ denote the dataset associated with training environment $i \in [N]$, where $[N] := \{1, 2, \ldots, N\}$ and N is the number of environments. IRM learns a representation model $\boldsymbol{\phi}$ such that there exists an invariant predictor $\mathbf{w}$ that is simultaneously optimal across all environments. This yields the invariant (environment-agnostic) model $\boldsymbol{\theta} := \mathbf{w} \circ \boldsymbol{\phi}$, where $\circ$ denotes model composition. Such a formulation ensures that $\boldsymbol{\phi}$ not only minimizes the empirical training objective but also supports optimal predictions on each environment $\mathcal{D}_i$. Formally, the IRM objective is given by

$$\begin{aligned} &\underset{\boldsymbol{\phi}}{\text{minimize}}\ \ell_{\text{ERM}}(\mathbf{w}^*(\boldsymbol{\phi}) \circ \boldsymbol{\phi}; \cup_i \mathcal{D}_i) \\ &\text{subject to } \mathbf{w}^*(\boldsymbol{\phi}) \in \arg\min_{\mathbf{w}} \ell_i(\mathbf{w} \circ \boldsymbol{\phi}; \mathcal{D}_i), \quad \forall i \in [N], \end{aligned} \tag{3.12}$$

where the upper-level objective $\ell_{\text{ERM}}(\cdot; \cup_i \mathcal{D}_i)$ represents empirical risk minimization (ERM) across all data to optimize $\boldsymbol{\phi}$ (e.g., $\ell_{\text{ERM}}(\cdot) = \sum_i \ell_i(\cdot; \mathcal{D}_i)$), ℓ_i denotes the individual training loss over $\mathcal{D}_i$, and $\mathbf{w}^*(\boldsymbol{\phi})$ denotes an invariant predictor (built upon $\boldsymbol{\phi}$) that is optimal to any specific training environment. The solution $\mathbf{w}^*(\boldsymbol{\phi})$ is explicitly expressed as a function of $\boldsymbol{\phi}$.

However, solving the original IRM problem (3.12) poses optimization challenges due to the need to compute the gradient $\frac{d\mathbf{w}^*(\boldsymbol{\phi})}{d\boldsymbol{\phi}}$. To circumvent that, problem (3.12) is typically relaxed to a single-level, regularized problem, referred to as IRMv1:

$$\underset{\boldsymbol{\theta}}{\text{minimize}}\ \underbrace{\ell_{\text{ERM}}(\boldsymbol{\theta})}_{\text{ERM}} + \underbrace{\lambda \sum_{i=1}^{N} \|\nabla_w \ell_i(w \circ \boldsymbol{\theta}; \mathcal{D}_i)\,|_{w=1}\,\|_2^2}_{\text{Invariance regularization}} \tag{3.13}$$

where the original predictor's parameters $\mathbf{w}$ are absorbed into the full model parameters $\boldsymbol{\theta}$, $\lambda > 0$ serves as a regularization parameter, $\|\cdot\|_2$ is the ℓ_2 norm, and $\nabla_w \ell_i$ represents the gradient of ℓ_i with respect to the (virtual) scalar predictor w, evaluated at $w = 1$. The invariance regularization in (3.13) enforces the necessary optimality condition for each lower-level problem in (3.12) by penalizing non-stationarity.

3.3.2 From IRM to Invariant LLM Unlearning (ILU)

The IRMv1 approach (3.13), while a relaxation of (3.12), facilitates the application of invariance regularization to LLM unlearning. First, composing $\boldsymbol{\theta}$ with a constant scalar predictor $w = 1$ ensures adaptability to a wide range of machine learning models, including LLMs. Second, if $\boldsymbol{\theta}$ represents the unlearned model obtained by replacing ERM with the unlearning objective ℓ_{u} in unlearning and $\mathcal{D}_i$ is a fine-tuning dataset, the invariance regularization in (3.13) enforces robustness of $\boldsymbol{\theta}$ against fine-tuning. ILU (invariant LLM unlearning) is thus introduced as an extension of IRMv1 (3.13), formulated as follows [9]:

$$\underset{\boldsymbol{\theta}}{\text{minimize}}\ \ell_{\text{u}}(\boldsymbol{\theta}) + \lambda \sum_{i=1}^{N} \|\nabla_{w|w=1} \ell_i(w \circ \boldsymbol{\theta}; \mathcal{D}_i)\|_2^2, \tag{3.14}$$

where ℓ_u is the unlearning objective, and $\mathcal{D}_i$ is a fine-tuning dataset that can be unrelated to the unlearning task, e.g., GSM8K or AGNews in Fig. 3.2.

A key question in (3.14) is whether the introduction of *multiple* fine-tuning sets $\{\mathcal{D}_i\}$ (i.e., $N > 1$) is necessary, as it assumes greater access to additional data and knowledge of potential fine-tuning tasks. An ideal ILU framework should minimize reliance on fine-tuning datasets while demonstrating generalization to *unseen* ones at test time. To explore this, only a single fine-tuning dataset ($\mathcal{D}$) is considered in (3.14):

$$\underset{\boldsymbol{\theta}}{\text{minimize}} \ \ell_u(\boldsymbol{\theta}) + \lambda \|\nabla_{w|w=1} \ell_i(w \circ \boldsymbol{\theta}; \mathcal{D})\|_2^2. \tag{3.15}$$

Figure 3.4 compares *ILU($\mathcal{D}$)*, i.e., ILU with the single fine-tuning dataset $\mathcal{D}$ in (3.15), with its multi-dataset variant, *ILU(Multi)*, which incorporates GSM8K, AGNews, and WinoGrande. As we can see, ILU consistently outperforms RMU as the number of fine-tuning epochs increases, demonstrating improved robustness to post-unlearning updates. Among the variants, ILU(GSM8K), which uses only a single dataset GSM8K unrelated to both the forget set WMDP and the evaluation fine-tuning set AGNews, achieves the strongest and most stable robustness, even as the model approaches full fine-tuning accuracy. In contrast, ILU(WMDP), which employs the forget set itself for invariance regularization, performs notably worse. This degradation is expected: using the same dataset for both unlearning and invariance regularization introduces conflicting objectives—unlearning aims to degrade performance on the forget set, whereas invariance regularization may inadvertently preserve it to enforce stationarity—thereby weakening robustness. Furthermore, although ILU(Multi) incorporates multiple fine-tuning datasets, it does not outperform the simpler ILU(GSM8K). A plausible explanation is that incorporating multiple datasets introduces additional optimization complexity: the unlearning direction must simultaneously align with multiple fine-tuning directions, without contradicting any of them, which makes the optimization more restrictive.

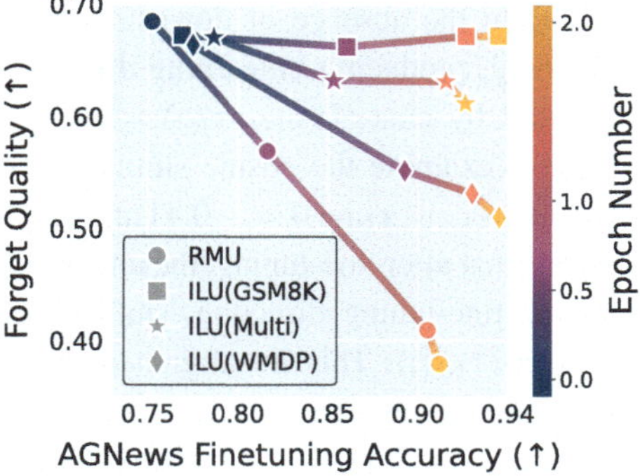

Fig. 3.4 A single fine-tuning dataset in ILU (3.15) suffices for enhancing unlearning robustness. Forget quality and fine-tuning accuracy of different unlearned models are presented against AGNews fine-tuning, following a similar setup and presentation format to Fig. 3.2

3.3.3 Understanding ILU via Task Vectors

To understand the effectiveness of ILU in building resilience against fine-tuning, the relationship between the 'unlearning direction' and the 'fine-tuning direction' is examined using task vector analysis [29].

A task vector defines a direction in the weight space for a specific task, where movement along this direction from the pre-trained model enhances performance on this task. Let $\boldsymbol{\theta}_{\mathrm{u}}$ and $\boldsymbol{\theta}_0$ denote the unlearned model (obtained via an unlearning approach) and the original pre-trained model, respectively. By the definition of task vector, the *unlearning direction* (i.e., unlearning task vector) is given by $\boldsymbol{\tau}_{\mathrm{u}} = \boldsymbol{\theta}_{\mathrm{u}} - \boldsymbol{\theta}_0$. Specifically, $\boldsymbol{\tau}_{\mathrm{ILU}}$ (or $\boldsymbol{\tau}_{\mathrm{NPO}}$) represents the unlearning direction resulting from ILU (or NPO), respectively. Similarly, the *fine-tuning direction* can be defined relative to $\boldsymbol{\theta}_0$ as $\boldsymbol{\tau}_{\mathrm{ft}} = \boldsymbol{\theta}_{\mathrm{ft}} - \boldsymbol{\theta}_0$, where $\boldsymbol{\theta}_{\mathrm{ft}}$ denotes the fine-tuned model obtained from $\boldsymbol{\theta}_0$. The fine-tuning direction lies in the *un*unlearning space, since fine-tuning alone does not achieve unlearning, particularly when applied to unrelated datasets [4]. Consequently, the unlearning direction is expected to oppose the fine-tuning direction, i.e.,

$$\cos(\angle(\boldsymbol{\tau}_{\mathrm{u}}, \boldsymbol{\tau}_{\mathrm{ft}})) = \frac{\boldsymbol{\tau}_{\mathrm{u}}^\top \boldsymbol{\tau}_{\mathrm{ft}}}{\|\boldsymbol{\tau}_{\mathrm{u}}\|_2 \, \|\boldsymbol{\tau}_{\mathrm{ft}}\|_2} < 0,$$

where $\angle$ and cos denote the angle between two vectors and its cosine, respectively.

The *post-unlearning fine-tuning direction* is given by $\boldsymbol{\tau}_{\mathrm{u}\to\mathrm{ft}} = \boldsymbol{\theta}_{\mathrm{u}}^{\mathrm{ft}} - \boldsymbol{\theta}_{\mathrm{u}}$, where $\boldsymbol{\theta}_{\mathrm{u}}^{\mathrm{ft}}$ is the fine-tuned model derived from $\boldsymbol{\theta}_{\mathrm{u}}$ through downstream fine-tuning. After unlearning, if $\boldsymbol{\tau}_{\mathrm{u}\to\mathrm{ft}}$ remains aligned with $\boldsymbol{\tau}_{\mathrm{u}}$, i.e., $\cos(\angle(\boldsymbol{\tau}_{\mathrm{u}\to\mathrm{ft}}, \boldsymbol{\tau}_{\mathrm{u}})) \geq 0$, the unlearning direction is preserved and the method is resilient to fine-tuning. Conversely, if $\boldsymbol{\tau}_{\mathrm{u}\to\mathrm{ft}}$ aligns more closely with $\boldsymbol{\tau}_{\mathrm{ft}}$, i.e., $\cos(\angle(\boldsymbol{\tau}_{\mathrm{u}\to\mathrm{ft}}, \boldsymbol{\tau}_{\mathrm{ft}})) \geq 0$, the model shifts toward the *un*unlearning space, indicating diminished robustness.

Figure 3.5 presents a 2D visualization of the task vector analysis, explaining the robustness advantage of ILU (implemented with the NPO-based unlearning objective and GSM8K-based invariance regularization) over the conventional NPO approach on WMDP. In the absence of downstream fine-tuning, both NPO and ILU are effective in unlearning, producing unlearning directions opposite to the fine-tuning direction. This is supported by $\cos(\angle(\boldsymbol{\tau}_{\mathrm{NPO}}, \boldsymbol{\tau}_{\mathrm{ft}})) = -0.92$ and $\cos(\angle(\boldsymbol{\tau}_{\mathrm{ILU}}, \boldsymbol{\tau}_{\mathrm{ft}})) = -0.64$. Focusing on NPO, we examine the cosine similarity between $\boldsymbol{\tau}_{\mathrm{NPO}\to\mathrm{ft}}$ and $\boldsymbol{\tau}_{\mathrm{NPO}}$ (or $\boldsymbol{\tau}_{\mathrm{ft}}$), we obtain $\cos(\angle(\boldsymbol{\tau}_{\mathrm{NPO}\to\mathrm{ft}}, \boldsymbol{\tau}_{\mathrm{NPO}})) = -0.41$ and $\cos(\angle(\boldsymbol{\tau}_{\mathrm{NPO}\to\mathrm{ft}}, \boldsymbol{\tau}_{\mathrm{ft}})) = 0.16$, i.e., ① in Fig. 3.5. This suggests that after fine-tuning, the unlearning task vector of NPO ($\boldsymbol{\tau}_{\mathrm{NPO}\to\mathrm{ft}}$) is more aligned with the fine-tuning direction, shifting towards the opposite to the original unlearning direction ($\boldsymbol{\tau}_{\mathrm{NPO}}$). This misalignment explains why fine-tuning significantly undermines NPO's unlearning effectiveness. In contrast, ILU yields $\cos(\angle(\boldsymbol{\tau}_{\mathrm{ILU}\to\mathrm{ft}}, \boldsymbol{\tau}_{\mathrm{ft}})) = 0.3554$ and $\cos(\angle(\boldsymbol{\tau}_{\mathrm{ILU}\to\mathrm{ft}}, \boldsymbol{\tau}_{\mathrm{ILU}})) = 0.09$, i.e., ② in Fig. 3.5. Comparing $\cos(\angle(\boldsymbol{\tau}_{\mathrm{ILU}\to\mathrm{ft}}, \boldsymbol{\tau}_{\mathrm{ILU}})) > 0$ with $\cos(\angle(\boldsymbol{\tau}_{\mathrm{NPO}\to\mathrm{ft}}, \boldsymbol{\tau}_{\mathrm{NPO}})) < 0$ shows that ILU maintains stronger alignment between the unlearning direction before and after fine-tuning. This alignment accounts for the substantially higher resilience of ILU to fine-tuning compared to NPO.

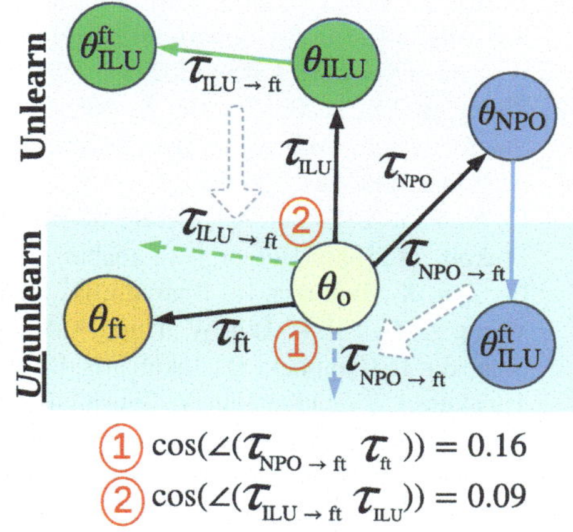

Fig. 3.5 Illustration of ILU's improved unlearning robustness compared to NPO through the relationships between unlearning and fine-tuning task vectors on the WMDP dataset

References

1. Deeb, A., Roger, F.: Do unlearning methods remove information from language model weights? arXiv preprint arXiv:2410.08827 (2024)
2. Zhang, Y., Jia, J., Chen, X., Chen, A., Zhang, Y., Liu, J., Ding, K., Liu, S.: To generate or not? safety-driven unlearned diffusion models are still easy to generate unsafe images... for now. In: European Conference on Computer Vision, pp. 385–403. Springer (2024)
3. Lynch, A., Guo, P., Ewart, A., Casper, S., Hadfield-Menell, D.: Eight methods to evaluate robust unlearning in llms. arXiv preprint arXiv:2402.16835 (2024)
4. Łucki, J., Wei, B., Huang, Y., Henderson, P., Tramèr, F., Rando, J.: An adversarial perspective on machine unlearning for AI safety. arXiv preprint arXiv:2409.18025 (2024)
5. Hu, S., Fu, Y., Wu, Z.S., Smith, V.: Jogging the memory of unlearned model through targeted relearning attack. arXiv preprint arXiv:2406.13356 (2024)
6. Fan, C., Jia, J., Zhang, Y., Ramakrishna, A., Hong, M., Liu, S.: Towards LLM unlearning resilient to relearning attacks: a sharpness-aware minimization perspective and beyond. In: Forty-Second International Conference on Machine Learning (2025)
7. Qi, X., Zeng, Y., Xie, T., Chen, P.Y., Jia, R., Mittal, P., Henderson, P.: Fine-tuning aligned language models compromises safety, even when users do not intend to! In: The Twelfth International Conference on Learning Representations (2024)
8. Jain, S., Kirk, R., Lubana, E.S., Dick, R.P., Tanaka, H., Rocktäschel, T., Grefenstette, E., Krueger, D.: Mechanistically analyzing the effects of fine-tuning on procedurally defined tasks. In: The Twelfth International Conference on Learning Representations (2024)
9. Wang, C., Zhang, Y., Jia, J., Ram, P., Wei, D., Yao, Y., Pal, S., Baracaldo, N., Liu, S.: Invariance makes LLM unlearning resilient even to unanticipated downstream fine-tuning. In: Forty-Second International Conference on Machine Learning (2025). https://openreview.net/forum?id=x2lm33kdrZ
10. Lo, M., Barez, F., Cohen, S.: Large language models relearn removed concepts. In: Findings of the Association for Computational Linguistics: ACL 2024, pp. 8306–8323. Association for Computational Linguistics (2024)

11. Zhang, Z., Wang, F., Li, X., Wu, Z., Tang, X., Liu, H., He, Q., Yin, W., Wang, S.: Does your llm truly unlearn? An embarrassingly simple approach to recover unlearned knowledge. arXiv preprint arXiv:2410.16454 (2024)
12. Zhang, R., Lin, L., Bai, Y., Mei, S.: Negative preference optimization: from catastrophic collapse to effective unlearning. In: First Conference on Language Modeling (2024). https://openreview.net/forum?id=MXLBXjQkmb
13. Li, N., Pan, A., Gopal, A., Yue, S., Berrios, D., Gatti, A., Li, J.D., Dombrowski, A.K., Goel, S., Mukobi, G., Helm-Burger, N., Lababidi, R., Justen, L., Liu, A.B., Chen, M., Barrass, I., Zhang, O., Zhu, X., Tamirisa, R., Bharathi, B., Herbert-Voss, A., Breuer, C.B., Zou, A., Mazeika, M., Wang, Z., Oswal, P., Lin, W., Hunt, A.A., Tienken-Harder, J., Shih, K.Y., Talley, K., Guan, J., Steneker, I., Campbell, D., Jokubaitis, B., Basart, S., Fitz, S., Kumaraguru, P., Karmakar, K.K., Tupakula, U., Varadharajan, V., Shoshitaishvili, Y., Ba, J., Esvelt, K.M., Wang, A., Hendrycks, D.: The WMDP benchmark: measuring and reducing malicious use with unlearning. In: Proceedings of the 41st International Conference on Machine Learning, Proceedings of Machine Learning Research, vol. 235, pp. 28525–28550. PMLR (2024)
14. Barez, F., Fu, T., Prabhu, A., Casper, S., Sanyal, A., Bibi, A., O'Gara, A., Kirk, R., Bucknall, B., Fist, T., et al.: Open problems in machine unlearning for AI safety. arXiv preprint arXiv:2501.04952 (2025)
15. Tamirisa, R., Bharathi, B., Phan, L., Zhou, A., Gatti, A., Suresh, T., Lin, M., Wang, J., Wang, R., Arel, R., Zou, A., Song, D., Li, B., Hendrycks, D., Mazeika, M.: Tamper-resistant safeguards for open-weight LLMs. In: The Thirteenth International Conference on Learning Representations (2025). https://openreview.net/forum?id=4FIjRodbW6
16. Madry, A., Makelov, A., Schmidt, L., Tsipras, D., Vladu, A.: Towards deep learning models resistant to adversarial attacks. In: International Conference on Learning Representations (2018). https://openreview.net/forum?id=rJzIBfZAb
17. Foret, P., Kleiner, A., Mobahi, H., Neyshabur, B.: Sharpness-aware minimization for efficiently improving generalization. In: International Conference on Learning Representations (2021)
18. Dauphin, Y., Agarwala, A., Mobahi, H.: Neglected hessian component explains mysteries in sharpness regularization. In: The Thirty-Eighth Annual Conference on Neural Information Processing Systems (2024)
19. Moosavi-Dezfooli, S.M., Fawzi, A., Uesato, J., Frossard, P.: Robustness via curvature regularization, and vice versa. In: Proceedings of the IEEE/CVF Conference on Computer Vision and Pattern Recognition, pp. 9078–9086 (2019)
20. Duchi, J.C., Bartlett, P.L., Wainwright, M.J.: Randomized smoothing for stochastic optimization. SIAM J. Optim. **22**(2), 674–701 (2012)
21. Cohen, J., Rosenfeld, E., Kolter, Z.: Certified adversarial robustness via randomized smoothing. In: International Conference on Machine Learning, pp. 1310–1320. PMLR (2019)
22. Finlay, C., Oberman, A.M.: Scaleable input gradient regularization for adversarial robustness. Mach. Learn. Appl. **3**, 100017 (2021)
23. Izmailov, P., Podoprikhin, D., Garipov, T., Vetrov, D., Wilson, A.G.: Averaging weights leads to wider optima and better generalization. arXiv preprint arXiv:1803.05407 (2018)
24. Chen, T., Zhang, Z., Liu, S., Chang, S., Wang, Z.: Robust overfitting may be mitigated by properly learned smoothening. In: International Conference on Learning Representations (2020)
25. Li, H., Xu, Z., Taylor, G., Studer, C., Goldstein, T.: Visualizing the loss landscape of neural nets. Adv. Neural Inform. Process. Syst. **31**, 11542 (2018)
26. Arjovsky, M., Bottou, L., Gulrajani, I., Lopez-Paz, D.: Invariant risk minimization. arXiv preprint arXiv:1907.02893 (2019)
27. Ahuja, K., Shanmugam, K., Varshney, K., Dhurandhar, A.: Invariant risk minimization games. In: International Conference on Machine Learning, pp. 145–155. PMLR (2020)

28. Zhang, Y., Sharma, P., Ram, P., Hong, M., Varshney, K.R., Liu, S.: What is missing in IRM training and evaluation? challenges and solutions. In: The Eleventh International Conference on Learning Representations (2023)
29. Ilharco, G., Ribeiro, M.T., Wortsman, M., Gururangan, S., Schmidt, L., Hajishirzi, H., Farhadi, A.: Editing models with task arithmetic. arXiv preprint arXiv:2212.04089 (2022)

Forget Data-Only Optimization for Unlearning 4

Yaxuan Wang

Abstract

Unlearning in Large Language Models (LLMs) is critical for ensuring ethical and responsible AI, particularly in mitigating privacy leaks, bias, safety concerns, and adapting to evolving regulations. Existing methods typically depend on retain data or a reference LLM, but they often struggle to balance unlearning effectiveness with preserving model utility. This difficulty arises because fine-tuning with explicit retain data or indirectly relying on a reference LLM's knowledge—tends to blur the boundary between forgotten and retained information, as similar queries frequently yield overlapping responses. To address this, this chapter proposes eliminating the reliance on retain data or a reference LLM for response calibration. Instead of directly applying gradient ascent on forget data, which commonly leads to instability and poor outcomes, this approach explicitly guides the LLM on what not to answer and, crucially, how to respond based on forget data. This chapter introduces **F**orget data only **L**oss **A**djustmen**T** (FLAT), a "flat" loss adjustment method that maximizes the f-divergence between template answers and forget answers solely with respect to forget data. The variational form of this f-divergence provides a principled way to adjust the loss by assigning different importance weights to template responses and to forgetting responses subject to unlearning. Empirical results show that FLAT achieves stronger unlearning performance than existing approaches while minimizing disruption to retained capabilities. This ensures high utility across a wide range of tasks, including copyrighted content unlearning on the Harry Potter dataset and MUSE benchmark, as well as entity unlearning on the TOFU dataset.

Y. Wang (✉)
University of California, Santa Cruz, Santa Cruz, CA, USA
e-mail: ywan1225@ucsc.edu

S. Liu et al. (eds.), *Machine Unlearning for Governance of Foundation Models*, Synthesis Lectures on Computer Vision, https://doi.org/10.1007/978-3-032-17282-2_4

4.1 Preliminaries of Forget Data-Only Loss Adjustment-Based Unlearning

The rapid adoption of LLMs in everyday applications has heightened concerns about their trustworthiness. Model outputs may expose sensitive, private, or even illegal content [1, 2], propagate societal biases [3, 4], or generate harmful instructions [5–7]. In particular, growing privacy concerns have prompted the introduction of regulations [8] requiring systems to support the removal of information embedded in training data upon user request. This has driven increasing attention to machine unlearning (MU) [9–13], a process designed to eliminate the influence of specific data points, categories, or higher-level concepts from trained models.

LLM unlearning [5, 13, 14] is a specialized branch of machine unlearning that aims to remove knowledge defined by a forget dataset, while preserving the model's ability to perform tasks unrelated to the unlearning target [13, 15]. Existing approaches can be broadly categorized into three types: input-based, data-based, and model-based methods. Input-based approaches [16, 17] rely on carefully designed instructions to guide the original LLM toward unlearning goals without modifying its parameters. Data-based methods [18] fine-tune the model on desirable responses constructed from prompts within the forget data distribution. Model-based strategies [5, 19] instead adjust the model's weights or architecture to achieve unlearning. Among these, the approach most closely related to FLAT [20] is fine-tuning the target LLM with a modified loss function. Such methods typically pursue two objectives: maximizing the loss on forget samples while minimizing (or preserving) the loss on retain samples.

Formulation. Let the forget dataset be D_f, the retain dataset be D_r, and the original LLM be θ_o. The goal of LLM unlearning is to fine-tune θ_o into an updated model θ that behaves as if it were trained without access to D_f. For a prompt-response pair (x, y), the fine-tuning loss on y is defined as: $\mathcal{L}(x, y; \theta) = \sum_{i=1}^{|y|} \ell(h_\theta(x, y_{<i}), y_i)$, where $\ell(\cdot)$ is the cross-entropy loss, and $h_\theta(x, y_{<i}) := \mathbb{P}(y_i | (x, y_{<i}); \theta)$ represents the probability assigned by model θ to token y_i, given the input prompt x and the previously generated tokens $y_{<i} := [y_1, ..., y_{i-1}]$.

Existing LLM Unlearning Paradigm. The predominant class of LLM unlearning methods fine-tunes the original model with respect to a tailored unlearning objective. While the specific designs differ across approaches, the underlying principle of loss adjustment in LLM unlearning can be generally formalized as follows:

$$L = L_{\mathbf{FG}} + L_{\mathbf{RT}} + L_{\mathbf{Custom}}. \tag{4.1}$$

The modified loss function comprises three main components:

- $L_{\mathbf{FG}}$ **(Forget Loss)**: This component encourages the model to forget undesired data or patterns by deliberately increasing the loss on the forget set. In doing so, the model's

performance on these specific examples is degraded, thereby reducing its reliance on them and minimizing their influence on future predictions.
- $L_{\mathbf{RT}}$ **(Retain Loss)**: This complementary objective ensures that the model preserves its overall performance and general knowledge on unaffected data. It is usually implemented by reusing the original training loss, or a modified variant applied to the retain dataset. In doing so, it safeguards the model's broader capabilities and prevents unintended degradation beyond the scope of the targeted unlearning objective.
- $L_{\mathbf{Custom}}$ **(Custom Loss)**: This additional component provides flexibility and customization in the unlearning process. It can incorporate regularization terms to constrain the magnitude of parameter updates or impose specific constraints to enforce desired unlearning behaviors. By doing so, it enables researchers to adapt the unlearning procedure to task-specific requirements or to integrate domain knowledge where appropriate.

The most direct approach to unlearning is Gradient Ascent (GA). GA adapts a trained model to forget or diminish the influence of particular data or patterns by inverting the typical training update. Formally, at step t, GA updates the model parameters by ascending the next-token prediction loss over the forget dataset: $\theta_{t+1} \leftarrow \theta_t + \lambda \nabla_{\theta_t} \mathcal{L}(x, y; \theta_t)$, where λ is the (un)learning rate. In practice, existing loss adjustment methods often combine one [21], two [22–24], or all three [5] of the aforementioned components to guide unlearning. The interplay among these terms enables controlled and targeted forgetting, ensuring that the model reduces reliance on undesired information while preserving its overall performance and utility.

However, as summarized in Table 4.1, existing loss adjustment methods suffer from several limitations. Many approaches rely on retain data [5, 22, 23], which may not be readily accessible in real-world settings [6]. Others depend on a reference model [5, 23–25] to maintain performance on the retain dataset, but this introduces substantial overhead, particularly when fine-tuning large-scale LLMs. Moreover, the use of explicit retain data or implicit knowledge from a reference model during fine-tuning can blur the boundary between forget and retain data, leading to a trade-off between unlearning quality and overall model utility. Designing an appropriate data mixing strategy further complicates this process when both retain and forget datasets are involved. To preserve model utility while improving forget quality, this chapter presents FLAT, a "flat" loss adjustment approach that operates solely on the forget data. FLAT guides the LLM not only in what to forget but also in how to respond, by optimizing the f-divergence between template and forget answers with respect to the forget data. The variational form of the f-divergence provides a principled mechanism for loss adjustment by assigning optimal importance weights—reinforcing learning from template responses while effectively suppressing responses that must be forgotten.

Table 4.1 Comparison of different loss adjustment-based baselines in terms of their requirement

Baselines	Forget data	Retain data	Reference model
Gradient ascent (GA) [23]	✓	✗	✗
Gradient difference (GD) [23]	✓	✓	✗
KL minimization (KL) [23]	✓	✓	✓
Preference optimization (PO) [23]	✓	✓	✗
Mismatch [16]	✓	✓	✗
Direct preference optimization (DPO) [25]	✓	✗	✓
Negative preference optimization (NPO) [24]	✓	✗	✓
Large language model unlearning (LLMU) [5]	✓	✓	✓
FLAT	✓	✗	✗

FLAT relies solely on forget data and available template responses, without using the retain data or a reference model for response calibration

4.2 Forget Data-Only Loss Adjustment (FLAT)

This section presents FLAT, a "flat" loss adjustment approach which adjusts the loss function using only the forget data, by leveraging f-divergence maximization towards the distance between the preferred template and original forget responses. It first derives the formulation of FLAT via f-divergence maximization (Sect. 4.2.1), followed by the presentation of the empirical alternative to FLAT (Sect. 4.2.2).

4.2.1 Loss-Adjustments via f-Divergence Maximization

For each training batch, FLAT assumes access only to a set of forget samples $(x_f, y_f) \in D_f$. Rather than directly applying gradient ascent on these samples, FLAT instead maximizes the divergence between exemplary responses and undesirable generations associated with the forget data. The key steps of this procedure are summarized as follows.

- **Step 1**: Equip example/template responses y_e for each forget sample x_f. Together FLAT denotes the paired samples as $\{(x_f^j, y_e^j)\}_{j\in[N]}$.

This can be achieved by leveraging open-source LLMs such as Llama 3.1 [26], or by manually defining responses as desired. The designated unlearning response may take the form of a rejection-based answer, such as "I don't know" (denoted as IDK), or an irrelevant response that contains no information related to the unlearning target.
Motivation: Step 1 generates exemplar responses for LLM fine-tuning, which serve as guidance on how the model should respond when presented with forget data. These exemplar responses not only provide positive instruction but also mitigate the risk of undesirable behaviors. Notably, prior studies have observed that certain unlearning methods may cause the LLM to produce hallucinated responses after unlearning, further underscoring the necessity of exemplar responses in guiding the unlearning process.

- **Step 2**: Loss adjustmens w.r.t. the sample pairs (x_f, y_e, y_f) through:

$$L(x_f, y_e, y_f; \theta) = \lambda_e \cdot L_e(x_f, y_e; \theta) - \lambda_f \cdot L_f(x_f, y_f; \theta), \tag{4.2}$$

where L_e, L_f are losses designed for the data sample (x_f, y_e) and (x_f, y_f), respectively. The corresponding closed form will be introduced in Sect. 4.2.2.
Motivation: Step 2. encourages the LLM to forget the undesired knowledge by penalizing bad responses on the forget data, while simultaneously reinforcing its ability to produce desirable responses, such as template answers on the same inputs. This dual process ensures that the model not only suppresses unwanted generations but also learns how to provide appropriate alternatives when confronted with forget data.

- **Step 3**: How to decide on the values of λ_e and λ_f?
FLAT leverages f-divergence to balance the two objectives $L_e(x_f, y_e; \theta)$ and $L_f(x_f, y_f; \theta)$. Specifically, let (x_f, y_e) be sampled from random variables (X_f, Y_e) following distribution $\mathcal{D}_e$, and (x_f, y_f) be sampled from (X_f, Y_f) following distribution $\mathcal{D}_f$. Intuitively, Step 2 can be viewed as maximizing the divergence between $\mathcal{D}_e$ and $\mathcal{D}_f$. The theoretical goal is thus to obtain a model that maximizes the f-divergence between the two distributions, formally expressed as: $f_{div}(\mathcal{D}_e||\mathcal{D}_f)$.
Instead of optimizing the f_{div} term directly, FLAT resolves to the variational form of it. Due to the Fenchel duality, FLAT would have:

$$f_{div}(\mathcal{D}_e||\mathcal{D}_f) = \sup_g \left[\mathbb{E}_{\mathcal{Z}_e \sim \mathcal{D}_e}\left[g(\mathcal{Z}_e)\right] - \mathbb{E}_{\mathcal{Z}_f \sim \mathcal{D}_f}\left[f^*(g(\mathcal{Z}_f))\right]\right] := \sup_g \mathrm{VA}(\theta, g), \tag{4.3}$$

Here, f^* is defined as the conjugate function of the chosen f-divergence. The function $\mathcal{Z}_e$ takes (x_f, y_e, θ) as input and measures the discrepancy between the model's response to x_f and the target y_e. Formally, this corresponds to the difference between $\theta(x_f)$, the output generated by the LLM parameterized by θ, and the exemplar response y_e. Similarly, $\mathcal{Z}_f$ quantifies the loss for (x_f, y_f, θ). The empirical estimation of Eq. (4.3) will be presented in the next section. For simplicity, FLAT defines $\mathrm{VA}(\theta, g^*) := \sup_g \mathrm{VA}(\theta, g)$, where g^* is the optimal variational function.

Hence, the objective of FLAT is to obtain: $\theta^* := \arg\max_\theta \mathrm{VA}(\theta, g^*)$.

Motivation: It is important to note that existing approaches often struggle to strike an effective balance between model performance on forget and retain data. Step 3 addresses this issue by providing a formal theoretical framework for FLAT's loss adjustment in Step 2. Specifically, under the objective of maximizing the f-divergence between $\mathcal{D}_e$ and $\mathcal{D}_f$, FLAT assigns optimal weighting functions $f(\cdot)$ and $g(f^*(\cdot))$ with respect to the joint data distributions $\mathcal{D}_e$ and $\mathcal{D}_f$.

4.2.2 Empirical Alternative of Loss Adjustment

It is worth noting that Eq. (4.3) can be interpreted as a distribution-level loss adjustment. In practice, however, when the model is provided with a set of forget data along with exemplar and undesirable responses (x_f, y_e, y_f), this objective admits a per-sample formulation. The resulting closed-form loss function, corresponding to Eq. (4.2), is given by:

$$\begin{aligned} L(x_f, y_e, y_f; \theta) &= -\left[\sup_g \left[g(\mathbb{P}(x_f, y_e; \theta)) - f^*(g(\mathbb{P}(x_f, y_f; \theta)))\right]\right] \\ &= -g^*(\mathbb{P}(x_f, y_e; \theta)) + f^*(g^*(\mathbb{P}(x_f, y_f; \theta))). \end{aligned} \tag{4.4}$$

Examples of commonly used f-divergence functions, together with their conjugate and variational forms [27, 28], are summarized in Table 4.2. To further clarify, FLAT demonstrates the idea through the following example.

Example 1: Total-Variation. For Total-Variation (TV), an example of f-divergence, $f^*(u) = u$, $g^*(v) = \frac{\tanh(v)}{2}$, hence, $g^*() - f^*(g^*(\mathbb{P}(x_f, y_f; \theta)) = \frac{\tanh()}{2} - \frac{\tanh(\mathbb{P}(x_f, y_f; \theta))}{2}$.

How to estimate, $\mathbb{P}(x_f, y_f; \theta)$? FLAT defines the following two quantities:

$$\mathbb{P}(x_f, y_e; \theta) := \frac{\sum_{i=1}^{|y_e|} P(\mathcal{M}_\theta(x_f, y_{e,<i}) = y_{e,i})}{|y_e|},$$

Table 4.2 f_{div}s, optimal variational g (g^*), conjugate functions (f^*)

Name	$g^*(v)$	dom_{f^*}	$f^*(u)$
Total variation	$\frac{1}{2}\tanh v$	$u \in [-\frac{1}{2}, \frac{1}{2}]$	u
Jensen-Shannon	$\log \frac{2}{1+e^{-v}}$	$u < \log 2$	$-\log(2 - e^u)$
Pearson	v	$\mathbb{R}$	$\frac{1}{4}u^2 + u$
KL	v	$\mathbb{R}$	e^{u-1}

$$\mathbb{P}(x_f, y_f; \theta) := \frac{\sum_{i=1}^{|y_f|} P(\mathcal{M}_\theta(x_f, y_{f,<i}) = y_{f,i})}{|y_f|}.$$

Here, $y_{e,i}$ and $y_{f,i}$ denote the i-th tokens of the exemplar response y_e and the forget response y_f, respectively, while $y_{e,<i}$ and $y_{f,<i}$ represent their corresponding prefixes. The lengths of the exemplar and forget responses are $|y_e|$ and $|y_f|$. Given a prompt x_f and a token prefix, $P(\mathcal{M}_\theta(x_f, y_{e,<i}) = y_{e,i})$ and $P(\mathcal{M}_\theta(x_f, y_{f,<i}) = y_{f,i})$ are the probabilities of correctly predicting the next token, where $\mathcal{M}_\theta(x_f, y_{e,<i})$, and $\mathcal{M}_\theta(x_f, y_{f,<i})$ are the predicted token using LLM θ given input prompt x_f and the already generated tokens $y_{e,<i}$ and $y_{f,<i}$. These two quantities represent the average probabilities of correctly generating tokens for the template and forget responses, respectively. To align Eq. (4.4) with Eq. (4.2), FLAT could define $\lambda_e = \lambda_f = 1$, $L_e(x_f, y_e; \theta) := -g^*(\mathbb{P}(x_f, y_e; \theta))$, and $L_f(x_f, y_f; \theta) := -f^*\left(g^*(\mathbb{P}(x_f, y_f; \theta))\right)$.

4.3 The Effectiveness of FLAT in Balancing Forget Quality and Model Utility

This section evaluates FLAT against baseline unlearning methods on two widely studied LLM unlearning tasks: copyrighted content unlearning using the Harry Potter (HP) series dataset [5] (Sect. 4.3.1), and entity unlearning using the TOFU dataset [23] (Sect. 4.3.2).

Baseline Methods. We evaluate the effectiveness of FLAT by benchmarking it against a suite of strong LLM unlearning baselines, with a particular focus on loss adjustment methods. Specifically, we compare FLAT with Gradient Ascent (GA) [5, 21], KL minimization (KL) [23], GradDiff (GD) [22], NPO [24], and Mismatch [16] across all tasks. For copyrighted content and entity unlearning, we additionally include Preference Optimization (PO) [23], Large Language Model Unlearning (LLMU) [5], and DPO [25].

4.3.1 FLAT Achieves Good Trade-Off in Copyrighted Content Unlearning

To evaluate the effectiveness of FLAT, Table 4.3 reports unlearning performance using three metrics: Forget Quality Gap (FQ Gap), perplexity (PPL) on Wikitext [29], and the zero-shot average accuracy across nine standard LLM benchmarks (Avg. Acc.). FLAT consistently ranks among the top two methods across all three metrics, demonstrating particularly strong performance when using the KL f-divergence function. Moreover, in terms of average accuracy on the nine benchmarks, FLAT achieves results comparable to the retained model, indicating minimal utility degradation.

Table 4.3 Performance of FLAT and baseline methods on the Harry Potter dataset using OPT-2.7B

Metric	FQ Gap(↓)	PPL(↓)	Avg.Acc.(↑)
Original LLM	1.5346	15.6314	0.4762
Retained LLM	0.0	14.3190	0.4686
GA	2.7301	1.0984e71	0.3667
KL	2.7301	16.1592	0.4688
GD	2.3439	16.1972	0.4690
PO	2.1601	**14.8960**	0.4583
Mismatch	1.4042	15.7507	0.4679
LLMU	2.4639	15.8398	0.4656
DPO	2.2152	16.8396	0.4621
NPO	**1.2611**	19.6637	0.4644
FLAT (TV)	1.4047	15.5512	0.4681
FLAT (KL)	**1.3238**	**15.5311**	**0.4694**
FLAT (JS)	1.4025	15.5499	**0.4693**
FLAT (Pearson)	1.4089	15.5543	0.4686

FLAT consistently ranks among the top two methods in terms of similarity to the retained model, as measured by Forget Quality Gap (FQ Gap). It further demonstrates strong generation ability, producing meaningful and diverse outputs as reflected by perplexity (PPL) and the average zero-shot accuracy across nine standard LLM benchmarks (Avg. Acc.). The top two results across the three main metrics are highlighted

4.3.2 FLAT Is Good at Preserving Model Utility and Achieving Top Forget Quality in Entity Unlearning

For the TOFU task, FLAT evaluates both forget quality and model utility using the two primary metrics introduced with the dataset: Forget Quality (FQ) and Model Utility (MU) [23]. Forget quality is measured via the p-value of a Kolmogorov-Smirnov test, which quantifies how closely the unlearned model's output distribution aligns with that of a model trained solely on retain data. A p-value above 0.01 indicates statistically significant forgetting. Model utility, in contrast, aggregates performance on held-out retain data, including fictional authors, real-world author profiles, and factual knowledge. In addition, ROUGE-L scores are reported for both the forget and retain sets. Importantly, for the forget set, a lower ROUGE-L score does not automatically imply better forgetting performance. Instead, FLAT emphasizes methods whose ROUGE-L scores closely match those of the retained model, as this alignment is considered a stronger indicator of effective and reliable unlearning.

FLAT preserves model utility most effectively. As shown in Table 4.4, FLAT exhibits almost no reduction in model utility compared to the original model. On LLaMA2-7B, although GA and KL achieve strong ROUGE-L scores on the retain dataset, their forgetting performance is weak. Similarly, on Phi-1.5B, GD achieves a strong ROUGE-L score on the retain dataset but fails to achieve significant forgetting (p-value ≤ 0.01).

FLAT consistently ranks among the top two in Forget Quality. On LLaMA2-7B and Phi-1.5B, FLAT achieves ROUGE-L scores on the forget dataset that are closest to those of

Table 4.4 Performance of FLAT and baseline methods on the TOFU dataset with three base LLMs: LLaMA2-7B, Phi-1.5B, and OPT-2.7B

Base LLM	Llama2-7B				Phi-1.5B				OPT-2.7B			
Metric	FQ	MU	F-RL(↓)	R-RL	FQ	MU	F-RL(↓)	R-RL	FQ	MU	F-RL(↓)	R-RL
Original LLM	4.4883e-06	0.6346	0.9851	0.9833	0.0013	0.5184	0.9607	0.9199	0.0013	0.5120	0.7537	0.7494
Retained LLM	1.0	0.6267	0.4080	0.9833	1.0	0.5233	0.4272	0.9269	1.0	0.5067	0.4217	0.7669
GA	0.0143	0.6333	0.4862	0.9008	0.0013	0.5069	0.5114	0.8048	**0.2657**	0.4639	**0.4748**	0.6387
KL	0.0068	0.6300	0.5281	**0.9398**	0.0030	0.5047	0.5059	0.8109	0.0286	0.4775	0.4810	0.6613
GD	0.0068	0.6320	0.4773	0.8912	0.0030	0.5110	0.4996	**0.8496**	0.0541	0.4912	**0.4521**	0.6603
PO	**0.0541**	0.6308	0.3640	0.8811	**0.0286**	0.5127	0.3170	0.7468	0.0068	0.4424	0.0589	0.4015
Mismatch	0.0143	0.6304	0.9406	**0.9741**	0.0030	0.5225	0.9612	**0.9194**	0.0030	0.5025	0.7525	**0.7475**
LLMU	**0.0541**	0.6337	0.4480	0.8865	**0.0286**	0.5110	0.3058	0.7270	0.0286	0.3296	0.0347	0.2495
DPO	**0.0541**	0.6359	0.5860	0.8852	0.0521	0.0519	0.3437	0.7349	0.0541	0.4264	0.0806	0.3937
NPO	0.0068	0.6321	0.4632	0.8950	0.0030	0.5057	0.5196	0.8000	0.0541	0.4788	0.4993	0.6490
FLAT (TV)	**0.0541**	0.6373	**0.4391**	0.8826	0.0143	**0.5168**	0.4689	0.8155	0.0068	**0.5086**	0.5217	**0.7067**
FLAT (KL)	0.0286	**0.6393**	0.5199	0.8750	0.0143	**0.5180**	**0.4524**	0.7850	0.0286	0.4838	0.4942	0.6974
FLAT (JS)	0.0541	0.6364	0.4454	0.8864	0.0068	0.5144	**0.4572**	0.8117	**0.0541**	0.4959	0.4938	0.7013
FLAT (Pearson)	0.0541	**0.6374**	**0.4392**	0.8857	0.0143	0.5175	0.4591	0.8099	0.0068	**0.5093**	0.5052	0.7059

Metrics include Forget Quality (FQ), Model Utility (MU), ROUGE-L on the retain dataset (R-RL), and ROUGE-L on the forget dataset (F-RL). Results for the original LLM and the retain LLM are also reported for reference. The top two results are highlighted

the retained model, indicating more reliable forgetting. While PO achieves the strongest forgetting efficiency on Phi-1.5B, its ROUGE-L scores on both forget and retain sets drop notably, reflecting degraded utility. A similar trade-off appears with GA: although it performs well on forgetting for OPT-2.7B, its model utility is considerably lower.

FLAT offers the best trade-off. Across all three models, FLAT ranks in the top two on the primary metrics, with the strongest overall performance in MU. Notably, the KL f-divergence variant achieves strong results in both FQ and MU on LLaMA2-7B and Phi-1.5B. Overall, all four f-divergence functions within FLAT maintain an effective balance between forgetting efficiency and model utility.

In summary, FLAT overcomes the key limitations of existing LLM unlearning methods, which often depend on retain data or a reference model for response calibration. By optimizing the f-divergence between template and forget responses, FLAT provides a principled and theoretically grounded framework for balancing forget quality with model utility. Extensive experiments on two representative unlearning tasks demonstrate that FLAT consistently delivers high unlearning efficiency while preserving overall model performance, thereby addressing both the practical and theoretical challenges of LLM unlearning.

References

1. Karamolegkou, A., Li, J., Zhou, L., Søgaard, A.: Copyright violations and large language models. arXiv preprint arXiv:2310.13771 (2023)
2. Patil, V., Hase, P., Bansal, M.: Can sensitive information be deleted from llms? Objectives for defending against extraction attacks. ICLR (2024)

3. Motoki, F., Pinho Neto, V., Rodrigues, V.: More human than human: measuring chatgpt political bias. Public Choice **198**(1), 3–23 (2024)
4. Yu, C., Jeoung, S., Kasi, A., Yu, P., Ji, H.: Unlearning bias in language models by partitioning gradients. In: Findings of the Association for Computational Linguistics: ACL 2023, pp. 6032–6048 (2023)
5. Yao, Y., Xu, X., Liu, Y.: Large language model unlearning. arXiv preprint arXiv:2310.10683 (2023)
6. Li, N., Pan, A., Gopal, A., Yue, S., Berrios, D., Gatti, A., Li, J.D., Dombrowski, A.K., Goel, S., Mukobi, G., Helm-Burger, N., Lababidi, R., Justen, L., Liu, A.B., Chen, M., Barrass, I., Zhang, O., Zhu, X., Tamirisa, R., Bharathi, B., Herbert-Voss, A., Breuer, C.B., Zou, A., Mazeika, M., Wang, Z., Oswal, P., Lin, W., Hunt, A.A., Tienken-Harder, J., Shih, K.Y., Talley, K., Guan, J., Steneker, I., Campbell, D., Jokubaitis, B., Basart, S., Fitz, S., Kumaraguru, P., Karmakar, K.K., Tupakula, U., Varadharajan, V., Shoshitaishvili, Y., Ba, J., Esvelt, K.M., Wang, A., Hendrycks, D.: The WMDP benchmark: measuring and reducing malicious use with unlearning. In: Proceedings of the 41st International Conference on Machine Learning, Proceedings of Machine Learning Research, vol. 235, pp. 28525–28550. PMLR (2024)
7. Barrett, C., Boyd, B., Bursztein, E., Carlini, N., Chen, B., Choi, J., Chowdhury, A.R., Christodorescu, M., Datta, A., Feizi, S., et al.: Identifying and mitigating the security risks of generative AI. Found. Trends Privacy Sec. **6**(1), 1–52 (2023)
8. Hoofnagle, C.J., van der Sloot, B., Borgesius, F.Z.: The European union general data protection regulation: what it is and what it means. Inform. Commun. Technol. Law **28**(1), 65–98 (2019)
9. Cao, Y., Yang, J.: Towards making systems forget with machine unlearning. In: Proceedings of the 2015 IEEE Symposium on Security and Privacy, pp. 463–480. IEEE (2015)
10. Liu, J., Ram, P., Yao, Y., Liu, G., Liu, Y., Sharma, P., Liu, S., et al.: Model sparsity can simplify machine unlearning. Adv. Neural Inform. Process. Syst. **36**, 552 (2024)
11. Fan, C., Liu, J., Zhang, Y., Wei, D., Wong, E., Liu, S.: Salun: Empowering machine unlearning via gradient-based weight saliency in both image classification and generation. In: International Conference on Learning Representations (2024)
12. Di, Z., Zhu, Z., Jia, J., Liu, J., Takhirov, Z., Jiang, B., Yao, Y., Liu, S., Liu, Y.: Label smoothing improves machine unlearning. arXiv preprint arXiv:2406.07698 (2024)
13. Liu, S., Yao, Y., Jia, J., Casper, S., Baracaldo, N., Hase, P., Yao, Y., Liu, C.Y., Xu, X., Li, H., Varshney, K.R., Bansal, M., Koyejo, S., Liu, Y.: Rethinking machine unlearning for large language models. arXiv preprint arXiv:2402.08787 (2024)
14. Eldan, R., Russinovich, M.: Who's harry potter? Approximate unlearning in llms. arXiv preprint arXiv:2310.02238 (2023)
15. Ji, J., Liu, Y., Zhang, Y., Liu, G., Kompella, R.R., Liu, S., Chang, S.: Reversing the forget-retain objectives: an efficient llm unlearning framework from logit difference. Adv. Neural. Inf. Process. Syst. **37**, 12581–12611 (2024)
16. Liu, C.Y., Wang, Y., Flanigan, J., Liu, Y.: Large language model unlearning via embedding-corrupted prompts. In: The Thirty-Eighth Annual Conference on Neural Information Processing Systems (2024). https://openreview.net/forum?id=e5icsXBD8Q
17. Pawelczyk, M., Neel, S., Lakkaraju, H.: In-context unlearning: language models as few-shot unlearners. In: Proceedings of the 41st International Conference on Machine Learning (2024)
18. Choi, M., Rim, D., Lee, D., Choo, J.: Snap: unlearning selective knowledge in large language models with negative instructions. arXiv preprint arXiv:2406.12329 (2024)
19. Chen, J., Yang, D.: Unlearn what you want to forget: efficient unlearning for llms. arXiv preprint arXiv:2310.20150 (2023)
20. Wang, Y., Wei, J., Liu, C.Y., Pang, J., Liu, Q., Shah, A.P., Bao, Y., Liu, Y., Wei, W.: Llm unlearning via loss adjustment with only forget data. arXiv preprint arXiv:2410.11143 (2024)

21. Jang, J., Yoon, D., Yang, S., Cha, S., Lee, M., Logeswaran, L., Seo, M.: Knowledge unlearning for mitigating privacy risks in language models. arXiv preprint arXiv:2210.01504 (2022)
22. Liu, B., Liu, Q., Stone, P.: Continual learning and private unlearning. In: Conference on Lifelong Learning Agents, pp. 243–254. PMLR (2022)
23. Maini, P., Feng, Z., Schwarzschild, A., Lipton, Z.C., Kolter, J.Z.: TOFU: a task of fictitious unlearning for LLMs. In: First Conference on Language Modeling (2024). https://openreview.net/forum?id=B41hNBoWLo
24. Zhang, R., Lin, L., Bai, Y., Mei, S.: Negative preference optimization: from catastrophic collapse to effective unlearning. In: First Conference on Language Modeling (2024). https://openreview.net/forum?id=MXLBXjQkmb
25. Rafailov, R., Sharma, A., Mitchell, E., Manning, C.D., Ermon, S., Finn, C.: Direct preference optimization: your language model is secretly a reward model. Adv. Neural Inform. Process. Syst. **36**, 63–78 (2024)
26. Dubey, A., Jauhri, A., Pandey, A., Kadian, A., Al-Dahle, A., Letman, A., Mathur, A., Schelten, A., Yang, A., Fan, A., et al.: The llama 3 herd of models. arXiv preprint arXiv:2407.21783 (2024)
27. Nowozin, S., Cseke, B., Tomioka, R.: f-gan: Training generative neural samplers using variational divergence minimization. Adv. Neural Inform. Process. Syst. **29**, 652 (2016)
28. Wei, J., Liu, Y.: When optimizing f-divergence is robust with label noise. In: International Conference on Learning Representations (2021). https://openreview.net/forum?id=WesiCoRVQ15
29. Merity, S., Xiong, C., Bradbury, J., Socher, R.: Pointer sentinel mixture models (2016)

Part II
Machine Unlearning: A Data-Centric Perspective

A Coreset Perspective for Machine Unlearning 5

Soumyadeep Pal

Abstract

This chapter explores coreset perspectives in machine unlearning across efficiency, robustness, and evaluation. We first reveal a coreset effect: small random subsets of the forget set can match full-data unlearning performance, with a coreset-training trade-off showing that fewer samples require proportionally more unlearning epochs. We then examine advanced coreset selection methods, e.g., gradient-, clustering-, and likelihood-based, finding only marginal gains over random selection across standard unlearning benchmarks. Next, we assess the faithfulness of coreset-unlearned models, showing they share the same optimal basin as full-forget-set models and exhibit similar robustness to adversarial attacks and downstream finetuning. We further introduce a utility-forget trade-off perspective, where pruning high-variance outliers mitigates collateral utility loss. Finally, using bi-level optimization, we identify worst-case forget sets that are hardest to erase yet least essential for generalization, linking coresets to adversarial unlearning evaluation. Overall, coresets provide a unifying lens for understanding data importance in unlearning, motivating principled selection methods and rigorous evaluation protocols for both vision and language models.

5.1 Preliminaries on Coreset Selection

Coreset selection aims to identify a small yet informative subset of the training data such that a model trained on this subset achieves generalization performance close to that of a model trained on the entire dataset. This problem has been extensively studied in computer

S. Pal (✉)
Michigan State University, East Lansing, MI, USA
e-mail: palsoum1@msu.edu

S. Liu et al. (eds.), *Machine Unlearning for Governance of Foundation Models*, Synthesis Lectures on Computer Vision, https://doi.org/10.1007/978-3-032-17282-2_5

vision, where numerous methods have been proposed to construct such subsets efficiently and effectively.

One major class of approaches assigns an importance score to each training sample and selects the top-scoring items. These scores are typically derived from training dynamics or model outputs. Examples include the expected gradient magnitude or error norm over training steps [1], the number of forgetting events where a sample's prediction switches from correct to incorrect during training [2], loss reduction on held-out test samples [3], or confidence- and logit-based measures [4].

A second line of work emphasizes representativeness, aiming to ensure that the selected subset captures the diversity of the full dataset. For example, clustering methods choose samples near cluster centers in a learned feature space [5], while coverage-based methods jointly optimize for representativeness and sample importance [6]. Recent work also uses message passing algorithms to balance sample diversity with training difficulty when constructing coresets [7].

A third class formulates coreset construction as an explicit optimization problem, solved via greedy algorithms, submodular optimization, or gradient-based proxies that approximate how removing or keeping a sample influences the training trajectory or final model parameters [8–11]. A comprehensive treatment of these methods can be found in [12].

Beyond vision, coreset selection for language modeling has also been explored for multiple purposes. Importance-based techniques have been adapted to loss-based scores for pruning and efficient training of language models [13], while classical scoring methods enable efficient downstream task-specific fine-tuning [14]. For instruction-following and alignment, specialized objectives score examples by gradient similarity to validation or alignment targets [15] or use large language models as judges to rank candidate training pairs [16, 17]. Moreover, work such as [18] shows that carefully curated, small-scale, high-quality datasets can achieve surprisingly strong alignment and fine-tuning performance, highlighting the importance of strategic data selection.

These developments naturally raise a central question: *what role can coresets play in machine unlearning, and more broadly, how does the quantity of forget data influence unlearning performance?*

5.2 Coreset-Based Unlearning

This section explores the intersection of coreset selection with the *training aspect* of machine unlearning. In standard machine learning, a coreset refers to a subset of training data whose use yields model performance close to that obtained when training on the full dataset. Extending this concept to unlearning, a natural question arises: *can a small subset of the forget set, termed a "forgetting coreset", achieve unlearning performance comparable to using the entire forget set?* Recent work by Pal et al. [19] demonstrates the prevalence of this so-called coreset effect in current unlearning benchmarks. Strikingly, their study shows that using an extremely small fraction of the forget set often suffices to achieve nearly the same unlearning effectiveness as the full set, suggesting the potential for substantial savings in data and computation without sacrificing forgetting fidelity.

Figure 5.1 illustrates this coreset effect across multiple unlearning benchmarks in large language models (LLMs), including WMDP [20] and MUSE [21], and multiple unlearning methods including RMU (representation misdirection unlearning) [20] and NPO (negative preference optimization) [22]; see Sect. 1.2 for details on these unlearning methods. The WMDP benchmark focuses on unlearning hazardous biomedical and cybersecurity knowledge, where unlearning effectiveness (UE) is defined as UE $= 100\% -$ (Accuracy on WMDP evaluation set). Thus, a higher UE reflects more successful forgetting. Model utility (UT) is measured by zero-shot accuracy on the MMLU benchmark [23]. In contrast, the MUSE benchmark targets unlearning memorized textual content from Harry Potter books (MUSE-Books) and news articles (MUSE-News). Here, UE includes three complementary metrics, with VerbMem (i.e., verbatim memorization accuracy) serving as the most distinctive one, as it measures exact reproduction from the forget set rather than knowledge-level extraction as in WMDP. UT for MUSE is the accuracy on its retain set.

As shown in Fig. 5.1, the coreset selection ratio ranges from 0% (i.e., no forget data used, corresponding to the original model before unlearning) to 100% (i.e., the full forget set used for unlearning). Across most settings, unlearning performance (UE) with small coreset ratios, such as 5 or 10%, is nearly indistinguishable from that obtained using the entire forget set. At the same time, model utility remains well preserved across all ratios, indicating no loss in general performance. Strikingly, in some cases, even a 1% coreset achieves forgetting effectiveness comparable to the full set, as observed in Fig. 5.1a, b. This pronounced coreset

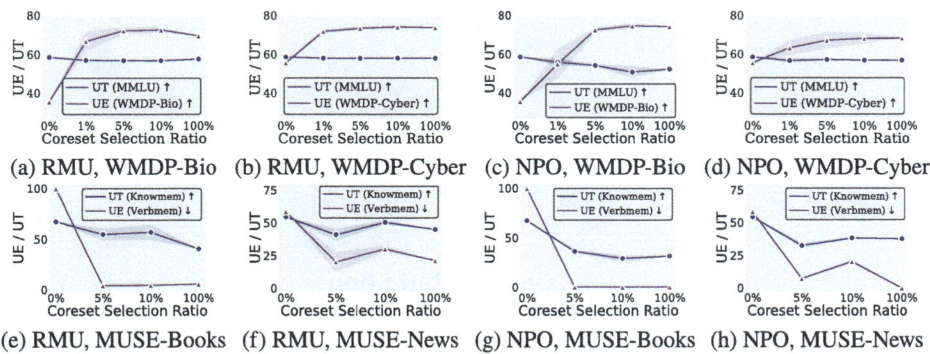

(a) RMU, WMDP-Bio (b) RMU, WMDP-Cyber (c) NPO, WMDP-Bio (d) NPO, WMDP-Cyber

(e) RMU, MUSE-Books (f) RMU, MUSE-News (g) NPO, MUSE-Books (h) NPO, MUSE-News

Fig. 5.1 Consistent unlearning performance in terms of UT (model utility) and UE (unlearning effectiveness) across varying coreset selection ratios. Results are averaged over five independent trials with random coreset selection, and shaded regions indicate performance variance. Panels a–d show results for different combinations of unlearning methods (RMU or NPO) and benchmark datasets (WMDP-Bio, WMDP-Cyber, MUSE-Books, MUSE-News). Following standard benchmark settings, unlearning was performed using Zephyr-7B-β on WMDP, LLaMA2-7B on MUSE-News, and ICLM-7B on MUSE-Books

effect in unlearning likely stems from the high redundancy of information within typical forget sets: once a small subset of samples captures the key semantic concepts to be erased, additional samples contribute little marginal forgetting signal. Moreover, modern deep models exhibit strong generalization from limited supervision, allowing them to propagate the forgetting effect induced by a small but representative subset across all semantically related instances in the forget set.

5.2.1 Revealing the Coreset Effect Through Extended Training

In standard practice, unlearning on benchmarks such as WMDP or MUSE is typically performed with only a small number of training epochs, often limited to a single epoch. However, recent work by Pal et al. [19] demonstrates that when unlearning is applied to only a small random subset of the forget set (i.e., a low coreset selection ratio), comparable performance to using the full forget set can still be achieved by proportionally increasing the number of unlearning epochs.

This uncovers a coreset-training trade-off: *fewer forget samples require proportionally more training iterations to reach the same unlearning efficacy.* Empirically, this trade-off exhibits a near-linear relationship between the number of required unlearning epochs and the coreset selection ratio in the log domain, as illustrated in Fig. 5.2. This phenomenon mirrors scaling laws widely observed in model training, suggesting that unlearning performance obeys predictable scaling behavior with respect to data quantity and optimization time.

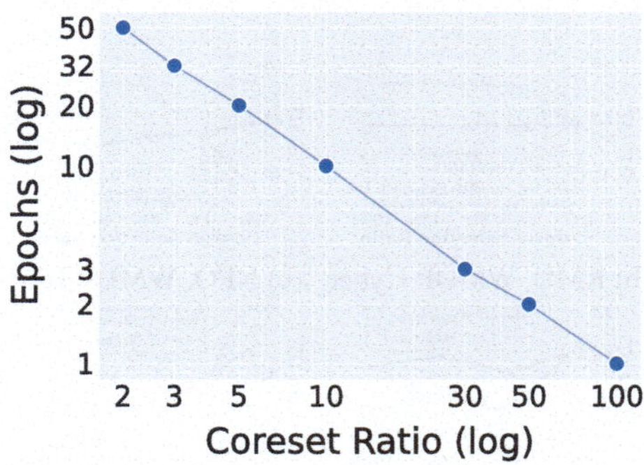

Fig. 5.2 Scaling law for unlearning: Relationship between the required number of unlearning epochs and the coreset selection ratio in the log domain for lossless unlearning. Experiments were conducted on WMDP-Bio using the RMU method with the pre-trained LLM Zephyr-7B-β

5.2.2 Do Sophisticated Coreset Selection Methods Improve Unlearning?

So far, our analysis has focused on unlearning performance when using *random* subsets of the forget set. While this was sufficient to reveal the *coreset effect*, it naturally raises the question: could more sophisticated coreset selection methods further improve unlearning efficiency? To address this, we discuss the coreset selection methods adapted to the unlearning setting by Pal et al. [19] and present empirical results comparing them to random selection.

GRAND. The core idea behind the coreset selection method GRAND [1] is to quantify the influence of each forget-sample z_f by the expected magnitude of the parameter gradient of the loss associated with that sample, where the expectation is taken over the unlearning trajectory of the model $\boldsymbol{\theta}_t$. Formally,

$$\chi(z_f) = \mathbb{E}_{\boldsymbol{\theta}_t} \|\nabla_{\boldsymbol{\theta}_t}[\ell_{\mathrm{f}}(z_f;\boldsymbol{\theta}_t) + \gamma\ell_{\mathrm{r}}(z_r;\boldsymbol{\theta}_t)]\|_2 \;;\; \text{where } z_f \sim \mathcal{D}_{\mathrm{f}},\;\; z_r \sim \mathcal{D}_{\mathrm{r}} \tag{5.1}$$

Here ℓ_f and ℓ_r denote the forget and retain losses defined in (1.1). Following Pal et al. [19], the expectation is computed over 10 unlearning epochs.

MODERATE. Originally introduced for classification tasks [5], the MODERATE method clusters samples using their feature representations and selects points nearest to the cluster centroids as the coreset, ensuring representativeness. In the unlearning setting of [19], where class labels are unavailable, forget samples are first partitioned into four clusters via K-means on penultimate-layer embeddings extracted from the reference model. Points closest to the cluster centers are then selected to form the coreset.

MIN-K% PROB. The MIN- K% PROB method [24] was initially proposed to test whether text appeared in the pretraining corpus. In [19], this idea was adapted for unlearning by selecting samples most likely to have influenced the model during training. For a sample z, the top $K = 40\%$ of tokens with the lowest log-likelihood under the model are identified as Min-K%(z). Coresets were obtained using the top scores of the samples in a forget dataset.

Table 5.1 shows unlearning performance across different datasets (WMDP, MUSE) and methods (RMU, NPO), where coreset selection ratios range from 0% (no unlearning) to 100% (full forget set). As we can see, all methods consistently demonstrate the *coreset effect*, that is, small subsets achieve performance close to the full forget set. However, sophisticated selection strategies rarely outperform random selection by large margins. In most cases, differences fall within the variance of random subsets, with occasional small gains for certain datasets and unlearning methods. This suggests that simple random selection remains surprisingly competitive for unlearning.

5.2.3 On the Faithfulness of Coreset-Unlearned Models

Although models unlearned using coresets achieve performance comparable to those unlearned with the full forget set under standard evaluation protocols, unlearning quality involves additional dimensions such as robustness to distribution shifts and adversarial

Table 5.1 Coreset unlearning performance (UE and UT, consistent with Fig. 5.1) using RMU and NPO on WMDP-Bio and WMDP-Cyber, evaluated using Zephyr-7B-β across varying coreset selection ratios (0, 5, 10, 100%) and selection methods (RANDOM, GRAND, MODERATE, MIN- K% PROB)

Coreset ratio (%)	Unlearning method	RMU on WMDP-Bio		NPO on WMDP-Bio		RMU on WMDP-Cyber		NPO on WMDP-Cyber	
		UE (↑)	UT (MMLU) (↑)	UE (↑)	UT (MMLU) (↑)	UE (↑)	UT (MMLU) (↑)	UE (↑)	UT (MMLU) (↑)
0	No unlearning	35.35	58.48	35.35	58.48	55.51	58.48	55.51	58.48
100	Full forget set	69.46	57.48	74.11	52.20	73.57	57.85	68.19	56.79
10	RANDOM	72.43 ± 1.34	56.66 ± 0.24	74.50 ± 1.22	50.69 ± 2.85	73.96 ± 0.74	57.95 ± 0.15	68.21 ± 2.69	56.88 ± 0.35
	GRAND	71.30	56.98	71.54	52.40	73.43	57.25	67.64	56.71
	MODERATE	70.86	56.66	75.92	52.36	73.53	57.20	69.65	56.69
	MIN- K% PROB	70.61	56.86	74.59	52.71	73.91	57.12	69.19	57.03
5	RANDOM	72.03 ± 1.78	56.69 ± 0.44	72.47 ± 1.23	54.12 ± 0.56	73.32 ± 0.79	57.89 ± 0.12	67.19 ± 4.17	57.23 ± 0.43
	GRAND	72.59	56.78	75.19	56.55	72.88	57.64	65.85	56.58
	MODERATE	73.78	56.90	71.72	56.18	73.56	57.52	67.47	55.72
	MIN- K% PROB	69.48	57.29	70.91	55.67	72.84	57.58	69.15	57.19

Here 0 and 100% refer to the pre-unlearning case (w/o using any forget data) and the standard unlearning case (w/ the full forget set), respectively. The performance for RANDOM-based selection were reported in the form $a \pm b$, where a is the mean and b is the standard deviation, computed over 5 independent trials. The performance of other data selection metrics were reported using the mean over 2 trials

perturbations. This raises an important question: *Are coreset-unlearned models faithful to their full forget set counterparts beyond standard utility and unlearning metrics?*

First, the recent work Pal et al. [19] attempted to answer this question by examining the structural relationship between coreset-unlearned and full-forget-set-unlearned models. Their key finding is that the two models exhibit *mode connectivity*, specifically, *linear mode connectivity* (LMC) [25]. In this setting, an interpolated model in parameter space, $\boldsymbol{\theta}(\alpha) := (\alpha\boldsymbol{\theta}_{\text{cu}} + (1-\alpha)\boldsymbol{\theta}_{\text{fu}})$, constructed between the coreset-unlearned model ($\boldsymbol{\theta}_{\text{cu}}$) and the full-forget-set-unlearned model ($\boldsymbol{\theta}_{\text{fu}}$), maintains approximately constant UE (unlearning efficacy) across interpolation coefficients $\alpha \in [0, 1]$. Figure 5.3 illustrates the LMC across different coreset-unlearned models on WMDP using both RMU and NPO methods. The results show that coreset-unlearned models consistently lie within the same optimal basin as their full-forget-set counterparts, providing strong evidence for the fidelity and representational faithfulness of coreset-based unlearning.

In addition, unlearned models are known to be vulnerable to input-level jailbreak attacks [26–28]. As shown in Table 5.2, coreset-unlearned models exhibit similar levels of vulnerability as their full-forget-set counterparts, reflected in comparable reductions in unlearning effectiveness under adversarial perturbations.

Finally, unlearned models are also known to be vulnerable to downstream finetuning, even when the finetuning tasks are unrelated to the forget set [29–31]. To examine whether this vulnerability extends to coreset-unlearned models, let us finetune models unlearned

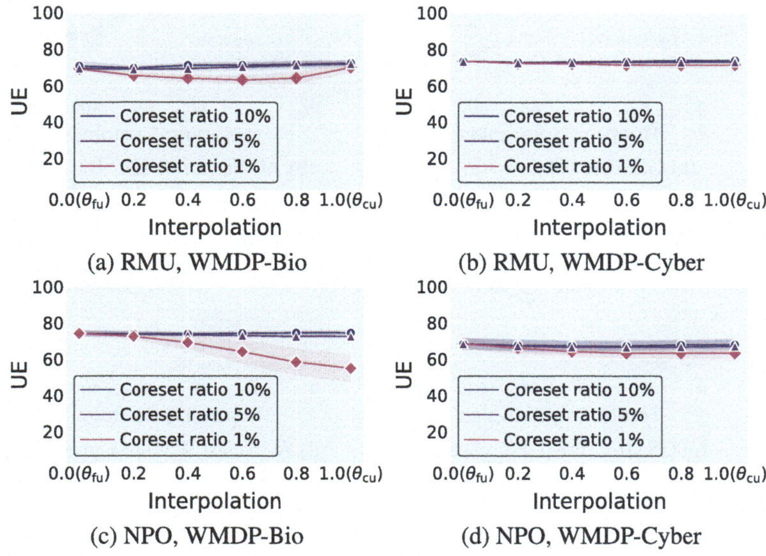

(a) RMU, WMDP-Bio (b) RMU, WMDP-Cyber

(c) NPO, WMDP-Bio (d) NPO, WMDP-Cyber

Fig. 5.3 LMC holds between coreset-unlearned model ($\boldsymbol{\theta}_{\text{cu}}$) and the full forget set-unlearned model ($\boldsymbol{\theta}_{\text{fu}}$), as evidenced by UE against the interpolation coefficient α (x-axis). Here the coreset-unlearned models are obtained using RANDOM-based coresets with the same setting as in Fig. 5.1a–d

Table 5.2 Robustness of models unlearned using RANDOM coreset selection on WMDP-Bio using RMU under Zephyr-7B-β, following the setting in Fig. 5.1a. Robustness is measured using the UE reduction after enhanced GCG attack

Coreset ratio (%)	U		EUE reduction
	Before attack	After attack	After attack
100	69.46	47.71	21.75
10	72.43 ± 1.34	53.39 ± 0.02	19.04
5	72.03 ± 1.78	51.29 ± 0.03	20.74

with NPO on two unrelated tasks: GSM8k [32] for arithmetic reasoning and AGNews [33] for news topic classification. As shown in Fig. 5.4, coreset-unlearned models exhibit levels of "relearning" comparable to their full-forget-set counterparts, indicating similar susceptibility to knowledge recovery under finetuning. However, for WMDP-Bio models finetuned on GSM8k, Fig. 5.4a reveals slightly higher relearning rates for coreset-unlearned models, suggesting a potentially increased vulnerability to downstream finetuning in certain settings.

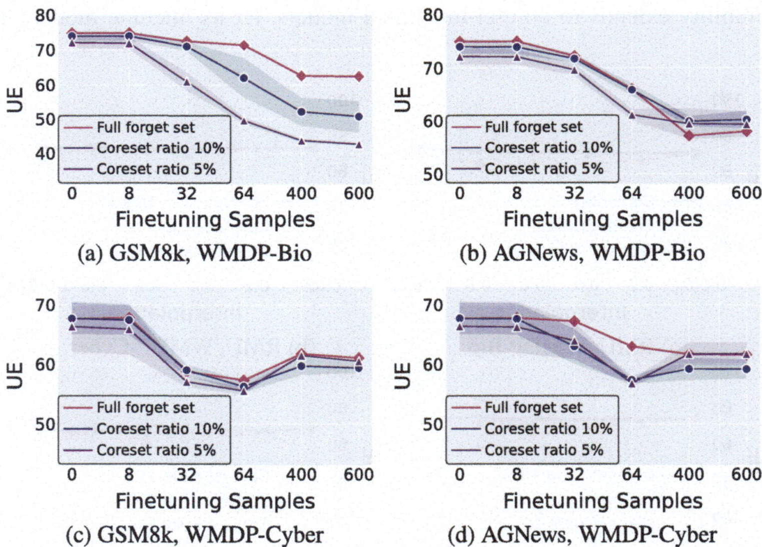

Fig. 5.4 Unlearning effectiveness (UE) of RANDOM-coreset unlearned models (using NPO under Zephyr-7B-β) against the number of fine-tuning samples. **a–d** Relearning using finetuning datasets (GSM8k, AGNews) for models unlearned on WMDP-Bio or WMDP-Cyber. The performance is averaged over 3 independent trials

5.2.4 Improving the Utility-Forget Tradeoff: A Different Coreset Perspective

One persistent challenge in machine unlearning is the collateral degradation to model utility on unrelated tasks (e.g., accuracy on MMLU) when forgetting is enforced. To address this issue, the coreset perspective was introduced in Patil et al. [34], leveraging coresets to balance effective forgetting with minimal utility degradation.

A key insight from Patil et al. [34] is that high variance in the representations of the forget set strongly correlates with collateral utility loss. Building on this observation, they employ the Isolation Forest algorithm [35] to detect outlier or anomalous samples within the forget set representations. By removing these outliers, the variance among the remaining samples is reduced, and the resulting corese,—consisting of more homogeneous, semantically representative data, is then used for unlearning. This pruning strategy effectively mitigates unnecessary utility loss while preserving forgetting performance. Crucially, this approach differs from the treatment of coreset-based unlearning in [19]. While the work Pal et al. [19] explores extreme pruning ratios to reveal the coreset effect, showing that even very small forget subsets can achieve comparable unlearning performance, Patil et al. [34] focuses on moderate pruning to explicitly optimize the utility-forget tradeoff. These perspectives highlight complementary roles of coreset selection: one for improving computational and data efficiency, and the other for safeguarding model utility during unlearning.

5.3 Challenging Forget "Coreset" in Worst-Case Unlearning Evaluation

A widely adopted protocol for evaluating machine unlearning methods, particularly in computer vision, is *random data forgetting*. In this setting, a subset of training examples is randomly sampled for removal, the candidate unlearning algorithm is applied, and the evaluation measures how effectively information about the removed examples has been erased. This protocol differs from many generative model unlearning benchmarks, where the forget set is predetermined by benchmark designers rather than sampled randomly. Fixed forget sets ensure reproducibility across methods, but random forgetting reveals average-case unlearning performance rather than adversarial or worst-case scenarios. To address this limitation, the work [36] introduced the notion of a *worst-case forget set*, termed "challenging forgets", a subset of training examples deliberately selected to be maximally challenging for unlearning algorithms.

Interestingly, it has been shown in [36] that the *complement of the worst-case forget set* naturally forms a *coreset* for the dataset. As illustrated in Fig. 5.5, experiments on CIFAR-10 and CIFAR-100 show that training on the complement of the worst-case forget set (i.e., challenging forgets) achieves testing accuracy (TA) comparable to, and in some cases exceeding, that of models trained on coresets selected by popular methods such as EL2N or GraNd [1],

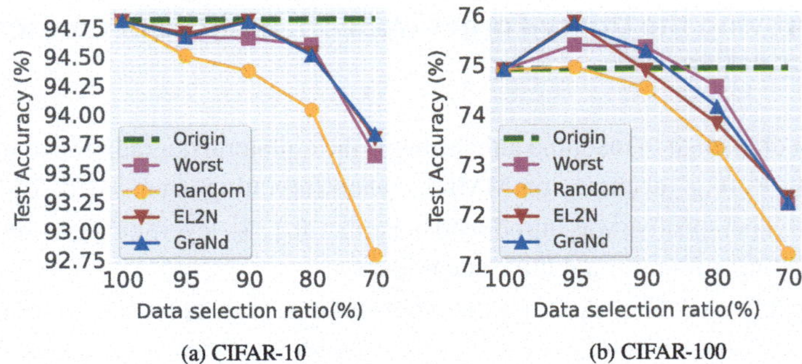

(a) CIFAR-10 (b) CIFAR-100

Fig. 5.5 Performance of ResNet-18 trained on coresets of **a** CIFAR-10 and **b** CIFAR-100, determined by different approaches, including the complement of the worst-case forget set (Worse), random select (Random), EL2N and GraNd, at varying coreset ratios. The dashed line represents the model's performance trained on the full dataset (Origin)

and even sometimes surpasses models trained on the full dataset (*Origin*). This finding establishes a surprising link between the worst-case unlearning evaluation and coreset construction, suggesting that the most challenging data to forget is often *least* essential for model generalization.

5.3.1 Identifying "Challenging Forgets" via Bi-Level Optimization

The problem of identifying the worst-case forget set (i.e., "challenging forgets") can be addressed using a *bi-level optimization* (BLO) framework [36]. In BLO, the solution to the *upper-level* objective depends on the outcome of a *lower-level* optimization problem [37].

For worst-case unlearning, the *lower-level optimization* trains the unlearning model for a given candidate forget set specified by a selection vector $\mathbf{w} \in \{0, 1\}^N$, where N is the total number of training samples in the training set $\mathcal{D}$. Specifically, $w_i = 1$ indicates that the i-th data point belongs to the forget set $\mathcal{D}_\mathrm{f} = \{\mathbf{z}_i \mid w_i = 1\}$, while $w_i = 0$ assigns it to the retain set. The lower-level problem then solves

$$\boldsymbol{\theta}_\mathrm{u}(\mathbf{w}) = \arg\min_{\boldsymbol{\theta}} \ \ell_\mathrm{MU}(\boldsymbol{\theta}; \mathbf{w}) := \sum_{\mathbf{z}_i \in \mathcal{D}} [w_i \ell_\mathrm{f}(\boldsymbol{\theta}; \mathbf{z}_i) + (1 - w_i)\ell_\mathrm{r}(\boldsymbol{\theta}; \mathbf{z}_i)], \tag{5.2}$$

where ℓ_f and ℓ_r are the loss terms on the forget and retain sets, respectively. In practice, the work [36] set $\ell_\mathrm{f} = -\ell_\mathrm{r}$, with both instantiated as the standard training loss ℓ (i.e., cross-entropy loss). This ensures that minimizing ℓ_MU simultaneously removes information from the forget set while retaining generalization on the rest of the data.

The *upper-level optimization* then chooses the forget set $\mathbf{w}$ to maximize the generalization performance of the resulting unlearned model $\boldsymbol{\theta}_{\mathrm{u}}(\mathbf{w})$, thereby constructing the most *adversarial* forget set: one that is hardest for the unlearning algorithm to erase while preserving accuracy elsewhere. Integrating both levels yields the full bi-level optimization framework for worst-case unlearning:

$$\min_{\mathbf{w}\in\mathcal{S}} \underbrace{\sum_{\mathbf{z}_i\in\mathcal{D}}[w_i\ell(\boldsymbol{\theta}_{\mathrm{u}}(\mathbf{w});\mathbf{z}_i)]+\gamma\|\mathbf{w}\|_2^2}_{\text{Upper-level objective}} := f(\mathbf{w},\boldsymbol{\theta}_{\mathrm{u}}(\mathbf{w}));\quad \text{subject to } \underbrace{\boldsymbol{\theta}_{\mathrm{u}}(\mathbf{w})=\arg\min_{\boldsymbol{\theta}}\ \ell_{\mathrm{MU}}(\boldsymbol{\theta};\mathbf{w})}_{\text{Lower-level optimization}}, \tag{5.3}$$

where $\mathbf{w}$ is the upper-level optimization variable subject to the data selection constraint set $\mathcal{S}$, e.g., $\mathcal{S}=\{\mathbf{w}|\mathbf{w}\in\{0,1\}^N, \mathbf{1}^\top\mathbf{w}=m\}$ with m being the forget set size, the lower-level objective function ℓ_{MU} has been defined in (5.2), and ℓ denotes the training loss. The ℓ_2 regularization with the regularization parameter $\gamma\geq 0$ encourages a sparse data selection scheme, while also enhancing the stability of BLO through a strongly convex regularizer.

Solving this bi-level optimization problem is challenging because the *implicit gradient* (IG), given by the full vector-wise derivative $\frac{d\boldsymbol{\theta}_{\mathrm{u}}(\mathbf{w})^\top}{d\mathbf{w}}$, is generally intractable. To address this issue, the work [36] adopted the *gradient unrolling* (GU) approach [37, 38], which approximates the lower-level solution $\boldsymbol{\theta}_{\mathrm{u}}(\mathbf{w})$ by performing a finite number of unrolled optimization steps. Interestingly, when the lower-level optimization within GU uses sign-based stochastic gradient descent (signSGD) [39], the IG term vanishes entirely, thereby eliminating the need for explicit IG computation and significantly simplifying the overall optimization.

We further investigate this "challenging forgets" finding through a case study on a biased dataset derived from CelebA [44], designed for hair color prediction (Blond vs. Non-blond) with a spurious correlation to the gender attribute (Male vs. Female) [40]. Figure 5.6 shows the composition of the identified worst-case forget set using BLO under a 10% forgetting ratio, where samples are categorized into four groups based on the combination of hair color

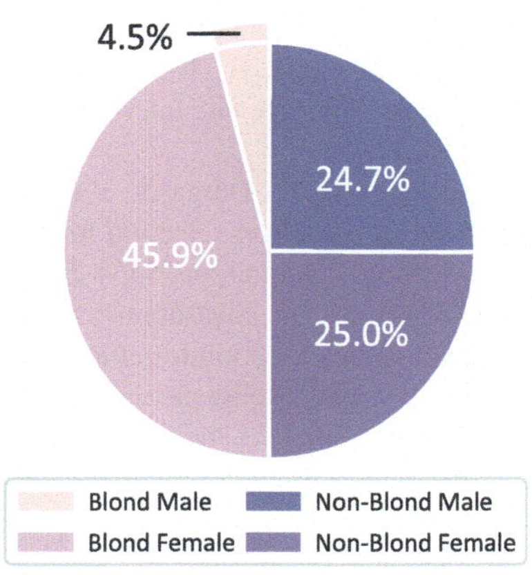

Fig. 5.6 Composition of the worst-case forget set under CelebA

and gender. Notably, the worst-case forget set contains a disproportionately large fraction of samples from the (Blond + Female) group. Since blonde hair is strongly correlated with the female attribute in CelebA, samples from this group are relatively easy to learn, i.e., harder to unlearn.

In summary, coresets offer a fresh perspective on machine unlearning beyond efficiency, revealing how small, strategically chosen subsets can match full-data unlearning performance, improve the forget-utility tradeoff, and expose weaknesses in current benchmarks. These findings highlight the need for more principled coreset selection methods and rigorous evaluation protocols to advance our understanding and practice of unlearning in both vision and language models.

References

1. Paul, M., Ganguli, S., Dziugaite, G.K.: Deep learning on a data diet: finding important examples early in training. In: Beygelzimer, A., Dauphin, Y., Liang, P., Vaughan, J.W. (eds.) Advances in Neural Information Processing Systems (2021). https://openreview.net/forum?id=Uj7pF-D-YvT
2. Toneva, M., Sordoni, A., des Combes, R.T., Trischler, A., Bengio, Y., Gordon, G.J.: An empirical study of example forgetting during deep neural network learning. In: International Conference on Learning Representations (2019). https://openreview.net/forum?id=BJlxm30cKm
3. Pruthi, G., Liu, F., Kale, S., Sundararajan, M.: Estimating training data influence by tracing gradient descent. Adv. Neural Inf. Process. Syst. **33**, 19920–19930 (2020)
4. Pleiss, G., Zhang, T., Elenberg, E., Weinberger, K.Q.: Identifying mislabeled data using the area under the margin ranking. Adv. Neural Inf. Process. Syst. **33**, 17044–17056 (2020)
5. Xia, X., Liu, J., Yu, J., Shen, X., Han, B., Liu, T.: Moderate coreset: a universal method of data selection for real-world data-efficient deep learning. In: The Eleventh International Conference on Learning Representations (2022)
6. Zheng, H., Liu, R., Lai, F., Prakash, A.: Coverage-Centric Coreset Selection for High Pruning Rates. arXiv preprint arXiv:2210.15809 (2022)
7. Maharana, A., Yadav, P., Bansal, M.: $\mathbf{D}^2$ pruning: message passing for balancing diversity & difficulty in data pruning. In: The Twelfth International Conference on Learning Representations (2024). https://openreview.net/forum?id=thbtoAkCe9
8. Mirzasoleiman, B., Bilmes, J., Leskovec, J.: Coresets for data-efficient training of machine learning models. In: International Conference on Machine Learning, pp. 6950–6960. PMLR (2020)
9. Killamsetty, K., Durga, S., Ramakrishnan, G., De, A., Iyer, R.: Grad-match: gradient matching based data subset selection for efficient deep model training. In: International Conference on Machine Learning, pp. 5464–5474. PMLR (2021)
10. Killamsetty, K., Sivasubramanian, D., Ramakrishnan, G., Iyer, R.: Glister: generalization based data subset selection for efficient and robust learning. In: Proceedings of the AAAI Conference on Artificial Intelligence, vol. 35, pp. 8110–8118 (2021)
11. Pooladzandi, O., Davini, D., Mirzasoleiman, B.: Adaptive second order coresets for data-efficient machine learning. In: International Conference on Machine Learning, pp. 17848–17869. PMLR (2022)

12. Guo, C., Zhao, B., Bai, Y.: Deepcore: a comprehensive library for coreset selection in deep learning. In: International Conference on Database and Expert Systems Applications, pp. 181–195. Springer (2022)
13. Azeemi, A., Qazi, I., Raza, A.: Data pruning for efficient model pruning in neural machine translation. In: Findings of the Association for Computational Linguistics: EMNLP 2023, pp. 236–246 (2023)
14. Zhang, X., Zhai, J., Ma, S., Shen, C., Li, T., Jiang, W., Liu, Y.: STAFF: speculative coreset selection for task-specific fine-tuning. In: The Thirteenth International Conference on Learning Representations (2025). https://openreview.net/forum?id=FAfxvdv1Dy
15. Xia, M., Malladi, S., Gururangan, S., Arora, S., Chen, D.: Less: Selecting Influential Data for Targeted Instruction Tuning. arXiv preprint arXiv:2402.04333 (2024)
16. Chen, L., Li, S., Yan, J., Wang, H., Gunaratna, K., Yadav, V., Tang, Z., Srinivasan, V., Zhou, T., Huang, H., Jin, H.: Alpagasus: training a better alpaca with fewer data. In: The Twelfth International Conference on Learning Representations (2024). https://openreview.net/forum?id=FdVXgSJhvz
17. Liu, W., Zeng, W., He, K., Jiang, Y., He, J.: What Makes Good Data for Alignment? A Comprehensive Study of Automatic Data Selection in Instruction Tuning. arXiv preprint arXiv:2312.15685 (2023)
18. Zhou, C., Liu, P., Xu, P., Iyer, S., Sun, J., Mao, Y., Ma, X., Efrat, A., Yu, P., Yu, L., et al.: Lima: less is more for alignment. Adv. Neural Inf. Process. Syst. **36**, 55006–55021 (2023)
19. Pal, S., Wang, C., Diffenderfer, J., Kailkhura, B., Liu, S.: LLM unlearning reveals a stronger-than-expected coreset effect in current benchmarks. In: Second Conference on Language Modeling (2025). https://openreview.net/forum?id=NMIqKUdDkw
20. Li, N., Pan, A., Gopal, A., Yue, S., Berrios, D., Gatti, A., Li, J.D., Dombrowski, A.K., Goel, S., Mukobi, G., Helm-Burger, N., Lababidi, R., Justen, L., Liu, A.B., Chen, M., Barrass, I., Zhang, O., Zhu, X., Tamirisa, R., Bharathi, B., Herbert-Voss, A., Breuer, C.B., Zou, A., Mazeika, M., Wang, Z., Oswal, P., Lin, W., Hunt, A.A., Tienken-Harder, J., Shih, K.Y., Talley, K., Guan, J., Steneker, I., Campbell, D., Jokubaitis, B., Basart, S., Fitz, S., Kumaraguru, P., Karmakar, K.K., Tupakula, U., Varadharajan, V., Shoshitaishvili, Y., Ba, J., Esvelt, K.M., Wang, A., Hendrycks, D.: The WMDP benchmark: measuring and reducing malicious use with unlearning. In: Proceedings of the 41st International Conference on Machine Learning. Proceedings of Machine Learning Research, vol. 235, pp. 28525–28550. PMLR (2024)
21. Shi, W., Lee, J., Huang, Y., Malladi, S., Zhao, J., Holtzman, A., Liu, D., Zettlemoyer, L., Smith, N.A., Zhang, C.: Muse: Machine Unlearning Six-Way Evaluation for Language Models. arXiv preprint arXiv:2407.06460 (2024)
22. Zhang, R., Lin, L., Bai, Y., Mei, S.: Negative preference optimization: from catastrophic collapse to effective unlearning. In: First Conference on Language Modeling (2024). https://openreview.net/forum?id=MXLBXjQkmb
23. Hendrycks, D., Burns, C., Basart, S., Zou, A., Mazeika, M., Song, D., Steinhardt, J.: Measuring Massive Multitask Language Understanding. arXiv preprint arXiv:2009.03300 (2020)
24. Shi, W., Ajith, A., Xia, M., Huang, Y., Liu, D., Blevins, T., Chen, D., Zettlemoyer, L.: Detecting pretraining data from large language models. In: The Twelfth International Conference on Learning Representations (2024). https://openreview.net/forum?id=zWqr3MQuNs
25. Frankle, J., Dziugaite, G.K., Roy, D., Carbin, M.: Linear mode connectivity and the lottery ticket hypothesis. In: International Conference on Machine Learning, pp. 3259–3269. PMLR (2020)
26. Łucki, J., Wei, B., Huang, Y., Henderson, P., Tramèr, F., Rando, J.: An Adversarial Perspective on Machine Unlearning for AI Safety. arXiv preprint arXiv:2409.18025 (2024)
27. Lynch, A., Guo, P., Ewart, A., Casper, S., Hadfield-Menell, D.: Eight Methods to Evaluate Robust Unlearning in LLMs. arXiv preprint arXiv:2402.16835 (2024)

28. Patil, V., Hase, P., Bansal, M.: Can Sensitive Information be Deleted from LLMs? Objectives for Defending Against Extraction Attacks. ICLR (2024)
29. Hu, S., Fu, Y., Wu, Z.S., Smith, V.: Jogging the Memory of Unlearned Model Through Targeted Relearning Attack. arXiv preprint arXiv:2406.13356 (2024)
30. Deeb, A., Roger, F.: Do Unlearning Methods Remove Information from Language Model Weights? arXiv preprint arXiv:2410.08827 (2024)
31. Lo, M., Barez, F., Cohen, S.: Large language models relearn removed concepts. In: Findings of the Association for Computational Linguistics: ACL 2024, pp. 8306–8323. Association for Computational Linguistics (2024)
32. Cobbe, K., Kosaraju, V., Bavarian, M., Chen, M., Jun, H., Kaiser, L., Plappert, M., Tworek, J., Hilton, J., Nakano, R., et al.: Training Verifiers to Solve Math Word Problems. arXiv preprint arXiv:2110.14168 (2021)
33. Zhang, X., Zhao, J., LeCun, Y.: Character-level convolutional networks for text classification. Adv. Neural Inf. Process. Syst. **28** (2015)
34. Patil, V., Stengel-Eskin, E., Bansal, M.: Upcore: Utility-Preserving Coreset Selection for Balanced Unlearning. arXiv preprint arXiv:2502.15082 (2025)
35. Liu, F.T., Ting, K.M., Zhou, Z.H.: Isolation forest. In: 2008 Eighth IEEE International Conference on Data Mining, pp. 413–422. IEEE (2008)
36. Fan, C., Liu, J., Hero, A., Liu, S.: Challenging forgets: unveiling the worst-case forget sets in machine unlearning. In: European Conference on Computer Vision (ECCV) (2024)
37. Zhang, Y., Khanduri, P., Tsaknakis, I., Yao, Y., Hong, M., Liu, S.: An Introduction to Bi-level Optimization: Foundations and Applications in Signal Processing and Machine Learning. arXiv preprint arXiv:2308.00788 (2023)
38. Shaban, A., Cheng, C.A., Hatch, N., Boots, B.: Truncated back-propagation for bilevel optimization. In: The 22nd International Conference on Artificial Intelligence and Statistics, pp. 1723–1732. PMLR (2019)
39. Bernstein, J., Wang, Y.X., Azizzadenesheli, K., Anandkumar, A.: signSGD: Compressed Optimisation for Non-convex Problems. ICML (2018)
40. Sagawa, S., Raghunathan, A., Koh, P.W., Liang, P.: An investigation of why overparameterization exacerbates spurious correlations. In: International Conference on Machine Learning, pp. 8346–8356. PMLR (2020)

Data Integrity for Machine Unlearning

6

Changsheng Wang and Yihua Zhang

Abstract

The reliability of machine unlearning critically depends on the integrity and fidelity of the data to be forgotten. In practice, forget sets may deviate from the original training distribution desired for unlearning due to noise, rewriting, or deliberate modifications such as watermarking. These variations raise important questions, *e.g., how do different types of data shift, from random perturbations to structured transformations, affect unlearning efficacy? When is forgetting robust to imperfect data, and when does performance degrade? Can perturbations, rather than being mere obstacles, be exploited to improve forgetting?* This chapter synthesizes research on data perturbationâŁ"centric machine unlearning, examining how perturbations at both test time and training time influence forgetting efficacy, model utility, and robustness. We begin with test-time data shifts, showing that while perturbations such as Gaussian noise, elastic transformations, and adversarial attacks degrade model utility, they often leave unlearning performance unaffected, or even enhanced, by amplifying forgetting signals. Extending to training-time shifts, we analyze incomplete, rewritten, and watermarked forget sets, finding that unlearning effectiveness depends primarily on the semantic fidelity rather than the surface form of forget data: as long as core concepts remain intact, forgetting proceeds reliably despite lexical or structural alterations. Finally, we demonstrate that watermarking can be transformed from a passive perturbation into an active unlearning signal through frameworks such as Water4MU, which deliberately embed

C. Wang (✉) · Y. Zhang
Michigan State University, East Lansing, MI, USA
e-mail: wangc168@msu.edu

Y. Zhang
e-mail: zhan1908@msu.edu

S. Liu et al. (eds.), *Machine Unlearning for Governance of Foundation Models*, Synthesis Lectures on Computer Vision, https://doi.org/10.1007/978-3-032-17282-2_6

semantically consistent yet separable patterns into forget data to improve forgetting accuracy, robustness against membership inference, and even safety in generative models, all while preserving model utility.

6.1 Machine Unlearning under Test-Time Data Shift

An important yet underexplored challenge in machine unlearning is its sensitivity to *data shift*–situations where the forget set or its underlying distribution deviates from that of the original training data. In real-world deployments, such shifts can arise from diverse sources, ranging from synthetic perturbations like Gaussian noise or elastic transformations to adversarial perturbations deliberately crafted to manipulate model predictions. In this section, we focus on *test-time data shift*, where the forget set encounters distributional distortions during evaluation rather than training. We present how these shifts influence unlearning effectiveness.

We focus on input perturbations at test time, where the forget set is subject to distributional distortions commonly studied in model generalization. Recent work on *forget vectors* [1] investigated whether unlearning remains effective when forget samples are perturbed by image corruptions such as zero-mean Gaussian noise and elastic transformations [2, 3], as well as worst-case adversarial perturbations [4, 5]. As demonstrated in Fig. 6.1, model utility (measured by test and retain accuracy) degrades as perturbation severity increases, which is consistent with the model generalization ability. Surprisingly, however, *unlearning accuracy* remains stable or even improves. Perturbations tend to reduce the model's predictive confidence on forget samples, thereby amplifying the forgetting signal and making them easier to erase.

The observation in Fig. 6.1 motivates a data-centric perspective on unlearning: rather than being treated solely as a threat, input-level shifts can be *leveraged proactively* to design more effective unlearning mechanisms. Building on this idea, the method "forget vectors" [1] introduced a data-driven approach based on a universal input perturbation to achieve unlearning without modifying model parameters. Formally, let $\boldsymbol{\delta}$ denote the universal perturbation applied to all forget samples $x \in \mathcal{D}_{\mathrm{f}}$, producing perturbed inputs:

$$x' = x + \boldsymbol{\delta}.$$

The forget vector in [1] is then obtained by solving

$$\underset{\boldsymbol{\delta}}{\text{minimize}} \quad \ell_{\mathrm{MU}}(\boldsymbol{\delta}; \boldsymbol{\theta}_0, \mathcal{D}_{\mathrm{f}}, \mathcal{D}_{\mathrm{r}}), \tag{6.1}$$

where $\boldsymbol{\theta}_0$ denotes the original model, $\mathcal{D}_{\mathrm{f}}$ the forget set, and $\mathcal{D}_{\mathrm{r}}$ the retain set.

The forget vector approach (6.1) effectively casts unlearning as a *data prompting* problem, drawing connections to visual prompting [7] and adversarial example generation [4]. Cru-

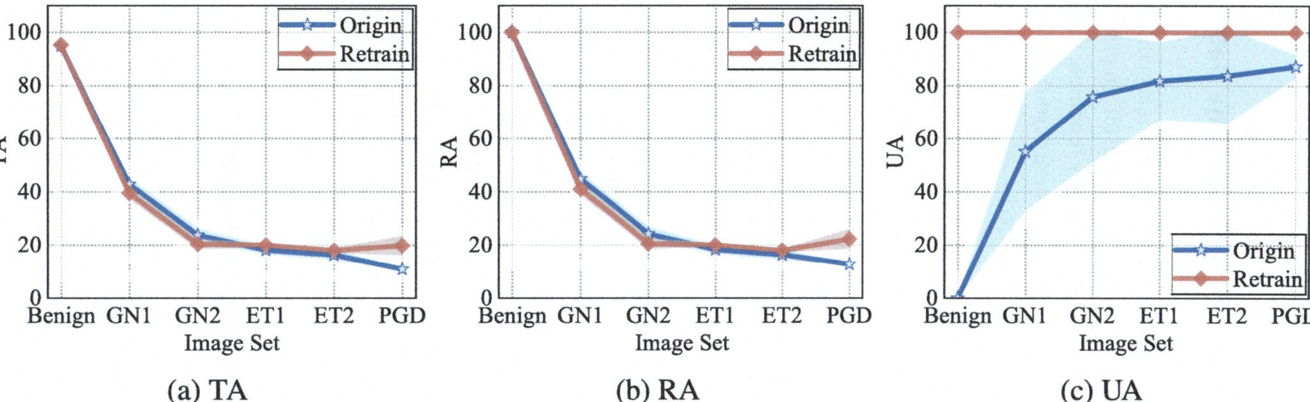

Fig. 6.1 The performance of class-wise forgetting for image classification under (ResNet-18, CIFAR-10) using the exact unlearning method *Retrain* versus the (pre-unlearning) original model performance (origin), evaluated on both benign evaluation sets (benign) and perturbed sets, which include (1) Gaussian noise (GN) with a standard deviation of 0.08 (termed GN1), (2) GN with a standard deviation of 0.2 (termed GN2), (3) Elastic transformation (ET) with parameters (488, 170.8, 24.4) regarding intensity, smoothing, and offset (termed ET1), (4) ET with parameters (488, 19.52, 48.8) (termed ET2), and (5) adversarial perturbations from a 7-step PGD attack [6] with strength $\epsilon = 8/255$. The unlearning performance metrics are reported as **a** TA (testing accuracy), **b** RA (retain accuracy), **c** UA (unlearning accuracy). The average performance is reported over 10 independent trials, where each trial focuses on forgetting one specific class from CIFAR-10. Shaded regions indicate the performance variance

cially, it enables efficient and parameter-free unlearning, offering a lightweight alternative to retraining-based methods while preserving the original model weights.

6.2 Machine Unlearning Under Train-Time Data Shift

While test-time data shift affects evaluation conditions, real-world unlearning requests often involve training-time perturbations to the forget set itself. In practice, forget data are rarely available in pristine form: they may be synthetically rewritten, partially redacted, or embedded with watermarking signals for provenance. These perturbations introduce a key question: *can unlearning methods reliably erase knowledge when the forget set is noisy, incomplete, or structurally altered?*

To formally capture the train-time data shift scenarios, the unlearning objective is extended to operate over a *perturbed forget set* $\mathcal{D}_\mathrm{f}'$ against its original version $\mathcal{D}_\mathrm{f}$:

$$\text{minimize}\,\boldsymbol{\theta} \quad \ell_\mathrm{f}(\boldsymbol{\theta}; \mathcal{D}_\mathrm{f}') + \gamma, \ell_\mathrm{r}(\boldsymbol{\theta}; \mathcal{D}_\mathrm{r}), \tag{6.2}$$

where ℓ_f denotes the unlearning loss on the perturbed forget set and ℓ_r evaluates model utility on benign data. In this section, we analyze three representative train-time data shift scenarios: incomplete forget sets with token-level masking, rewritten forget sets, and watermarked forget sets.

A key takeaway is that unlearning performance depends primarily on the *semantic fidelity* of the forget set rather than its exact lexical form [8, 9]. As long as the core concepts remain intact, forgetting proceeds effectively even when surface tokens differ significantly. To illustrate this point, Fig. 6.2 presents representative noisy forget set scenarios from the WMDP dataset [10] and summarizes the resulting unlearning performance. As shown in Fig. 6.2(right), models unlearned with perturbed forget data achieve performance nearly identical to those trained on the original clean forget set, both in terms of unlearning efficacy and general utility. Notably, all unlearned models exhibit substantially improved unlearning efficacy–reflected by lower accuracy on sensitive content–compared to the original model before unlearning.

6.2.1 Unlearning from Incomplete Forget Sets

A representative scenario involves the *token-level masking* of forget data in LLMs, which arises when sensitive information cannot be disclosed in full or when only partial document excerpts are legally or ethically available. While this setting is loosely related to "coreset"-based unlearning [12], there is an important difference in granularity. Coreset methods typically select a small but representative subset of entire forget samples, whereas the masked setting considered here retains *all* forget samples but with incomplete content, i.e., incompleteness occurs *within* samples rather than across them.

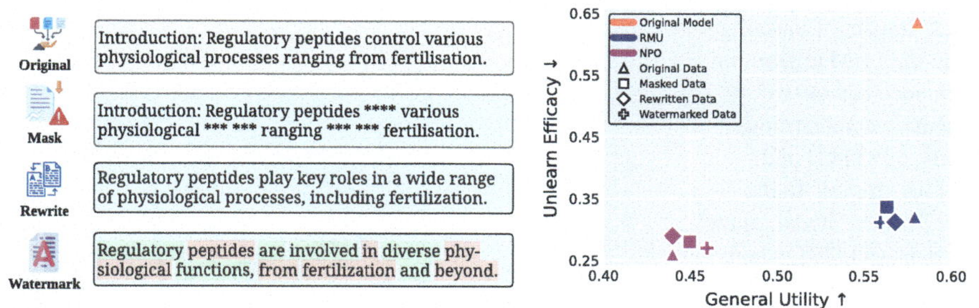

Fig. 6.2 Illustrative examples of "noisy" forget data used during LLM unlearning training (left), and the performance (unlearning efficacy and utility) of the unlearned model evaluated on clean test data (right). (Left) Different perturbation types applied to forget data during unlearning training. These include: *Mask*, where partial or missing content is simulated (masked tokens are indicated by *); *Rewrite*, where LLMs are prompted to generate semantically equivalent variants; and *Watermark*, where identifiable signals are embedded while preserving semantic meaning (tokens containing watermark signals are highlighted in red). **(Right)** Performance evaluation of two representative unlearning methods, NPO [11] and RMU [10], applied to the Zephyr-7b-beta model on the WMDP dataset [10]. The forget data used for unlearning contains different types of perturbations (Mask, Rewrite, Watermark). *Unlearn Efficacy* is reflected by the WMDP evaluation accuracy, where lower values indicate better unlearning performance. *General Utility* reflects MMLU accuracy, where higher values indicate better retention of general model utility. Compared with unlearning on the original forget data format, different perturbation types have minimal impact on unlearning performance

Formally, let $\textsc{Mask}_k(x)$ denote a token-level masking function that randomly masks a proportion k of tokens in each forget sample $x \in \mathcal{D}_\mathrm{f}$:

$$\mathcal{D}'_\mathrm{f} = \left\{\textsc{Mask}_k(x) |\; x \in \mathcal{D}_\mathrm{f}\right\}. \tag{6.3}$$

Empirical results in Fig. 6.3 demonstrate that unlearning remains highly effective under moderate masking. As we can see, the performance of LLM unlearning remains stable until the masking ratio exceeds roughly 30%, after which a sharp drop is observed. This robustness suggests that unlearning algorithms can generalize forgetting behavior even when only partially observed evidence of the forget set is available.

6.2.2 Unlearning Under Rewritten Forget Data

Another important dimension of data shift is scenarios where the forget set is inherently low-quality or synthetically rewritten. A common example occurs when the forget data are *rewritten by LLMs themselves*. This typically happens when the original information exists only in high-level conceptual form or when underspecified content must be regenerated for unlearning purposes. In these cases, complete forget samples are synthesized from coarse descriptions, and the unlearning procedure is applied to these generated variants.

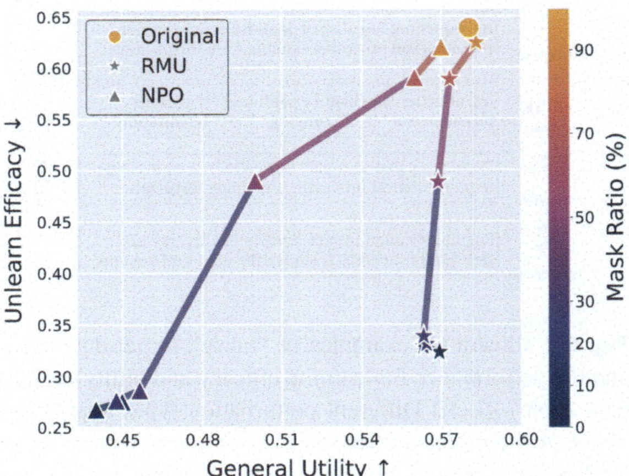

Fig. 6.3 Impact of masking ratio on the LLM unlearning performance across two representative unlearning methods, NPO [11] and RMU [10], applied to the Zephyr-7b-beta model on the WMDP dataset [10], where the masking ratio (k) varies from 0 to 90%. Performance is stable up to 30% masking, after which a sharp drop is observed

To simulate this setting, we prompt the target LLM (the model before unlearning) to produce semantically faithful rewrites of each forget example. This setup enables us to analyze how rewriting-induced variation influences unlearning performance. Formally, let $\textsc{Rewrite}(\cdot)$ denote a function that generates a paraphrased variant of a forget sample x while preserving its original semantics. The rewritten forget set is then given by

$$\mathcal{D}'_{\mathrm{f}} = \{\textsc{Rewrite}(x) \mid x \in \mathcal{D}_{\mathrm{f}}\}. \tag{6.4}$$

The rewriting process follows structured prompting schemes inspired by back-translation and controlled paraphrasing, ensuring both semantic consistency and fidelity to the original intent–two properties widely leveraged in text generation tasks to maintain meaning preservation.

Empirical evaluations demonstrate that unlearning performance remains robust even under such rewriting-induced perturbations. Table 6.1 reports the performance of the RMU unlearning method [10] on the WMDP benchmark using the Zephyr-7B-beta model under three forget set conditions: clean, randomly masked (with 30% masking ratio), and semantically rewritten. Two key metrics are used: unlearning effectiveness (UE), which measures the extent to which the model forgets the specified data (lower values indicate better forgetting), and model utility (UT), which measures retained performance on benign tasks unrelated to the forget set (higher values indicate better utility preservation).

As shown in Table 6.1, the rewritten condition yields a UE score nearly identical to that obtained using the clean forget set, with only negligible changes in UT. These results demonstrate that unlearning algorithms such as RMU can reliably erase semantically equivalent data even when the lexical surface form of the forget examples differs significantly from the original, emphasizing that semantic fidelity rather than surface identity governs unlearning effectiveness.

Table 6.1 Performance of RMU [10] unlearning on perturbed forget data using Zephyr-7B-beta

Metric	No unlearning	Clean	Mask	Rewrite
UE ↓	0.6386	0.3229	0.3382	0.3142
UT ↑	0.5805	0.5692	0.5632	0.5680

Comparison of unlearning efficacy and general utility on the WMDP benchmark under various forget data conditions: no unlearning (i.e., original model before unlearning on forget data), clean, randomly masked (incomplete, $k = 30\%$), semantically rewritten (prompt-based)

6.2.3 Unlearning Under Watermarked Forget Sets

Watermarking introduces a unique and increasingly important dimension to machine unlearning. Unlike random corruption or unintentional distributional shifts, watermarking is a deliberate process embedded to protect intellectual property, trace data provenance, or ensure accountability in the use of generative models. While we defer a detailed discussion of watermarking for unlearning to Sect. 6.3, we briefly highlight a few key results for unlearning under watermarked data below.

In the context of LLMs, watermarking equips models with decoding processes deliberately modified to embed imperceptible patterns into generated text, enabling ownership verification and content attribution [13]. These techniques are increasingly adopted by LLM providers to deter misuse in content-generation scenarios. However, they introduce a new and largely unexplored challenge for machine unlearning: when synthetic or rewritten forget data are produced by a watermarked LLM, the resulting text inherits watermark artifacts, making the forget set $\mathcal{D}_f$ inherently "noisy."

To examine the above challenge, we consider the scenario where the forget data is rewritten using a watermarked LLM. The watermarking process introduces imperceptible signals into the rewritten text due to the modified decoding behavior, yielding a watermarked forget set

$$\mathcal{D}_{\mathrm{f}}' = \left\{ \text{WATERMARK}(x) \mid \forall x \in \mathcal{D}_{\mathrm{f}} \right\}, \tag{6.5}$$

where $\text{WATERMARK}(x)$ denotes the output of a watermark-enabled LLM decoding process applied to the original forget sample x. The resulting forget set thus combines both semantic transformations from rewriting and hidden watermark patterns from the generation process.

Two representative strategies for embedding watermarks are logits-based watermarking (e.g., KGW [14]) and sampling-based watermarking (e.g., SynthID [13]). Figure 6.2-(Left) shows an example of KGW-based watermarking, whose strength is measured by the proportion of red-list tokens in the generated text. Increasing the so-called hardness parameter δ forces a larger fraction of tokens into the red list, making the watermark signal stronger and more detectable. By contrast, SynthID embeds watermarks by modifying the sampling process via multi-layer tournament selection guided by a hidden key, rather than perturbing token logits directly. Raising the number of tournament layers enhances watermark strength.

Table 6.2 Performance of RMU-based LLM unlearning [10] on perturbed forget data using the model Zephyr-7B-beta

Metric	No unlearning	Clean	Mask	Rewrite	WM (KGW)	WM (SynID)
UE ↓	0.6386	0.3229	0.3382	0.3142	0.3134	0.3221
UT ↑	0.5805	0.5692	0.5632	0.5680	0.5694	0.5684

Comparison of unlearning efficacy and general utility on the WMDP benchmark under various forget data conditions: no unlearning (i.e., original model before unlearning on forget data), clean, randomly masked (incomplete), semantically rewritten (prompt-based), and watermarked (KGW [14] and SynthID [13])

As shown in Table 6.2, these watermarking schemes do not compromise unlearning effectiveness: on the WMDP benchmark, models unlearned with watermarked data achieve utility and forgetting scores nearly identical to those obtained using the original, unperturbed data.

6.3 Watermarking for Unlearning: From Passive Perturbation to Active Unlearning Signal

Beyond passive robustness, watermarking can be deliberately leveraged to facilitate forgetting by creating a clearer separation between forget and retain sets. Recent work in the vision domain introduces Water4MU [15], a bi-level optimization framework where watermark encoders and decoders are trained to minimize unlearning difficulty while preserving downstream utility [9]. Under this framework, watermarked data amplify the unlearning signal during model training without corrupting semantic content or generalization ability.

Empirical results in Table 6.3 demonstrate that across diverse classification tasks, including CIFAR-10, CIFAR-100, SVHN, and ImageNet, Water4MU [15] consistently enhances unlearning accuracy (UA) and membership inference attack efficacy (MIA) while preserving high retain accuracy (RA) and test accuracy (TA). Notably, in class-wise forgetting on CIFAR-10, Water4MU improves UA by over five percentage points compared to standard fine-tuning baselines. A similar pattern emerges on ImageNet, where Water4MU achieves substantial UA gains under GA-based unlearning, accompanied by only a marginal reduction in RA.

Moreover, the benefits of watermark-based unlearning extend beyond classification. In the scenario of diffusion model unlearning on the UnlearnCanvas dataset [16], watermarked forget sets outperform erasure baselines such as ESD [17] and UCE [18] on multiple metrics, including in-domain and cross-domain retain accuracy (IRA, CRA) and Fréchet Inception Distance (FID). For instance, as shown in Table 6.4, Water4MU achieves lower FID scores, indicating higher image quality, while simultaneously strengthening forgetting efficacy relative to ESD and UCE. Figure 6.4 showcases the visualizations of the generations produced by

Table 6.3 Performance of different unlearning methods under both unwatermarked forget/retain sets (Original) and Water4MU-induced watermarked forget/retain sets

Metric	Retrain			GA			FT			Sparse			IU		
	Original	WATER4MU	Diff	Original	WATER4MU	Diff.	Original	WATER4MU	Diff.	Original	WATER4MU	Diff.	Original	WATER4MU	Diff.
10% random data forgetting, CIFAR-10															
UA↑	6.78$_{\pm0.63}$	10.01$_{\pm0.69}$	3.23▲	0.80$_{\pm0.32}$	1.92$_{\pm0.25}$	1.12▲	1.85$_{\pm0.69}$	4.93$_{\pm0.76}$	3.08▲	6.11$_{\pm0.72}$	7.50$_{\pm0.64}$	1.39▲	0.64$_{\pm0.21}$	2.62$_{\pm0.22}$	1.98▲
MIA↑	16.06$_{\pm0.07}$	19.33$_{\pm0.05}$	3.27▲	1.89$_{\pm0.02}$	5.67$_{\pm0.02}$	3.78▲	5.60$_{\pm0.22}$	8.26$_{\pm0.12}$	2.66▲	13.08$_{\pm0.08}$	14.70$_{\pm0.02}$	1.62▲	1.53$_{\pm0.01}$	3.67$_{\pm0.01}$	2.14▲
RA↑	100.00$_{\pm0.00}$	99.93$_{\pm0.01}$	0.07▼	99.42$_{\pm0.27}$	99.18$_{\pm0.34}$	0.24▼	99.66$_{\pm0.06}$	98.75$_{\pm0.10}$	0.91▼	97.76$_{\pm0.32}$	97.22$_{\pm0.23}$	0.54▼	99.43$_{\pm0.22}$	98.98$_{\pm0.36}$	0.45▼
TA↑	92.02$_{\pm0.02}$	90.63$_{\pm0.02}$	1.39▼	92.19$_{\pm0.52}$	92.39$_{\pm0.38}$	0.20▲	93.54$_{\pm0.27}$	92.11$_{\pm0.41}$	1.43▼	91.61$_{\pm0.46}$	91.65$_{\pm0.55}$	0.04▲	94.51$_{\pm0.87}$	92.87$_{\pm0.74}$	1.64▼
Class-wise forgetting, CIFAR-10															
UA↑	100.00$_{\pm0.00}$	100.00$_{\pm0.00}$	0.00-	41.84$_{\pm7.35}$	49.03$_{\pm5.46}$	7.19▲	37.29$_{\pm8.18}$	53.25$_{\pm6.55}$	15.96▲	87.09$_{\pm2.19}$	94.60$_{\pm2.02}$	7.51▲	92.03$_{\pm2.54}$	98.49$_{\pm1.12}$	6.46▲
MIA↑	100.00$_{\pm0.00}$	100.00$_{\pm0.00}$	0.00-	55.11$_{\pm8.32}$	61.42$_{\pm7.75}$	6.31▲	55.96$_{\pm6.49}$	69.24$_{\pm9.17}$	13.28▲	90.62$_{\pm2.13}$	98.71$_{\pm1.14}$	8.09▲	99.32$_{\pm0.21}$	100.00$_{\pm0.00}$	0.68▲
RA↑	100.00$_{\pm0.00}$	99.84$_{\pm0.03}$	0.16▼	99.20$_{\pm0.12}$	99.07$_{\pm0.23}$	0.13▼	99.31$_{\pm0.07}$	98.76$_{\pm0.12}$	0.55▼	99.58$_{\pm0.05}$	95.27$_{\pm0.33}$	4.31▼	92.27$_{\pm0.26}$	92.17$_{\pm0.41}$	0.10▼
TA↑	91.20$_{\pm0.11}$	89.95$_{\pm0.21}$	1.25▼	87.87$_{\pm0.07}$	88.62$_{\pm0.18}$	0.75▲	90.67$_{\pm0.22}$	90.61$_{\pm0.14}$	0.06▼	90.48$_{\pm0.31}$	90.24$_{\pm0.10}$	0.24▼	89.50$_{\pm0.07}$	88.67$_{\pm0.05}$	0.83▼
10% random data forgetting, CIFAR-100															
UA↑	24.88$_{\pm0.13}$	28.32$_{\pm0.67}$	3.44▲	0.05$_{\pm0.01}$	2.29$_{\pm0.12}$	2.24▲	0.98$_{\pm0.06}$	3.11$_{\pm0.27}$	2.13▲	9.12$_{\pm0.21}$	13.31$_{\pm0.11}$	4.19▲	3.19$_{\pm0.09}$	5.43$_{\pm0.33}$	2.24▲
MIA↑	49.83$_{\pm0.17}$	49.96$_{\pm0.25}$	0.13▲	2.40$_{\pm0.21}$	5.49$_{\pm0.52}$	3.09▲	2.29$_{\pm0.29}$	5.89$_{\pm0.32}$	3.60▲	12.56$_{\pm0.72}$	19.64$_{\pm1.07}$	7.08▲	6.71$_{\pm0.40}$	10.78$_{\pm0.33}$	4.07▲
RA↑	99.99$_{\pm0.01}$	99.38$_{\pm0.21}$	0.61▼	99.97$_{\pm0.02}$	99.58$_{\pm0.09}$	0.39▼	99.98$_{\pm0.01}$	99.91$_{\pm0.03}$	0.07▼	96.32$_{\pm0.07}$	96.71$_{\pm0.11}$	0.39▲	97.86$_{\pm0.02}$	97.11$_{\pm0.03}$	0.75▼
TA↑	76.35$_{\pm0.13}$	73.31$_{\pm0.22}$	3.04▼	75.91$_{\pm0.19}$	74.43$_{\pm0.35}$	1.48▼	75.96$_{\pm0.27}$	74.83$_{\pm0.14}$	1.13▼	70.43$_{\pm0.49}$	69.77$_{\pm0.26}$	0.66▼	72.75$_{\pm0.25}$	72.53$_{\pm0.16}$	0.22▼
10% random data forgetting, SVHN															
UA↑	7.49$_{\pm0.16}$	8.79$_{\pm0.29}$	1.30▲	0.02$_{\pm0.01}$	2.18$_{\pm0.17}$	2.16▲	1.98$_{\pm0.23}$	4.31$_{\pm0.20}$	2.33▲	5.01$_{\pm0.17}$	7.66$_{\pm0.12}$	2.65▲	4.94$_{\pm0.12}$	7.02$_{\pm0.47}$	2.08▲
MIA↑	17.78$_{\pm0.14}$	17.67$_{\pm0.15}$	0.11▼	0.30$_{\pm0.06}$	3.67$_{\pm0.22}$	3.37▲	5.35$_{\pm0.19}$	11.69$_{\pm0.55}$	6.34▲	10.62$_{\pm0.15}$	13.81$_{\pm0.38}$	3.19▲	7.23$_{\pm0.09}$	8.96$_{\pm0.25}$	1.73▲
RA↑	100.00$_{\pm0.00}$	99.95$_{\pm0.02}$	0.05▼	99.97$_{\pm0.01}$	99.00$_{\pm0.13}$	0.97▼	98.98$_{\pm0.07}$	99.59$_{\pm0.11}$	0.61▲	99.14$_{\pm0.05}$	98.57$_{\pm0.12}$	0.57▼	99.46$_{\pm0.02}$	99.20$_{\pm0.04}$	0.26▼
TA↑	93.71$_{\pm0.42}$	93.63$_{\pm0.28}$	0.08▼	92.59$_{\pm0.57}$	92.51$_{\pm0.18}$	0.08▼	92.89$_{\pm0.13}$	93.40$_{\pm0.37}$	0.51▲	93.70$_{\pm0.08}$	93.22$_{\pm0.17}$	0.48▼	91.13$_{\pm0.15}$	90.46$_{\pm0.11}$	0.67▼
Class-wise forgetting, ImageNet (10% classes)															
UA↑	74.33$_{\pm0.62}$	76.45$_{\pm0.49}$	2.12▲	61.10$_{\pm0.45}$	67.23$_{\pm1.36}$	6.13▲	45.52$_{\pm0.86}$	53.71$_{\pm0.59}$	8.19▲	80.25$_{\pm0.59}$	85.09$_{\pm0.37}$	4.84▲	50.34$_{\pm0.56}$	53.71$_{\pm0.86}$	3.37▲
MIA↑	98.30$_{\pm0.17}$	98.69$_{\pm0.08}$	0.39▲	98.71$_{\pm0.16}$	99.64$_{\pm0.22}$	0.93▲	96.35$_{\pm0.34}$	99.56$_{\pm0.27}$	3.21▲	98.20$_{\pm0.20}$	99.94$_{\pm0.33}$	3.21▲	95.78$_{\pm0.18}$	95.88$_{\pm0.13}$	0.10▲
RA↑	65.80$_{\pm0.31}$	64.73$_{\pm0.29}$	1.07▼	63.37$_{\pm0.76}$	63.05$_{\pm0.52}$	0.32▼	65.86$_{\pm0.46}$	65.68$_{\pm0.19}$	0.18▼	64.34$_{\pm0.83}$	62.41$_{\pm0.08}$	1.93▼	66.63$_{\pm0.36}$	66.24$_{\pm0.12}$	0.39▼
TA↑	65.48$_{\pm0.63}$	64.21$_{\pm0.77}$	1.27▼	63.22$_{\pm0.47}$	63.10$_{\pm0.40}$	0.12▼	64.89$_{\pm0.18}$	64.37$_{\pm0.52}$	0.52▼	63.69$_{\pm0.34}$	63.22$_{\pm0.23}$	0.47▼	65.80$_{\pm0.24}$	65.09$_{\pm0.25}$	0.71▼

Results are reported on (CIFAR-10, ResNet-18), as well as on additional datasets (CIFAR-100, SVHN under random data forgetting, and IMAGENET under class-wise forgetting). The result format is $a_{\pm b}$ with mean a and standard deviation b over 10 independent trials. We present results under random data forgetting (10% forgetting ratio) and class-wise forgetting (entire classes, or 10% of ImageNet classes). The performance difference is provided in Diff. The results in this table are sourced from [15]

Table 6.4 Performance overview of WATER4MU evaluated on style and object unlearning for diffusion models on the dataset UNLEARNCANVAS

Metric	Style unlearning			Object unlearning			FID
	UA↑	IRA↑	CRA↑	UA↑	IRA↑	CRA↑	
ESD	98.58	80.97	93.96	92.15	55.78	44.23	65.55
FMN	88.48	56.77	46.60	45.64	90.63	73.46	131.37
UCE	98.40	60.22	47.71	94.31	39.35	34.67	182.01
WATER4MU	93.49	92.75	94.01	82.24	85.60	81.52	86.62

The performance metrics include UA (unlearning accuracy), IRA (in-domain retain accuracy), CRA (cross-domain retain accuracy), and FID. Results are averaged over all the style and object unlearning cases. The results in this table is sourced from [15]

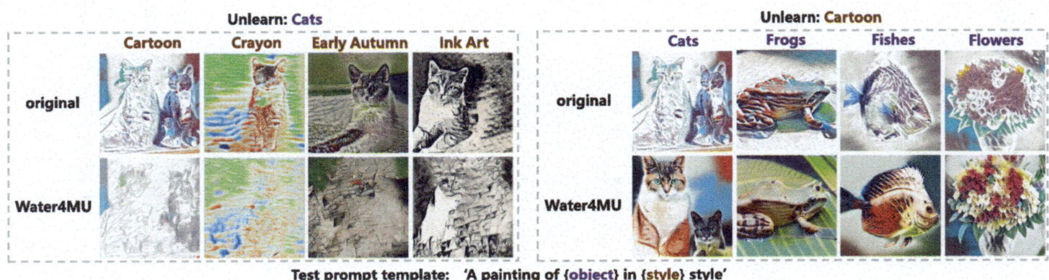

Fig. 6.4 Examples of generated images by SD w/ and w/o MU with Water4MU. Concept types are distinguished by color: styles in brown and objects in purple. The first row serves as a reference for comparison before unlearning

unlearned SD with Water4MU. As observed, Water4MU achieves effective concept erasure in both style and object unlearning.

Watermarking for unlearning also improves safety in generative models. On the I2P (Inappropriate-to-Prompt) benchmark for unsafe image generation, Water4MU reduces the number of detected unsafe outputs relative to competing erasure methods such as ESD [17], FMN [19], and UCE [18]. For example, as shown in Table 6.5, Water4MU generates fewer instances of explicit content across multiple nudity-related body part categories, highlighting its potential for aligning unlearning with broader safety goals.

The above findings support a unified, data-centric perspective: watermarking should be viewed not merely as a perturbation to be tolerated, but as a design lever for reliable and efficient unlearning. By embedding semantically consistent yet separable signals into the forget set, watermarking enables stronger forgetting with minimal cost to utility, and, in some cases, with measurable safety gains.

Table 6.5 Number of nude body parts detected by Nudenet [20] on I2P dataset [21] with threshold 0.6. The results in this table are sourced from [15]

Method	Breast	Genitalia	Buttocks	Feet	Belly	Armpits	Total
SD v1.4	229	31	44	42	171	129	646
ESD	**22**	**6**	5	24	31	33	121
FMN	172	17	12	56	116	42	415
UCE	50	14	11	20	55	36	186
WATER4MU	29	15	**5**	**10**	**29**	**21**	**109**

References

1. Sun, C., Wang, R., Zhang, Y., Jia, J., Liu, J., Liu, G., Yan, Y., Liu, S.: Forget vectors at play: Universal input perturbations driving machine unlearning in image classification (2024). arXiv:2412.16780
2. Hendrycks, D., Basart, S., Mu, N., Kadavath, S., Wang, F., Dorundo, E., Desai, R., Zhu, T., Parajuli, S., Guo, M., et al.: The many faces of robustness: a critical analysis of out-of-distribution generalization. In: Proceedings of the IEEE/CVF International Conference on Computer Vision, pp. 8340–8349 (2021)
3. Hendrycks, D., Dietterich, T.: Benchmarking neural network robustness to common corruptions and perturbations (2019). arXiv:1903.12261
4. Goodfellow, I.J., Shlens, J., Szegedy, C.: Explaining and harnessing adversarial examples (2014). arXiv:1412.6572
5. Madry, A., Makelov, A., Schmidt, L., Tsipras, D., Vladu, A.: Towards deep learning models resistant to adversarial attacks (2017). arXiv:1706.06083
6. Madry, A., Makelov, A., Schmidt, L., Tsipras, D., Vladu, A.: Towards deep learning models resistant to adversarial attacks. In: International Conference on Learning Representations (2018). https://openreview.net/forum?id=rJzIBfZAb
7. Chen, A., Yao, Y., Chen, P.Y., Zhang, Y., Liu, S.: Understanding and improving visual prompting: A label-mapping perspective. In: Proceedings of the IEEE/CVF Conference on Computer Vision and Pattern Recognition, pp. 19133–19143 (2023)
8. Wang, C., Zhang, Y., Wei, D., Jia, J., Chen, P.Y., Liu, S.: LLM unlearning on noisy forget sets: a study of incomplete, rewritten, and watermarked data. In: ACM CCS AISec'25 (2025)
9. Wang, C., Zhang, Y., Jia, J., Wei, D., Liu, S.: Noisy but forgotten: LLM unlearning are robust against perturbed data in the wild. In: ICML 2025 Workshop on Machine Unlearning for Generative AI
10. Li, N., Pan, A., Gopal, A., Yue, S., Berrios, D., Gatti, A., Li, J.D., Dombrowski, A.K., Goel, S., Mukobi, G., Helm-Burger, N., Lababidi, R., Justen, L., Liu, A.B., Chen, M., Barrass, I., Zhang, O., Zhu, X., Tamirisa, R., Bharathi, B., Herbert-Voss, A., Breuer, C.B., Zou, A., Mazeika, M., Wang, Z., Oswal, P., Lin, W., Hunt, A.A., Tienken-Harder, J., Shih, K.Y., Talley, K., Guan, J., Steneker, I., Campbell, D., Jokubaitis, B., Basart, S., Fitz, S., Kumaraguru, P., Karmakar, K.K., Tupakula, U., Varadharajan, V., Shoshitaishvili, Y., Ba, J., Esvelt, K.M., Wang, A., Hendrycks, D.: The WMDP benchmark: measuring and reducing malicious use with unlearning. In: Proceedings

of the 41st International Conference on Machine Learning, Proceedings of Machine Learning Research, vol. 235, pp. 28525–28550. PMLR (2024)

11. Zhang, R., Lin, L., Bai, Y., Mei, S.: Negative preference optimization: From catastrophic collapse to effective unlearning. In: First Conference on Language Modeling (2024). https://openreview.net/forum?id=MXLBXjQkmb
12. Pal, S., Wang, C., Diffenderfer, J., Kailkhura, B., Liu, S.: LLM unlearning reveals a stronger-than-expected coreset effect in current benchmarks. In: Second Conference on Language Modeling (2025). https://openreview.net/forum?id=NMIqKUdDkw
13. Dathathri, S., See, A., Ghaisas, S., Huang, P.S., McAdam, R., Welbl, J., Bachani, V., Kaskasoli, A., Stanforth, R., Matejovicova, T., et al.: Scalable watermarking for identifying large language model outputs. Nature **634**(8035), 818–823 (2024)
14. Kirchenbauer, J., Geiping, J., Wen, Y., Katz, J., Miers, I., Goldstein, T.: A watermark for large language models. In: International Conference on Machine Learning, pp. 17061–17084. PMLR (2023)
15. Sun, Y., Zhang, Y., Liu, G., Xie, H., Liu, S.: Invisible watermarks, visible gains: steering machine unlearning with bi-level watermarking design. In: International Conference on Computer Vision (2025)
16. Zhang, Y., Fan, C., Zhang, Y., Yao, Y., Jia, J., Liu, J., Zhang, G., Liu, G., Kompella, R.R., Liu, X., Liu, S.: UnlearnCanvas: stylized image dataset for enhanced machine unlearning evaluation in diffusion models. In: The Thirty-Eight Conference on Neural Information Processing Systems Datasets and Benchmarks Track (2024). https://openreview.net/forum?id=t9aThFL1lE
17. Gandikota, R., Materzynska, J., Fiotto-Kaufman, J., Bau, D.: Erasing concepts from diffusion models. In: Proceedings of the IEEE/CVF International Conference on Computer Vision, pp. 2426–2436 (2023)
18. Gandikota, R., Orgad, H., Belinkov, Y., Materzyńska, J., Bau, D.: Unified concept editing in diffusion models (2023). arXiv:2308.14761
19. Zhang, E., Wang, K., Xu, X., Wang, Z., Shi, H.: Forget-me-not: learning to forget in text-to-image diffusion models (2023). arXiv:2303.17591
20. notAI-tech: NudeNet: lightweight nudity detection. GitHub repository (2019). https://github.com/notAI-tech/NudeNet. Accessed 2025 Aug 24
21. Schramowski, P., Brack, M., Deiseroth, B., Kersting, K.: Safe latent diffusion: Mitigating inappropriate degeneration in diffusion models. In: Proceedings of the IEEE/CVF Conference on Computer Vision and Pattern Recognition, pp. 22522–22531 (2023)

Split, Unlearn, Merge: Leveraging Data Attributes for More Effective Unlearning in LLMs

7

Swanand Ravindra Kadhe, Farhan Ahmed, Dennis Wei, Nathalie Baracaldo and Inkit Padhi

Abstract

This chapter considers the use of attributes of unlearning data to enhance unlearning performance. It studies a framework called "**SP**lit, **UN**learn, Mer**GE**" (SPUNGE), which splits unlearning data into subsets based on the values of a selected attribute, unlearns each subset separately, and merges the unlearned models. SPUNGE can be used with any unlearning method to amplify its effectiveness. We empirically demonstrate such improvements for two recent unlearning methods, reducing undesirable behaviors and hazardous knowledge in two popular LLMs.

7.1 Introduction

Most unlearning methods do not take advantage of characteristics of the data. In this chapter, we explore how leveraging *attributes* associated with unlearning data can significantly

S. R. Kadhe · F. Ahmed · D. Wei · N. Baracaldo (✉) · I. Padhi
IBM Research, San Jose, CA, USA
e-mail: baracald@us.ibm.com

S. R. Kadhe
e-mail: swanand.kadhe@ibm.com

F. Ahmed
e-mail: Farhan.Ahmed@ibm.com

D. Wei
e-mail: dwei@us.ibm.com

I. Padhi
e-mail: inkit.padhi@ibm.com

S. Liu et al. (eds.), *Machine Unlearning for Governance of Foundation Models*, Synthesis Lectures on Computer Vision, https://doi.org/10.1007/978-3-032-17282-2_7

improve the effectiveness of unlearning. Making use of such characteristics of unlearning *data* [1, 2] is a direction complementary to most work on unlearning, which focuses on the efficiency and effectiveness of unlearning methods. We specifically study a simple yet effective framework called SPUNGE: "**SP**lit, **UN**learn, then mer**GE**," which operates in three steps (see Fig. 7.1):

1. the unlearning data is split into subsets based on the values of a selected attribute;
2. each subset is separately used to unlearn a subtype of the undesired behavior, resulting in multiple unlearned LLMs;
3. the unlearned LLMs are *merged* to obtain the final unlearned LLM.

In the rest of this chapter, we show that by simply performing these steps, techniques shown in previous chapters can improve their performance. We focus our assessment in three scenarios where unlearning has been shown to reduce undesirable behaviors and knowledge: toxicity and hate speech [3–7]; social bias; and hazardous scientific knowledge [8]. For toxicity and hate speech, we configure SPUNGE to leverage demographic information (e.g., gender, ethnicity, religion, etc.), while for hazardous knowledge, we use the scientific domain (e.g., biosecurity, cybersecurity).

In the next section, we provide background on social bias mitigation techniques. Then, we show how SPUNGE can be combined with popular unlearning techniques such as Task Vector Negation [3] and Representation Misdirection Unlearning [8]. Finally, we empirically demonstrate applying SPUNGE maintains the general capabilities of the LLMs as measured on 10 standard benchmarks while substantially reducing harmful content.

7.2 Background on Social Bias and Biosecurity Risks

The rapid improvement and increasing adoption of large language models (LLMs) has been accompanied by their downsides, notably their potential harmful behaviors [9]. LLMs are known to generate harmful content such as toxic, hateful, or biased language [10–12]. LLMs also contain hazardous knowledge of sensitive topics such as biosecurity, which can be (mis)used to empower malicious actors [13, 14]. A widely adopted way to safeguard against harmful or objectionable responses is to *align* LLMs via safety tuning [15–18]. However, safety tuning of LLMs has shown to be vulnerable to adversarial or *jailbreak* attacks where adversarial prompts break through alignment and re-invoke harmful responses [19–21]. Even subsequent benign fine-tuning can degrade alignment [22].

Early works in reducing toxicity in language models [23–25] have focused on small to moderated sized models and restrict to explicit toxicity. Detoxification techniques primarily employ controlled text generation methods, which incurs heavy inference overhead and it is difficult to measure model performance on benchmark tasks.

In parallel, machine *unlearning* has emerged as a promising paradigm for more targeted and efficient sociotechnical harm reduction. It has been shown that unlearning can reduce toxicity and other harmful responses [3–7] and erase hazardous scientific knowledge [8]. Unlearning can be considered a complementary safety tool to alignment techniques and can be used before or after alignment [2].

However, as noted earlier, existing approaches rarely leverage the specific characteristics of the domain in which unlearning is applied. Most prior work has focused on developing unlearning methods, evaluation metrics, and benchmarks, without considering the underlying attributes of the data being unlearned. In contrast, our proposed Spunge framework explicitly incorporates data attributes to strengthen the effectiveness of any unlearning method.

7.3 Spunge Framework

The proposed SPUNGE framework is illustrated in Fig. 7.1 and in Algorithm 1. We focus on unlearning behaviors or bodies of knowledge (as opposed to smaller, discrete units of information) from a given LLM with parameters θ_{init}; this is represented by a dataset D consisting of examples of the undesired behavior or knowledge. We consider scenarios in which the dataset can be partitioned into subsets corresponding to different values $a_1, \ldots, a_n$ of an attribute a in the data which can often be identified. For example, the attribute for the

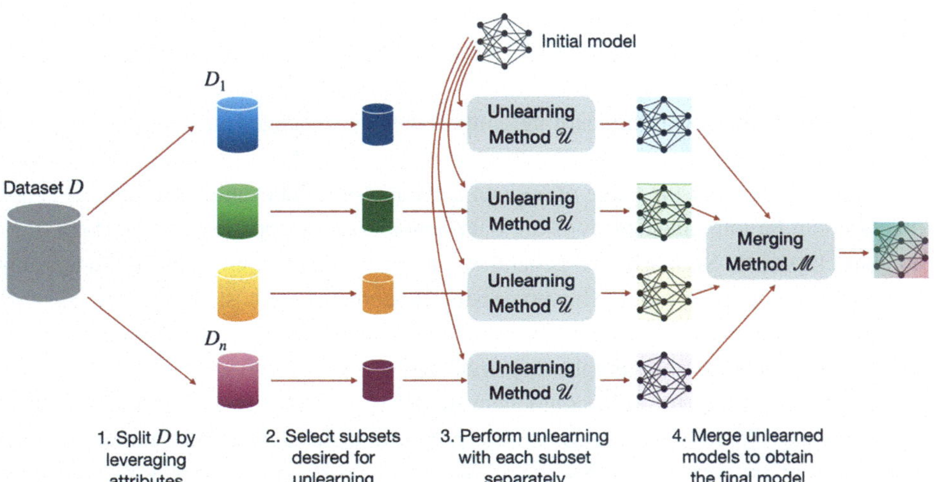

Fig. 7.1 An overview of the SPlit, UNlearn, then merGE (SPUNGE) framework. SPUNGE splits the unlearning dataset into subsets based on selected attribute values, unlearns each subset separately, and then merges the unlearned models

Algorithm 1 SPUNGE Framework

Input: Initial model parameters θ_{init}, Unlearning dataset D, Attribute with values $a_1, \ldots, a_n$, Processing pipeline `proc`, Unlearning method $\mathcal{U}$, Merging method $\mathcal{M}$
Output: Unlearned model θ^u
for $t = 1$ **to** n **do**
  Select subset associated with data attribute value a_t as
    $D_t = \{\mathbf{x} \in D \mid \texttt{attr}(\mathbf{x}) = a_t\}$
  Process subset for unlearning
    $D_t^u = \{\texttt{proc}(\mathbf{x}) \mid \mathbf{x} \in D_t\}$
  Perform unlearning $\theta_t^u \leftarrow \mathcal{U}(\theta_{\text{init}}, D_t^u)$
end for
Perform merging $\theta^u \leftarrow \mathcal{M}(\theta_1^u, \ldots, \theta_n^u)$

case of unlearning toxicity could be the demographic group (e.g., women, Muslims) targeted by the toxic text.

Given a dataset and attribute as described above, the SPUNGE framework consists of the following steps: (1) Split the dataset into subsets D_t for $t = 1, \ldots, n$ based on the attribute. (2) Perform unlearning separately on each subset D_t, all starting from the given LLM, θ_{init}, and yielding n different unlearned LLMs, θ_t^u. (3) Merge the unlearned LLMs into a single final unlearned LLM, θ^u. (We provide more details about merging later.)

SPUNGE can be instantiated with any unlearning method $\mathcal{U}(\theta_{\text{init}}, D_t^u)$ and merging method $\mathcal{M}(\theta_1^u, \ldots, \theta_n^u)$, where the unlearning method updates model parameters from θ_{init} to θ_t^u using data subset D_t^u, and the merging method combines these independent parameters $\theta_1^u, \ldots, \theta_n^u$ into one θ^u. We present details of the methods we use for instantiating SPUNGE in Sect. 7.4.

It is frequently the case for unlearning samples to have associated attributes. SPUNGE can be applied to a variety of attributes. For this reason, in Algorithm 1, we consider a function `attr(·)` that can output the value of a given attribute for a data sample. In practice, such a function can be implemented by using data annotations or appropriate classifiers (e.g., a domain classifier). Similarly, we generalize any processing required by the unlearning method with function `proc(·)`. This processing function abstracts steps such as selecting representative samples with undesirable behavior or knowledge; it can also augment the unlearning set with desirable samples to be retained.

Note that unlearning for each component model θ_t^u is performed on the subset D_t of the original data. When $D_1, \ldots, D_n$ are the partition of the unlearning data D, the total number of gradient steps in SPUNGE is the same as applying the unlearning method $\mathcal{U}$ on the entire data D without using SPUNGE. Additional computation for using SPUNGE on top of an unlearning method $\mathcal{U}$ comes from the merging step, and model merging methods are computationally efficient [26–28].

7.4 Details on Unlearning and Merging Methods

We describe the specific unlearning and merging methods used in this work in the following.

7.4.1 Task Vector Negation (TVN)

TVN uses the notion of *task vector arithmetic* for unlearning [3]. Let $\theta_{\text{init}} \in \mathbb{R}^d$ denote the initial model weights and $\theta_{\text{ft}} \in \mathbb{R}^d$ the corresponding weights after fine-tuning the model on unlearning dataset D. The task vector used for unlearning is computed as $\tau = \theta_{\text{ft}} - \theta_{\text{init}}$. TVN obtains the unlearned model as $\theta^u = \theta_{\text{init}} - \lambda\tau$ where $\lambda \geq 0$ is a scaling parameter. Following [4], we employ Parameter-Efficient Fine-Tuning (PEFT) instead of full fine-tuning and compute the task vector using Parameter Efficient Modules (PEMs). In our experiments, we use a state-of-the-art PEFT method, LoRA [29], and perform negation using LoRA modules with $\lambda = 1$.

7.4.2 Representation Misdirection Unlearning (RMU)

This method from [8] randomizes model activations on unlearning data while preserving model activations on data to be kept. Specifically, RMU uses a two-part loss function: (1) a forget loss to bring the model activations on unlearning data close to a scaled uniform random vector, and (2) a retain loss to preserve model activations on data to be retained. Here, let D denote the unlearning dataset and D' denote the retain set (containing samples with desirable behavior or knowledge). Let $f_\theta(\cdot)$ and $f_{\theta_{\text{init}}}(\cdot)$ denote the hidden states of the model being unlearned and the initial model, respectively, at some layer ℓ. Then, the forget loss is $\mathcal{L}_u = \mathbb{E}_{\mathbf{x}_u \sim D}\left[\frac{1}{|\mathbf{x}_u|}\sum_{\text{token } t \in \mathbf{x}_u} \|f_\theta(t) - c \cdot \mathbf{u}\|_2^2\right]$, where $\mathbf{u}$ is a random unit vector with entries sampled independently, and uniformly at random from $[0, 1)$, and c is a hyperparameter. The retain loss is $\mathcal{L}_r = \mathbb{E}_{\mathbf{x}_r \sim D'}\left[\frac{1}{|\mathbf{x}_r|}\sum_{\text{token } t \in \mathbf{x}_r} \left\|f_\theta(t) - f_{\theta_{\text{init}}}(t)\right\|_2^2\right]$. The model parameters are updated to minimize the combined loss $\mathcal{L} = \mathcal{L}_u + \alpha\mathcal{L}_r$, where $\alpha > 0$ is a hyperparameter. The loss is typically computed only on layer ℓ and gradients are updated only on layers $\ell - 2$, $\ell - 1$, and ℓ.

7.4.3 TIES-Merging

This method from [28] allows one to merge multiple model parameters using task vector arithmetic. Given a set of model weights $\theta_1^u, \ldots, \theta_n^u$ along with the initial weights θ_{init}, TIES-Merging computes a task vector for each model as $\tau_t = \theta_t^u - \theta_{\text{init}}$. Then, it operates in three steps: (i) trim each task vector by selecting the parameters with largest magnitudes, (ii) resolve sign conflicts by creating an aggregate elected sign vector, and (iii) average only the parameters whose signs are the same as the aggregated elected sign.

Algorithm 2 presents the instantiation of SPUNGE with RMU and TIES.

Algorithm 2 SPUNGE Framework Instantiated with RMU [8] and TIES-Merging [28]

Input: Initial model parameters θ_{init}, Dataset D for unlearning, Retain dataset D^r (as needed by RMU), Data attributes $a_1, \ldots, a_n$, Parameters for RMU c, α, Parameters for TIES-merging λ, k
Output: Unlearned model θ_u
for $t = 1$ **to** n **do**
  Select subset associated with data attribute a_t as $D_t = \{\mathbf{x} \in D \mid \texttt{attr}(\mathbf{x}) = a_t\}$
  Process subset for unlearning $D_t^u = \{\texttt{proc}(\mathbf{x}) \mid \mathbf{x} \in D_t\}$
  Perform unlearning $\theta_i^u \leftarrow \text{RMU}(\theta_{\text{init}}, D_t^u, D^r, c, \alpha)$
end for
Perform merging $\theta^u \leftarrow \texttt{TIES}(\theta_1^u, \ldots, \theta_n^u, \theta_{\text{init}}, \lambda)$

Function RMU$(\theta, D^u, D^r, c, \alpha)$
Sample unit vector $\mathbf{u}$ with entries drawn independently, and uniformly at random from $[0, 1)$
for data points $\mathbf{x}_u \sim D^u, \mathbf{x}_r \sim D^r$ **do**
  Set $\mathcal{L}_u = \frac{1}{L} \sum_{t \in \mathbf{x}_u} \| f_\theta(t) - c \cdot \mathbf{u} \|_2^2$, where $\mathbf{x}_u$ contains L tokens
  Set $\mathcal{L}_r = \frac{1}{L} \sum_{t \in \mathbf{x}_r} \left\| f_\theta(t) - f_{\theta_{\text{init}}}(t) \right\|_2^2$, where $\mathbf{x}_r$ contains L tokens
  Update parameters θ using $\mathcal{L} = \mathcal{L}_u + \alpha \cdot \mathcal{L}_r$
end for
return θ

Function TIES$(\theta_1, \ldots, \theta_n, \theta_{\text{init}}, \lambda, k)$
for $t = 1$ **to** n **do**
  Create task vector $\tau_t = \theta_t^u - \theta_{\text{init}}$
  Sparsify the task vector to keep only largest k elements to obtain $\hat{\tau}_t$
  Collect signs for components $\hat{\gamma}_t \leftarrow \text{sign}(\hat{\tau}_t)$
  Collect magnitudes for components $\hat{\mu} \leftarrow |\hat{\tau}_t|$
end for
Elect final signs as $\gamma_u \leftarrow \text{sign}\left(\sum_{t=1}^n \hat{\tau}_t\right)$
for $p = 1$ **to** d **do**
  $\mathcal{A}^p = \{t \in [n] \mid \hat{\gamma}_t^p = \gamma^p\}$
  $\tau_u^p = \frac{1}{|\mathcal{A}^p|} \sum_{t \in \mathcal{A}^p} \hat{\tau}_t^p$
end for
$\theta_u \leftarrow \theta_{\text{init}} + \lambda \tau_u$
return θ_u

7.5 Evaluation of Spunge

In the following, we evaluate SPUNGE on three unlearning scenarios: (i) unlearning toxicity and hate speech, (ii) unlearning social bias, and (iii) unlearning hazardous knowledge. For each scenario, we take an unlearning method that has been shown to be effective in the literature, apply SPUNGE on top of it, and evaluate how SPUNGE impacts the performance of the baseline unlearning method. We measure the effectiveness of unlearning by using scenario-specific metrics. To measure the general capability of the model, we consider 10 standard academic benchmarks, including all 6 benchmarks from the Open LLM Leaderboard v1 [30] (see Sect. 7.6.1 for details). We perform benchmark evaluations the LM Evaluation Harness framework [31].

7.5.1 Unlearning Toxicity and Hate Speech

We apply SPUNGE on top of Task Vector Negation (TVN) [3], which has been shown to reduce toxicity in LLMs [4].

7.5.1.1 Experiment Set Up

Unlearning via TVN: To unlearn toxicity via TVN, we first fine-tune the model on a subset of toxic sentences from ToxiGen [32]. Then, we compute tasks vectors by subtracting the base model weights from the fine-tuned toxic model. Finally, we negate the task vectors and add them to the base model to detoxify the base model.

Specifically, we select toxic sentences from ToxiGen [32], which contains 8.96k samples designed to measure implicit toxicity and hate speech across 13 demographic groups (e.g., African Americans, women, Mexicans, etc.). ToxiGen benchmark contains, for each prompt, the target demographic group and the toxicity level evaluated by human annotators. While ToxiGen encompasses 13 demographic groups, for our experiments, we choose the following 5 representative demographic groups: Nationality (Mexican), Gender and Sex (Women), Religion (Muslim), Sexual Orientation (LGBTQ), and Health Condition (Physical Disability). We perform TVN using the ToxiGen training samples with toxicity scores ≥ 3, restricted to the five demographic groups.

SPUNGE + TVN: We instantiate SPUNGE by leveraging the demographic information in the ToxiGen unlearning set as attributes. Specifically, we choose the following 5 representative demographic groups out of 13 demographic groups in ToxiGen: Nationality (Mexican), Gender and Sex (Women), Religion (Muslim), Sexual Orientation (LGBTQ), and Health Condition (Physical Disability). SPUNGE first splits the unlearning set ToxiGen into 5 subsets—$D_1, \ldots, D_5$—based on the 5 demographic groups. Next, from each set D_t, we select a subset of samples with toxicity score[1] ≥ 3 to get five unlearning subsets $D_1^u, \ldots, D_5^u$. SPUNGE then

[1] The range of toxicity annotation scores is from 1 to 5.

performs TVN on the base model θ_{init} with each D_t^u to obtain $\theta_1^u, \ldots, \theta_5^u$. Finally, we use TIES-merging (Sect. 7.4.3) to merge the five unlearned models.

Evaluation Set Up: For toxicity evaluation, we consider a similar experimental setup to [33, 34]. To evaluate the amount of toxicity and hate speech in model generations, we use the ToxiGen benchmark [32]. ToxiGen is designed to measure implicit toxicity and hate speech across 13 demographic groups (e.g., African Americans, women, Mexicans, etc.). Following [33], to measure the toxicity of the model completions, we use a RoBERTA model fine-tuned on ToxiGen [32]. In our evaluation, we prompt the model for completions, with toxic and benign examples from the test subset of ToxiGen, and measure the toxicity of the model completions using a RoBERTA model fine-tuned on ToxiGen [32]. We use greedy decoding and compute the percentage of completions that are deemed toxic by the classifier as *toxicity*. We also assess how unlearning impacts the fluency of the model, similar to [5, 24], by computing the perplexity of the model completions with an independent, larger model, LLAMA2- 13B.

7.5.1.2 Experimental Results

As shown in Table 7.1, SPUNGE boosts the performance of TVN for both ZEPHYR- 7B- BETA and LLAMA2- 7B. For ZEPHYR- 7B- BETA, SPUNGE reduces the toxicity percentage of TVN by 31% (from 5.65 to 3.88), while maintaining the fluency of generations as measured by the perplexity computed with LLAMA- 13B. Notably, SPUNGE maintains general capabilities of the model as measured by the average accuracy on the benchmarks. Similarly, for LLAMA2-7B, SPUNGE reduces the toxicity percentage of TVN by 30% (from 4.26 to 2.96) while maintaining the average accuracy on benchmarks within 1% of the base model (Tables 7.2 and 7.3).

Table 7.1 Evaluation of toxicity unlearning on ToxiGen

Model	ToxiGen		Average
+ method	Toxicity (↓)	PPL (↓)	Acc. (↑)
ZEPHYR- 7B- BETA	20.48	7.62	65.72
+ TVN	5.65	8.36	65.67
+ SPUNGE-TVN	**3.88**	8.66	65.53
LLAMA2- 7B	15.95	5.97	56.29
+ TVN	4.26	8.42	56.35
+ SPUNGE-TVN	**2.96**	7.88	55.72

SPUNGE boosts the reduction in toxicity, while maintaining benchmark performance similar to the base model

Table 7.2 Accuracy on the benchmarks for the ZEPHYR- 7B- BETA model and the models after performing unlearning on ToxiGen

Benchmark	ZEPHYR- 7B- BETA	TVN	SPUNGE
Arc-C (↑)	63.90	64.50	63.73
Arc-E (↑)	84.89	83.96	83.37
HellaSwag (↑)	84.21	84.41	84.28
MMLU (↑)	59.75	58.14	58.52
Winogrande (↑)	77.42	78.05	77.82
GSM8K (↑)	34.42	34.79	33.43
MathQA (↑)	38.05	36.88	36.71
PIQA (↑)	82.69	82.26	82.42
PubmedQA (↑)	76.80	76.60	77.00
TruthfulQA (↑)	55.12	57.20	58.01
Average (↑)	65.72	65.67	65.52

Table 7.3 Accuracy on the benchmarks for the LLAMA2- 7B model and the models after performing unlearning on ToxiGen

Benchmark	LLAMA2- 7B	TVN	SPUNGE
Arc-C (↑)	53.32	53.32	52.04
Arc-E (↑)	81.48	81.64	81.69
HellaSwag (↑)	78.57	77.44	74.39
MMLU (↑)	45.99	44.74	44.22
Winogrande (↑)	72.45	73.71	74.11
GSM8K (↑)	15.01	8.11	9.47
MathQA (↑)	29.41	29.31	29.14
PIQA (↑)	79.37	79.97	79.65
PubmedQA (↑)	68.40	71.00	69.80
TruthfulQA (↑)	38.97	44.34	42.72
Average (↑)	56.29	56.35	55.72

Toxicity per Demographic Group: We compare the toxicity percentage for each demographic and show that SPUNGE strengthens TVN. We focus on the same 5 demographic groups used during unlearning: Nationality (Mexican), Gender and Sex (Women), Religion (Muslim), Sexual Orientation (LGBTQ), and Health Condition (Physical Disability). In Figs. 7.2 and 7.3, we present radar plots for toxicity percentage per demographic group. The plots show results for the base model, TVN, and SPUNGE used with TVN. SPUNGE reduces the toxi-

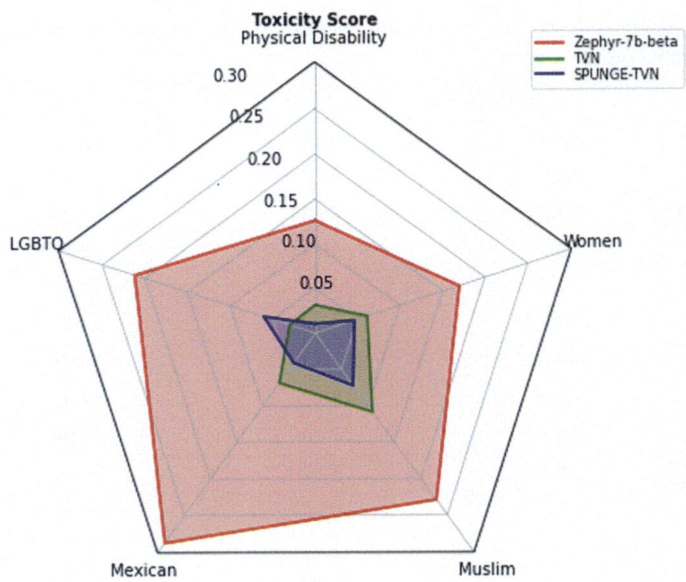

Fig. 7.2 Toxicity scores per demographic group on ToxiGen test set for the ZEPHYR- 7B- BETA base model, after unlearning with TVN, and after unlearning with SPUNGE used with TVN.

city for every demographic group for LLAMA2- 7B (Fig. 7.3) whereas for ZEPHYR- 7B- BETA, SPUNGE cuts down toxicity percentage for most demographic groups (Fig. 7.2).

7.5.1.3 Spunge Leveraging Type of Toxicity

We consider the goal of unlearning implicit as well as explicit toxicity from LLMs. Explicit toxicity is a conventional form of toxicity containing profanity, slurs, swearwords, and offensive language. On the other hand, implicit toxicity does not include such terms in contrast to explicit toxicity and can even be positive in sentiment [32]. Examples of implicit toxicity include stereotyping and microaggressions. The ToxiGen dataset [32] is focused on implicit and subtly toxic samples. There are datasets that contains samples with explicit toxicity such as Civil Comments [35].

As a baseline, we perform unlearning on LLAMA2- 7B with TVN using a dataset consisting of samples with implicit as well as explicit toxicity. To represent implicit toxicity, we take samples from the (annotated) train set of ToxiGen with human toxicity level of 5 (highest level). To represent explicit toxicity, we take samples from Civil Comments with severe toxicity score greater than 0.35.

For comparison, we instantiate SPUNGE to leverage type of toxicity. Specifically, we separate the unlearning set into two subsets: examples with implicit toxicity (D_1) and exam-

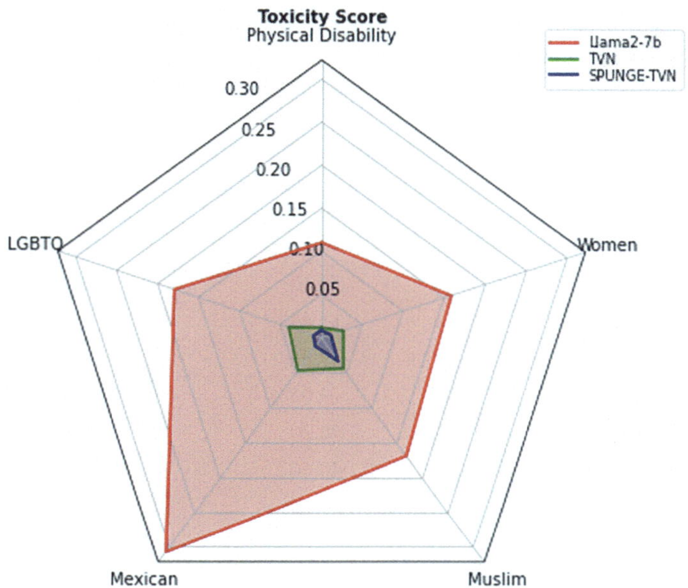

Fig. 7.3 Toxicity scores per demographic group on ToxiGen test set for the LLAMA2- 7B base model, after unlearning with TVN, and after unlearning with SPUNGE used with TVN.

ples with explicit toxicity (D_2). We separately unlearn the two subsets, and then merge the unlearning models with TIES-merging (Table 7.4).

Experimental Results: Table 7.6 compares TVN and its SPUNGE-enhanced version. In addition to computing toxicity on the ToxiGen test set (which contains implicitly toxic and benign samples), we also compute toxicity on Real Toxicity Prompts (RTP) [11] (which contains explicitly toxic and benign samples). We see that SPUNGE amplifies the performance of TVN on both ToxiGen and RTP, while maintaining the performance on benchmark tasks. We present the accuracy results on benchmark tasks in Table 7.5.

Table 7.4 Evaluation of toxicity unlearning on ToxiGen and RealToxicityPrompts (RTP)

Model	Toxicity		Average
+ method	ToxiGen (↓)	RTP (↓)	Acc. (↑)
LLAMA2- 7B	15.95	6.40	56.29
+ TVN	8.42	3.17	**56.14**
+ SPUNGE-TVN	**4.81**	**1.97**	55.23

We consider LLAMA2- 7B with TVN. Toxicity is the percentage of toxic generations and Average Acc. is the average performance on the 10 benchmarks. SPUNGE is configured to leverage type of toxicity: implicit versus explicit toxicity

Table 7.5 Accuracy on the benchmarks for the LLAMA2- 7B model and the models after performing unlearning on Civil Comments and ToxiGen

Benchmark	LLAMA2- 7B	TVN	SPUNGE
Arc-C (↑)	53.32	53.75	53.24
Arc-E (↑)	81.48	81.35	79.33
HellaSwag (↑)	78.57	78.41	77.82
MMLU (↑)	45.99	44.32	44.16
Winogrande (↑)	72.45	73.16	73.16
GSM8K (↑)	15.01	11.44	4.16
MathQA (↑)	29.41	29.34	29.41
PIQA (↑)	79.37	79.05	79.65
PubmedQA (↑)	68.40	70.20	70.20
TruthfulQA (↑)	38.97	40.40	41.23
Average (↑)	56.29	56.14	55.23

7.5.2 Unlearning Social Bias

Unlearning methods, especially Task Vector Negation (TVN), have been shown to effectively mitigate social bias in LLMs that is characterized by deliberate or unintentional discrimination towards individuals, groups, or specific ideas and beliefs, resulting in unfair treatment [36, 37].

7.5.2.1 Experiment Set Up

Unlearning via TVN: Following [37], we first fine-tune the model using biased samples from StereoSet [38]. Then, we compute tasks vectors by subtracting the base model weights from the fine-tuned biased model. Finally, we negate the task vectors and add them to the base model to debias the base model.

We select biased samples from the StereoSet [38], which consists of sentences that measures model preferences across gender, race, religion, and profession. *intersentence* subset of StereoSet. Each row consists of the *context* and 3 *sentences*, which are stereotypical, anti-stereotypical, and unrelated with regards to the context. For each row in the subset, we concatenate the context and the stereotyped sequence from the sentences to generate a biased sentence, which is used for fine tuning.

SPUNGE + TVN: We instantiate SPUNGE by using the *bias domain* information in the StereoSet dataset. StereoSet samples measure stereotypical biases in four target domains: gender, profession, race, and religion. SPUNGE first splits the unlearning set StereoSet into 4 subsets – $D_1, \ldots, D_4$ – based on the 4 bias domains: gender, profession, race, and religion. Next, from each set D_t, we concatenate the context and the stereotyped sequence from

Table 7.6 Evaluation of social bias unlearning on CrowS-Pairs

Model	CrowS-Pairs	Average
ZEPHYR- 7B- BETA	0.649	65.72
+ TVN	0.556	65.76
+ SPUNGE-TVN	**0.534**	65.84
LLAMA2- 7B	0.677	56.29
+ TVN	0.565	56.55
+ SPUNGE-TVN	**0.540**	56.67

SPUNGE mitigates the bias further without sacrificing benchmark performance

the sentences to generate 4 unlearning subsets $D_1^u, \ldots, D_4^u$. SPUNGE then performs TVN on the base model θ_{init} with each D_i^u to obtain $\theta_1^u, \ldots, \theta_4^u$. Finally, we use TIES-merging (Sect. 7.4.3) to merge the five unlearned models.

Evaluation Set Up: For evaluating bias, we use the CrowS-Pairs benchmark [39], similar to [36]. Each sample in CrowS-Pairs consists of two sentences: one that is more stereotyping and another that is less stereotyping. *CrowS-Pair bias score* of a model is the percentage of more-stereotypical sentences that are rated as more likely by the model than the non-stereotypical sentences. Ideally, for an unbiased model, the bias score should be closer to 0.5.

7.5.2.2 Experimental Results

As shown in Table 7.6, SPUNGE strengthens the performance of TVN for both ZEPHYR-7B- BETA and LLAMA2- 7B. In particular, SPUNGE reduces the bias of TVN by ~4% (from 0.556 to 0.534 for ZEPHYR- 7B- BETA, and from 0.565 to 0.540 for LLAMA2- 7B). Notably, SPUNGE maintains general capabilities of the models as measured by the average accuracy on the benchmarks (Tables 7.7 and 7.8).

7.5.3 Unlearning Hazardous Knowledge

We focus on reducing the model's ability to answer questions about hazardous knowledge (e.g., cultivating virus) while maintaining the ability to answer questions about non-hazardous knowledge (e.g., properties of fungi). Reference [8] designed Representation Misdirection Unlearning (RMU) for unlearning hazardous knowledge from LLMs, and showed its superiority to several unlearning methods. We demonstrate that SPUNGE enhances the performance of RMU.

Table 7.7 Accuracy on the benchmarks for the ZEPHYR- 7B- BETA model and the models after performing unlearning on StereoSet.

Benchmark	ZEPHYR- 7B- BETA	TVN	SPUNGE
Arc-C (↑)	63.90	63.13	63.23
Arc-E (↑)	84.89	84.34	83.71
HellaSwag (↑)	84.12	85.05	85.13
MMLU (↑)	59.75	59.65	59.66
Winogrande (↑)	77.42	76.16	76.48
GSM8K (↑)	34.42	34.04	34.42
MathQA (↑)	38.05	36.95	36.78
PIQA (↑)	82.69	82.26	81.66
PubmedQA (↑)	76.80	77.0	77.2
TruthfulQA (↑)	55.12	59.01	60.14
Average (↑)	65.72	65.76	65.84

Table 7.8 Accuracy on the benchmarks for the LLAMA2- 7B model and the models after performing unlearning on StereoSet

Benchmark	LLAMA2- 7B	TVN	SPUNGE
Arc-C (↑)	53.32	53.49	52.81
Arc-E (↑)	81.48	82.28	81.90
HellaSwag (↑)	78.57	78.24	78.54
MMLU (↑)	45.99	44.74	45.49
Winogrande (↑)	72.45	72.13	71.51
GSM8K (↑)	15.01	12.43	13.04
MathQA (↑)	29.41	29.21	29.31
PIQA (↑)	79.37	80.08	79.97
PubmedQA (↑)	68.40	72.8	72.0
TruthfulQA (↑)	38.97	40.08	42.16
Average (↑)	56.29	56.55	56.67

7.5.3.1 Experiment Set Up

Unlearning via RMU: Given an unlearning dataset and a retain dataset, RMU randomizes model activations on unlearning data while preserving model activations on data to be kept (Sect. 7.4.2). As unlearning datasets, we use the *bio corpora* and *cyber corpora* – training documents specially collected by [8] for performing hazardous knowledge unlearning. The bio corpora consist of a selected subset of PubMed papers that are related to the topics appearing in WMDP-Bio questions. The cyber corpora consist of passages scraped from GitHub via keyword search on topics related to WMDP-Cyber questions. These corpora are specially collected training sets for performing hazardous knowledge unlearning, and are separate from the WMDP benchmark, which is designed for evaluation [8]. We use a subset of WikiText [40] as the retain dataset.

SPUNGE + RMU: We instantiate SPUNGE to leverage the scientific domain attribute in the unlearning set. As mentioned in the previous section, the unlearning dataset is a combination of bio and cyber corpora. In other words, $n = 2, a_1 =$ bio, and $a_2 =$ cyber. Given a document $\mathbf{x}$ from the corpora, the function $\texttt{attr}(\mathbf{x})$ outputs the scientific domain of $\mathbf{x}$, whether it is cyber or bio. Thus, SPUNGE splits the data by domain to separate bio corpora (D_1) and cyber corpora (D_2). SPUNGE performs unlearning separately on each of them to obtain two unlearned LLMs: one with biosecurity hazardous knowledge removed θ_1^u and the other with cybersecurity hazardous knowledge removed θ_2^u. SPUNGE then merges θ_1^u and θ_2^u using TIES-merging (Sect. 7.4). Note that, in contrast to SPUNGE + RMU, the vanilla RMU (and other baselines) in [8] use the bio and cyber corpora together during unlearning—in particular, RMU alternates between one batch from the bio corpora and one from the cyber corpora during unlearning. Algorithm 2 presents SPUNGE instantiated with RMU and TIES-merging.

Evaluation Set Up: To evaluate hazardous knowledge removal, we use the Weapons of Mass Destruction Proxy (WMDP) benchmark [8] which consists of 3.6k multiple-choice questions on biosecurity (WMDP-Bio), cybersecurity (WMDP-Cyber), and chemistry (WMDP-Chem).[2] To evaluate general-knowledge question answering, we use the MMLU benchmark [41]. Similar to [8], we conduct unlearning evaluation only on the challenging subsets WMDP-Bio and WMDP-Cyber.

7.5.3.2 Experimental Results

Table 7.9 shows that SPUNGE fortifies the performance of RMU in removing hazardous knowledge while maintaining general-knowledge capabilities. In particular, SPUNGE reduces WMDP-Bio accuracy by 11.8% (from 31.26 to 27.57) and WMDP-Cyber accuracy by 4% (from 27.62 to 26.47), while maintaining MMLU accuracy within 1% of RMU.

7.6 Additional Details

7.6.1 Benchmarks Used for Evaluation

We use the following 10 benchmarks for evaluating the general capability of models. We use all six benchmarks from the Open LLM Leaderboard v1 [30]. We use the same few-shot prompting evaluation method used by the Open LLM Leaderboard and select the same

[2] This benchmark is different than bio and cyber corpora used for unlearning [8].

Table 7.9 Evaluation of hazardous knowledge unlearning on WMDP

Model + method	WMDP-Bio (↓)	WMDP-cyber (↓)	MMLU (↑)
ZEPHYR- 7B- BETA	63.55	43.63	58.15
+ RMU	31.26	27.62	56.48
+ SPUNGE-RMU	**27.57**	**26.47**	55.83

SPUNGE strengthens the performance of RMU, while preserving general knowledge on MMLU

number of shots as prescribed for each benchmark. For the remaining four benchmarks, we choose those commonly in literature and perform 5-shot prompting for each. We perform benchmark evaluations the Language Model Evaluation Harness framework [31].

1. AI2 Reasoning Challenge (ARC-Challenge and ARC-Easy) [42] (25-shot)
2. HellaSwag [43] (10-shot)
3. MMLU [41] (5-shot)
4. TruthfulQA [44] (0-shot)
5. Winogrande [45] (5-shot)
6. GSM8K [46] (5-shot)
7. MathQA [47] (5-shot)
8. PIQA [48] (5-shot)
9. PubMedQA [49] (5-shot)

7.6.2 Training Parameters

TVN with ZEPHYR-7B-BETA: For TVN with ZEPHYR- 7B- BETA [50], we set the LoRA rank to 16, α associated with LoRA to 16, LoRA dropout to 0.01, and the target modules as the default modules in the HuggingFace PEFT library. We use the Adam optimizer with a learning rate of 2×10^{-5} and a cosine learning rate schedule to train for 1 epoch. For social bias unlearning with SPUNGE, we use the same learning rate. For toxicity unlearning with SPUNGE, since the unlearning subsets are substantially smaller, we perform training with a learning rate of 1×10^{-4} for 1 epoch. All experiments are performed on one NVIDIA V100 GPU with 32GB memory.

TVN with LLAMA2-7B: For TVN with LLAMA2- 7B [33], we set the LoRA rank to 64, α associated with LoRA to 64, LoRA dropout to 0.01, and the target modules as `key`, `value`, `query`, `up`, `down`, and `gate` projections. We use the Adam optimizer with a learning rate of 1×10^{-4} and a cosine learning rate scheduling. All experiments are performed on one NVIDIA V100 GPU with 32GB memory.

RMU with ZEPHYR-7B-BETA: For RMU with ZEPHYR- 7B- BETA [50], we use the hyperparameters from [8]. In particular, we use $c = 6.5$ and $\alpha = 1200$. We use the Adam optimizer with a learning rate of 5×10^{-5} and a batch size of 150. We select layer 7 to perform the unlearning loss and layers 5, 6, and 7 to update gradients. When performing separate unlearning with SPUNGE, the unlearning subsets are substantially smaller. Thus, we perform training for 2 epochs with early stopping if the cosine similarity between the activations of the unlearned model and the initial model drops below 0.5. All experiments are performed on one NVIDIA A100 GPU with 80GB memory.

7.7 Limitations and Ethical Considerations

We demonstrated the performance gains of SPUNGE for scenarios wherein unlearning samples had associated attributes. For nuanced datasets with less clearly defined attributes, it is possible to apply SPUNGE by splitting the data based on clustering with LLM embeddings. Evaluating the performance of SPUNGE in unlearning scenarios when data attributes are less clearly defined is an exciting future direction.

If attribute selection is incorrect or noisy, then it may potentially lead to ineffective unlearning. An important future work is to investigate the impact of the accuracy or noise in attribute selection on the performance of SPUNGE.

Our evaluation of SPUNGE is limited to unlearning undesirable behaviors (toxicity and social bias) and hazardous knowledge. Unlearning is often applied in other scenarios such as data unlearning (e.g., copyrighted or licensed data) and reducing harmfulness (e.g., harmful responses to provocative prompts). It will be interesting to investigate how much benefits SPUNGE provides for such diverse scenarios.

Due to compute limitations, we restricted our experiments to two unlearning methods on models of size 7B. Exploring SPUNGE with larger and newer models and different unlearning is potential future direction.

Unlearning undesirable behaviors and hazardous knowledge from LLMs often involves the use of offensive, toxic, biased, or malicious data samples. As in the case of training datasets of LLMs, data used for unlearning may also include personally identifiable information. There might be ethical implications related to how data used for unlearning are obtained and used. It is crucial to carefully consider such ethical implications when unlearning is employed to mitigate undesirable behaviors and reduce hazardous knowledge from LLMs, irrespective of whether our framework SPUNGE is used to enhance the performance of unlearning methods.

7.8 Conclusion

Unlearning procedures can boost their performance by adapting how data is fed. In this chapter, we presented SPUNGE, a novel unlearning framework that takes advantage of attributes associated with the data to be unlearned. SPUNGE leverages attributes using a *split-unlearn-then-merge* approach, and can be applied on top of any unlearning method. We empirically demonstrated that SPUNGE significantly improves the effectiveness of unlearning methods for reducing undesirable behaviors and hazardous knowledge. An interesting future work is to explore using SPUNGE for data unlearning (e.g., copyrighted or licensed data).

References

1. Xu, H., Zhu, T., Zhang, L., Zhou, W., Yu, P.S.: Machine unlearning: a survey. ACM Comput. Surv. **56**(1) (2023)
2. Liu, S., Yao, Y., Jia, J., Casper, S., Baracaldo, N., Hase, P., Yao, Y., Liu, C.Y., Xu, X., Li, H., Varshney, K.R., Bansal, M., Koyejo, S., Liu, Y.: Rethinking machine unlearning for large language models. arXiv preprint arXiv:2402.08787 (2024)
3. Ilharco, G., Ribeiro, M.T., Wortsman, M., Schmidt, L., Hajishirzi, H., Farhadi, A.: Editing models with task arithmetic. In: The Eleventh International Conference on Learning Representations. https://openreview.net/forum?id=6t0Kwf8-jrj (2023)
4. Zhang, J., Chen, S., Liu, J., He, J.: Composing parameter-efficient modules with arithmetic operations. arXiv preprint arXiv:2306.14870 (2023)
5. Lu, X., Welleck, S., Hessel, J., Jiang, L., Qin, L., West, P., Ammanabrolu, P., Choi, Y.: Quark: controllable text generation with reinforced unlearning. Adv. Neural. Inf. Process. Syst. **35**, 27591–27609 (2022)
6. Yao, Y., Xu, X., Liu, Y.: Large language model unlearning. Adv. Neural. Inf. Process. Syst. **37**, 105425–105475 (2024)
7. Liu, Z., Dou, G., Tan, Z., Tian, Y., Jiang, M.: Towards safer large language models through machine unlearning (2024)
8. Li, N., Pan, A., Gopal, A., Yue, S., Berrios, D., Gatti, A., Li, J.D., Dombrowski, A.K., Goel, S., Mukobi, G., Helm-Burger, N., Lababidi, R., Justen, L., Liu, A.B., Chen, M., Barrass, I., Zhang, O., Zhu, X., Tamirisa, R., Bharathi, B., Herbert-Voss, A., Breuer, C.B., Zou, A., Mazeika, M., Wang, Z., Oswal, P., Lin, W., Hunt, A.A., Tienken-Harder, J., Shih, K.Y., Talley, K., Guan, J., Steneker, I., Campbell, D., Jokubaitis, B., Basart, S., Fitz, S., Kumaraguru, P., Karmakar, K.K., Tupakula, U., Varadharajan, V., Shoshitaishvili, Y., Ba, J., Esvelt, K.M., Wang, A., Hendrycks, D.: The WMDP benchmark: measuring and reducing malicious use with unlearning. In: Proceedings of the 41st International Conference on Machine Learning, Proceedings of Machine Learning Research, vol. 235, pp. 28525–28550. PMLR (2024)
9. Weidinger, L., Uesato, J., Rauh, M., Griffin, C., Huang, P.S., Mellor, J., Glaese, A., Cheng, M., Balle, B., Kasirzadeh, A., Biles, C., Brown, S., Kenton, Z., Hawkins, W., Stepleton, T., Birhane, A., Hendricks, L.A., Rimell, L., Isaac, W., Haas, J., Legassick, S., Irving, G., Gabriel, I.: Taxonomy of risks posed by language models. In: Proceedings of the 2022 ACM Conference on Fairness, Accountability, and Transparency, pp. 214–229. Association for Computing Machinery (2022)
10. Sheng, E., Chang, K.W., Natarajan, P., Peng, N.: The woman worked as a babysitter: on biases in language generation. In: Inui, K., Jiang, J., Ng, V., Wan, X. (eds.) Proceedings of the 2019 Conference on Empirical Methods in Natural Language Processing and the 9th International Joint Conference on Natural Language Processing (EMNLP-IJCNLP), pp. 3407–3412. Association for Computational Linguistics, Hong Kong, China (2019). https://doi.org/10.18653/v1/D19-1339
11. Gehman, S., Gururangan, S., Sap, M., Choi, Y., Smith, N.A.: Realtoxicityprompts: evaluating neural toxic degeneration in language models. arXiv preprint arXiv:2009.11462 (2020)

12. Gallegos, I.O., Rossi, R.A., Barrow, J., Tanjim, M.M., Kim, S., Dernoncourt, F., Yu, T., Zhang, R., Ahmed, N.K.: Bias and fairness in large language models: a survey. Comput. Linguist. **50**(3), 1097–1179 (2024)
13. Sandbrink, J.B.: Artificial intelligence and biological misuse: differentiating risks of language models and biological design tools. arXiv preprint arXiv:2306.13952 (2023)
14. Fang, R., Bindu, R., Gupta, A., Zhan, Q., Kang, D.: LLM agents can autonomously hack websites. arXiv preprint arXiv:2402.06664 (2024)
15. Ouyang, L., Wu, J., Jiang, X., Almeida, D., Wainwright, C., Mishkin, P., Zhang, C., Agarwal, S., Slama, K., Ray, A., et al.: Training language models to follow instructions with human feedback. Adv. Neural. Inf. Process. Syst. **35**, 27730–27744 (2022)
16. Bai, Y., Kadavath, S., Kundu, S., Askell, A., Kernion, J., Jones, A., Chen, A., Goldie, A., Mirhoseini, A., McKinnon, C., et al.: Constitutional AI: harmlessness from AI feedback. arXiv preprint arXiv:2212.08073 (2022)
17. Korbak, T., Shi, K., Chen, A., Bhalerao, R., Buckley, C.L., Phang, J., Bowman, S.R., Perez, E.: Pretraining language models with human preferences. ICML'23 (2023)
18. Glaese, A., McAleese, N., Trebacz, M., Aslanides, J., Firoiu, V., Ewalds, T., Rauh, M., Weidinger, L., Chadwick, M., Thacker, P., Campbell-Gillingham, L., Uesato, J., Huang, P.S., Comanescu, R., Yang, F., See, A., Dathathri, S., Greig, R., Chen, C., Fritz, D., Elias, J.S., Green, R., Mokr, S., Fernando, N., Wu, B., Foley, R., Young, S., Gabriel, I., Isaac, W., Mellor, J., Hassabis, D., Kavukcuoglu, K., Hendricks, L.A., Irving, G.: Improving alignment of dialogue agents via targeted human judgements (2022)
19. Wei, A., Haghtalab, N., Steinhardt, J.: Jailbroken: How does LLM safety training fail? In: Thirty-Seventh Conference on Neural Information Processing Systems (2023)
20. Zou, A., Wang, Z., Kolter, J.Z., Fredrikson, M.: Universal and transferable adversarial attacks on aligned language models. arXiv preprint arXiv:2307.15043 (2023)
21. Carlini, N., Nasr, M., Choquette-Choo, C.A., Jagielski, M., Gao, I., Koh, P.W., Ippolito, D., Tramèr, F., Schmidt, L.: Are aligned neural networks adversarially aligned? In: Thirty-Seventh Conference on Neural Information Processing Systems (2023)
22. Qi, X., Zeng, Y., Xie, T., Chen, P.Y., Jia, R., Mittal, P., Henderson, P.: Fine-tuning aligned language models compromises safety, even when users do not intend to! In: The Twelfth International Conference on Learning Representations (2024)
23. Krause, B., Gotmare, A.D., McCann, B., Keskar, N.S., Joty, S., Socher, R., Rajani, N.F.: GeDi: generative discriminator guided sequence generation. In: Moens, M.F., Huang, X., Specia, L., Yih, S.W.T. (Eds.) Findings of the Association for Computational Linguistics: EMNLP 2021, pp. 4929–4952. Association for Computational Linguistics, Punta Cana, Dominican Republic (2021). https://doi.org/10.18653/v1/2021.findings-emnlp.424
24. Liu, A., Sap, M., Lu, X., Swayamdipta, S., Bhagavatula, C., Smith, N.A., Choi, Y.: DExperts: Decoding-time controlled text generation with experts and anti-experts. In: C. Zong, F. Xia, W. Li, R. Navigli (eds.) Proceedings of the 59th Annual Meeting of the Association for Computational Linguistics and the 11th International Joint Conference on Natural Language Processing (Volume 1: Long Papers), pp. 6691–6706. Association for Computational Linguistics, Online (2021). https://doi.org/10.18653/v1/2021.acl-long.522
25. Dathathri, S., Madotto, A., Lan, J., Hung, J., Frank, E., Molino, P., Yosinski, J., Liu, R.: Plug and play language models: a simple approach to controlled text generation. In: International Conference on Learning Representations (2020)
26. Matena, M.S., Raffel, C.A.: Merging models with fisher-weighted averaging. In: Koyejo, S., Mohamed, S., Agarwal, A., Belgrave, D., Cho, K., Oh, A. (eds.) Advances in Neural Information Processing Systems, vol. 35, pp. 17703–17716. Curran Associates, Inc. (2022)
27. Choshen, L., Venezian, E., Slonim, N., Katz, Y.: Fusing finetuned models for better pretraining (2022)

28. Yadav, P., Tam, D., Choshen, L., Raffel, C.A., Bansal, M.: Ties-merging: resolving interference when merging models. In: Oh, A., Naumann, T., Globerson, A., Saenko, K., Hardt, M., Levine, S. (eds.) Advances in Neural Information Processing Systems, vol. 36, pp. 7093–7115. Curran Associates, Inc. (2023)
29. Hu, E.J., Shen, Y., Wallis, P., Allen-Zhu, Z., Li, Y., Wang, S., Wang, L., Chen, W.: Lora: Low-rank adaptation of large language models. In: ICLR (2022)
30. Beeching, E., Fourrier, C., Habib, N., Han, S., Lambert, N., Rajani, N., Sanseviero, O., Tunstall, L., Wolf, T.: Open LLM leaderboard. https://huggingface.co/spaces/open-llm-leaderboard/open_llm_leaderboard (2023)
31. Gao, L., Tow, J., Abbasi, B., Biderman, S., Black, S., DiPofi, A., Foster, C., Golding, L., Hsu, J., Le Noac'h, A., Li, H., McDonell, K., Muennighoff, N., Ociepa, C., Phang, J., Reynolds, L., Schoelkopf, H., Skowron, A., Sutawika, L., Tang, E., Thite, A., Wang, B., Wang, K., Zou, A.: A framework for few-shot language model evaluation (2023)
32. Hartvigsen, T., Gabriel, S., Palangi, H., Sap, M., Ray, D., Kamar, E.: ToxiGen: a large-scale machine-generated dataset for adversarial and implicit hate speech detection. In: S. Muresan, P. Nakov, A. Villavicencio (eds.) Proceedings of the 60th Annual Meeting of the Association for Computational Linguistics (Volume 1: Long Papers), pp. 3309–3326. Association for Computational Linguistics, Dublin, Ireland (2022). https://doi.org/10.18653/v1/2022.acl-long.234
33. Touvron, H., Martin, L., Stone, K., Albert, P., Almahairi, A., Babaei, Y., Bashlykov, N., Batra, S., Bhargava, P., Bhosale, S., et al.: Llama 2: open foundation and fine-tuned chat models. arXiv preprint arXiv:2307.09288 (2023)
34. Mukherjee, S., Mitra, A., Jawahar, G., Agarwal, S., Palangi, H., Awadallah, A.: Orca: progressive learning from complex explanation traces of GPT-4. arXiv:2306.02707 (2023)
35. Borkan, D., Dixon, L., Sorensen, J., Thain, N., Vasserman, L.: Nuanced metrics for measuring unintended bias with real data for text classification. CoRR arXiv:1903.04561 (2019)
36. Dige, O., Singh, D., Yau, T.F., Zhang, Q., Bolandraftar, B., Zhu, X., Khattak, F.K.: Mitigating social biases in language models through unlearning. https://arxiv.org/abs/2406.13551 (2024)
37. Dige, O., Arneja, D., Yau, T.F., Zhang, Q., Bolandraftar, M., Zhu, X., Khattak, F.K.: Can machine unlearning reduce social bias in language models? In: F. Dernoncourt, D. Preoţiuc-Pietro, A. Shimorina (eds.) Proceedings of the 2024 Conference on Empirical Methods in Natural Language Processing: Industry Track, pp. 954–969. Association for Computational Linguistics, Miami, Florida, US (2024). https://doi.org/10.18653/v1/2024.emnlp-industry.71. https://aclanthology.org/2024.emnlp-industry.71/
38. Nadeem, M., Bethke, A., Reddy, S.: StereoSet: measuring stereotypical bias in pretrained language models. In: Zong, C., Xia, F., Li, W., Navigli, R. (eds.) Proceedings of the 59th Annual Meeting of the Association for Computational Linguistics and the 11th International Joint Conference on Natural Language Processing (Volume 1: Long Papers), pp. 5356–5371. Association for Computational Linguistics, Online (2021). https://doi.org/10.18653/v1/2021.acl-long.416. https://aclanthology.org/2021.acl-long.416/
39. Nangia, N., Vania, C., Bhalerao, R., Bowman, S.R.: CrowS-pairs: a challenge dataset for measuring social biases in masked language models. In: Webber, B., Cohn, T., He, Y., Liu, Y. (eds.) Proceedings of the 2020 Conference on Empirical Methods in Natural Language Processing (EMNLP), pp. 1953–1967. Association for Computational Linguistics, Online (2020). https://doi.org/10.18653/v1/2020.emnlp-main.154. https://aclanthology.org/2020.emnlp-main.154/
40. Merity, S., Xiong, C., Bradbury, J., Socher, R.: Pointer sentinel mixture models. In: International Conference on Learning Representations (2017). https://openreview.net/forum?id=Byj72udxe
41. Hendrycks, D., Burns, C., Kadavath, S., Arora, A., Basart, S., Tang, E., Song, D., Steinhardt, J.: Measuring mathematical problem solving with the MATH dataset. In: NeurIPS Datasets and Benchmarks (2021)

42. Clark, P., Cowhey, I., Etzioni, O., Khot, T., Sabharwal, A., Schoenick, C., Tafjord, O.: Think you have solved question answering? Try ARC, the AI2 reasoning challenge. https://arxiv.org/abs/1803.05457 (2018)
43. Zellers, R., Holtzman, A., Bisk, Y., Farhadi, A., Choi, Y.: Hellaswag: can a machine really finish your sentence? arXiv preprint arXiv:1905.07830 (2019)
44. Lin, S., Hilton, J., Evans, O.: TruthfulQA: measuring how models mimic human falsehoods. In: Muresan, S., Nakov, P., Villavicencio, A. (eds.) Proceedings of the 60th Annual Meeting of the Association for Computational Linguistics (Volume 1: Long Papers), pp. 3214–3252. Association for Computational Linguistics, Dublin, Ireland (2022). https://doi.org/10.18653/v1/2022.acl-long.229
45. Sakaguchi, K., Bras, R.L., Bhagavatula, C., Choi, Y.: Winogrande: an adversarial winograd schema challenge at scale. Commun. ACM **64**(9), 99–106 (2021)
46. Cobbe, K., Kosaraju, V., Bavarian, M., Chen, M., Jun, H., Kaiser, L., Plappert, M., Tworek, J., Hilton, J., Nakano, R., et al.: Training verifiers to solve math word problems. arXiv preprint arXiv:2110.14168 (2021)
47. Amini, A., Gabriel, S., Lin, S., Koncel-Kedziorski, R., Choi, Y., Hajishirzi, H.: MathQA: towards interpretable math word problem solving with operation-based formalisms. In: Burstein, J., Doran, C., Solorio, T. (eds.) Proceedings of the 2019 Conference of the North American Chapter of the Association for Computational Linguistics: Human Language Technologies, Volume 1 (Long and Short Papers), pp. 2357–2367. Association for Computational Linguistics, Minneapolis, Minnesota (2019). https://doi.org/10.18653/v1/N19-1245
48. Bisk, Y., Zellers, R., Bras, R.L., Gao, J., Choi, Y.: PIQA: reasoning about physical commonsense in natural language. In: AAAI Conference on Artificial Intelligence (2019)
49. Jin, Q., Dhingra, B., Liu, Z., Cohen, W., Lu, X.: PubmedQA: a dataset for biomedical research question answering. In: Proceedings of the 2019 Conference on Empirical Methods in Natural Language Processing and the 9th International Joint Conference on Natural Language Processing (EMNLP-IJCNLP), pp. 2567–2577 (2019)
50. Tunstall, L., Beeching, E., Lambert, N., Rajani, N., Rasul, K., Belkada, Y., Huang, S., von Werra, L., Fourrier, C., Habib, N., Sarrazin, N., Sanseviero, O., Rush, A.M., Wolf, T.: Zephyr: direct distillation of LM alignment (2023)

Label Smoothing Improves Machine Unlearning

8

Zonglin Di

Abstract

The goal of machine unlearning (MU) is to remove previously learned data from a model. However, existing MU techniques struggle to balance computational cost with performance. Inspired by the effects of label smoothing on model confidence and by differential privacy, this chapter introduces UGradSL, a simple, gradient-based MU method that applies an inverse form of label smoothing. UGradSL is plug-and-play and theoretically grounded, with analyses showing that appropriately applying label smoothing enhances MU effectiveness. Extensive experiments across datasets of varying sizes and modalities confirm its robustness and efficiency. Notably, UGradSL exhibits a strong connection to improvements in local differential privacy. It achieves consistent MU performance gains with only minimal additional computation. For example, UGradSL improves unlearning accuracy by 66% over the gradient ascent baseline without reducing efficiency. A self-adaptive version is also presented to enable straightforward parameter selection.

8.1 Introduction

Machine unlearning (MU) aims to remove specific learned data points from a model to enhance fairness, robustness, and privacy [1, 2]. Interest in MU has grown alongside privacy-focused legislation [3, 4] and concerns over information leakage [5–7].

Although retraining a model from scratch guarantees privacy protection, it is computationally prohibitive. As a result, most existing approaches [8–15] focus on approximate MU to balance unlearning effectiveness with efficiency. Gradient ascent (GA) [12] is a natural

Z. Di (✉)
University of California Santa Cruz, 1156 High St, Santa Cruz, CA 95064, USA
e-mail: zdi@ucsc.edu

S. Liu et al. (eds.), *Machine Unlearning for Governance of Foundation Models*, Synthesis Lectures on Computer Vision, https://doi.org/10.1007/978-3-032-17282-2_8

candidate, since MU can be viewed as the inverse of ML. However, GA often falls short because the gradients of well-memorized samples diminish (corresponding to near-zero loss), limiting its unlearning power.

Our approach is inspired by label smoothing [16], a regularization technique shown to improve generalization and reduce overfitting [17, 18]. While most work focuses on positive label smoothing, [19] explored negative label smoothing with a broader feasible domain. Building on this, we introduce GA with "negative" label smoothing, which applies standard label smoothing in a descending manner to quickly reduce the model's confidence in forgotten data. We call this method *UGradSL* (Unlearning using Gradient-based Smoothed Labels), a plug-and-play approach that consistently enhances gradient-based MU performance without degrading accuracy on the remaining or testing datasets.

In addition to gradient-based methods, related work has investigated probabilistic unlearning [2, 20], differential privacy techniques [21, 22], and influence functions [8] for tracing the impact of training samples. MU applications now extend across diverse learning paradigms, including federated learning [23], graph learning [24], and representation learning [25].

The core contributions of this chapter are summarized as follows:

- It introduces a lightweight tool to improve MU by joining the label smoothing and gradient ascent.
- It theoretically analyzes the role of gradient ascent in MU and how negative label smoothing is able to boost MU performance.
- Extensive experiments in seven datasets in different modalities and several unlearning paradigms regarding different MU metrics show the robustness and generalization of the method.
- It investigates the relationship between label smoothing and label differential privacy (LDP), showing that label smoothing can aid LDP.

8.2 Label Smoothing Enables Fast and Effective Unlearning

This section sets up the analysis and shows that properly performing label smoothing enables fast and effective unlearning. The key ingredients of the proposed approach are gradient ascent (GA) and label smoothing (LS). This section starts with understanding how GA helps with unlearning and then moves on to show the power of LS. At the end of the section, the algorithm is formally presented.

8.2.1 Preliminary

Consider a K-class classification problem over a training data distribution $\mathcal{D}_{tr} = (\mathcal{X} \times \mathcal{Y})$, where $\mathcal{X}$ and $\mathcal{Y}$ denote the feature and label spaces, respectively. Owing to privacy regulations, there exists a forgetting data distribution $\mathcal{D}_f$ that the model is required to unlearn. $\boldsymbol{\theta}_{tr}$ refers to the model trained on $\mathcal{D}tr$ and $\boldsymbol{\theta}_u$ refers to the model excluding the influence of $\mathcal{D}_f$. The goal of machine unlearning (MU) is how to generate $\boldsymbol{\theta}_u$ from $\boldsymbol{\theta}_{tr}$ (Fig. 8.1).

Label Smoothing In a K-class classification task, let $\boldsymbol{y}_i$ denote the one-hot encoded vector of $y_i \in \mathcal{Y}$. Following [19], positive label smoothing (PLS) and negative label smoothing (NLS) can be unified under the framework of generalized label smoothing (GLS). The smoothed label $\boldsymbol{y}_i^{\text{GLS},\alpha}$, parameterized by a smoothing rate $\alpha \in (-\infty, 1]$, is defined as $\boldsymbol{y}_i^{\text{GLS},\alpha} = (1-\alpha)\cdot \boldsymbol{y}_i + \frac{\alpha}{K}\cdot \mathbf{1} = [\frac{\alpha}{K}, \cdots, \frac{\alpha}{K}, (1+\frac{1-K}{K}\alpha), \frac{\alpha}{K}, \cdots, \frac{\alpha}{K}]$, where the element $1+\frac{1-K}{K}\alpha$ corresponds to the y_i-th entry of the encoded label vector. Notably, GLS reduces to NLS when $\alpha < 0$.

8.2.2 Gradient Ascent Can Help Gradient-Based Machine Unlearning

Three sets of model parameters are given in the MU problem: (1) $\boldsymbol{\theta}_{tr}^*$, the optimal parameters trained from $D_{tr} \sim \mathcal{D}_{tr}$, (2) $\boldsymbol{\theta}_r^*$, the optimal parameters trained from $D_r \sim \mathcal{D}_r$, such that $D_r = D_{tr} \backslash D_f$ and (3) $\boldsymbol{\theta}_f^*$, the optimal parameters unlearned using gradient ascent (GA) on $D_f \sim \mathcal{D}_f$. Note $\boldsymbol{\theta}_r^*$ can be viewed as the *exact* MU model. The definitions of $\boldsymbol{\theta}_{tr}^*$ and $\boldsymbol{\theta}_r^*$ follow the standard empirical risk minimization as

$$\boldsymbol{\theta}^* = \arg\min_{\boldsymbol{\theta}} \frac{1}{n} \sum_{z \in D} \ell\left(h_{\boldsymbol{\theta}}, z\right). \tag{8.1}$$

and by using the influence function, $\boldsymbol{\theta}_f^*$ is

$$\boldsymbol{\theta}_f^* = \arg\min_{\boldsymbol{\theta}} \{R_{tr}(\boldsymbol{\theta}) + \varepsilon \sum_{z^f \in D_f} \ell(h_{\boldsymbol{\theta}}, z^f)\}$$

where $R_{tr}(\boldsymbol{\theta}) = \sum_{z^{tr} \in D_{tr}} \ell(h_{\boldsymbol{\theta}}, z^{tr})$ and $R_f(\boldsymbol{\theta}) = \sum_{z^f \in D_f} \ell(h_{\boldsymbol{\theta}}, z^f)$ are the empirical risk on D_{tr} and D_f, respectively. Notations $\ell(h_{\boldsymbol{\theta}}, z)$ is used to specify the loss of an example $z = (x, y)$ in the dataset. $h_{\boldsymbol{\theta}}$ is a function h parameterized by $\boldsymbol{\theta}$. ε is the weight of D_f compared with D_{tr}. The optimal parameter can be found when the gradient is 0:

$$\nabla_{\boldsymbol{\theta}} R_{tr}(\boldsymbol{\theta}_f^*) + \varepsilon \sum_{z^f \in D_f} \nabla_{\boldsymbol{\theta}} \ell(h_{\boldsymbol{\theta}_f^*}, z^f) = 0. \tag{8.2}$$

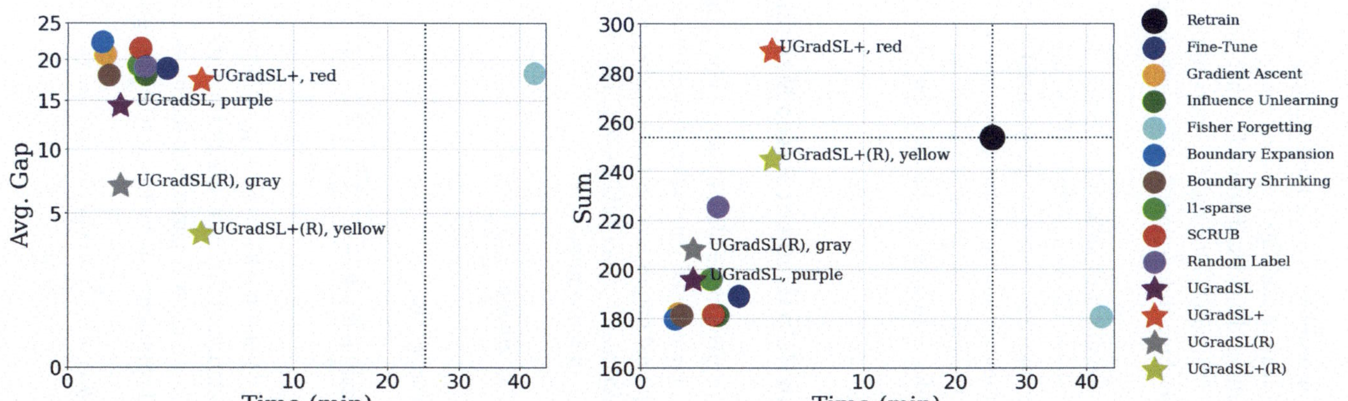

Fig. 8.1 Summary of the method and baselines (CIFAR-100, random forgetting across all classes) in terms of the two different MU metrics and speed. The proposed method with a following *(R)* means the method is optimized according to the retrained model while the method without a *(R)* means the method is optimized according to the *complete unlearning* [26]. For the plot of **Average Gap**, the **lower left** indicates better performance. Average Gap is to evaluate how close the approximate MU method is to the retrained model. The smaller the average gap is, the better MU performance is. The retrained model has no average gap as reference in this plot. The **bold** vertical dash shows the speed. When we target to make the approximate MU method close to the retrained model, UGradSL (★), UGradSL (R, ★) and UGradSL+ (R, ★) outperform the other methods. For the plot of **Sum**, the **upper left** indicates better performance. Sum is to evaluate the comprehensive performance of the approximate MU model. The higher the sum is, the better performance of the MU method is. The black dot represents the retrained model. When we target to improve the comprehensive performance of the approximate MU, UGradSL+ (★) and UGradSL+ (R, ★) outperform the other methods without too much delay. Moreover, in both plots, the methods are always the top methods. The methods show the robustness and generalization capability in both MU metrics

Expanding Eq. (8.2) at $\boldsymbol{\theta} = \boldsymbol{\theta}_{tr}^*$ using the Taylor series, there is

$$\boldsymbol{\theta}_f^* - \boldsymbol{\theta}_{tr}^* \approx -\Big[\sum_{z^{tr}\in D_{tr}} \nabla_{\boldsymbol{\theta}}^2 \ell(h_{\boldsymbol{\theta}_{tr}^*}, z^{tr}) + \varepsilon \sum_{z^f\in D_f} \nabla_{\boldsymbol{\theta}}^2 \ell(h_{\boldsymbol{\theta}_{tr}^*}, z^f)\Big]^{-1} \Big(\varepsilon \sum_{z^f\in D_f} \nabla_{\boldsymbol{\theta}} \ell(h_{\boldsymbol{\theta}_{tr}^*}, z^f)\Big). \qquad (8.3)$$

Similarly, $\nabla_{\boldsymbol{\theta}} R_{tr}(\boldsymbol{\theta}_{tr}^*)$ can be expanded at $\boldsymbol{\theta} = \boldsymbol{\theta}_r^*$ and $\boldsymbol{\theta}_r^* - \boldsymbol{\theta}_{tr}^*$ is derived as

$$\boldsymbol{\theta}_r^* - \boldsymbol{\theta}_{tr}^* \approx \left[\sum_{z^{tr}\in D_{tr}} \nabla_{\boldsymbol{\theta}}^2 \ell(h_{\boldsymbol{\theta}_r^*}, z^{tr})\right]^{-1} \left(\sum_{z^{tr}\in D_{tr}} \nabla_{\boldsymbol{\theta}} \ell(h_{\boldsymbol{\theta}_r^*}, z^{tr})\right).$$

The average operation in the original definition of the influence function is ignored for computation convenience because the sizes of D_{tr} or D_f are fixed. For GA, let $\varepsilon = -1$ in Eq. 8.3 and there is

$$\boldsymbol{\theta}_r^* - \boldsymbol{\theta}_f^* = \boldsymbol{\theta}_r^* - \boldsymbol{\theta}_{tr}^* - (\boldsymbol{\theta}_f^* - \boldsymbol{\theta}_{tr}^*) = \Delta\boldsymbol{\theta}_r - \Delta\boldsymbol{\theta}_f \qquad (8.4)$$

$(-\Delta\boldsymbol{\theta}_r)$ represents the learning gap from $\boldsymbol{\theta}_r^*$ to $\boldsymbol{\theta}_{tr}^*$ while vector $\Delta\boldsymbol{\theta}_f$ represents how much the model unlearns (backtracked progress) between $\boldsymbol{\theta}_f^*$ and $\boldsymbol{\theta}_{tr}^*$.

Ideally, when $\Delta\boldsymbol{\theta}_r$ and $\Delta\boldsymbol{\theta}_f$ are identical vectors, GA guides the model toward the optimal retrained solution, as $\boldsymbol{\theta}_r = \boldsymbol{\theta}_f$ holds in this case. However, this condition is hard to satisfy in practice. Thus, GA cannot always help MU. It is summarized in Theorem 8.1.

Theorem 8.1 *Given the approximation in Eq. (8.4), GA achieves exact MU if and only if*

$$\sum_{z^f\in D_f} \nabla_{\boldsymbol{\theta}} \ell(h_{\boldsymbol{\theta}_r^*}, z^f) \approx -\boldsymbol{H}(\boldsymbol{\theta}_r^*, \boldsymbol{\theta}_{tr}^*) \cdot \sum_{z^f\in D_f} \nabla_{\boldsymbol{\theta}} \ell(h_{\boldsymbol{\theta}_{tr}^*}, z^f),$$

$\boldsymbol{H}(\boldsymbol{\theta}_r^*, \boldsymbol{\theta}_{tr}^*) = \left[\sum_{z^{tr}\in D_{tr}} \nabla_{\boldsymbol{\theta}}^2 \ell(h_{\boldsymbol{\theta}_r^*}, z^{tr})\right]\left[\sum_{z^r\in D_r} \nabla_{\boldsymbol{\theta}}^2 \ell(h_{\boldsymbol{\theta}_{tr}^*}, z^r)\right]^{-1}$. *Otherwise, there exist* $\boldsymbol{\theta}_r^*, \boldsymbol{\theta}_{tr}^*$ *such that GA can not help MU, i.e.,* $\|\boldsymbol{\theta}_r^* - \boldsymbol{\theta}_f^*\| > \|\boldsymbol{\theta}_r^* - \boldsymbol{\theta}_{tr}^*\|$.

8.2.3 Label Smoothing Improves MU

In practice, GA cannot be guaranteed to consistently improve MU, as shown in Theorem 8.1. To mitigate the potential adverse effects of GA, this chapter proposes incorporating label smoothing as a plug-in module. Taking the cross-entropy loss as an example, the loss under GLS is computed as

$$\ell(h_{\boldsymbol{\theta}}, z^{\mathrm{GLS},\alpha}) = \left(1 + \frac{1-K}{K}\alpha\right) \cdot \ell(h_{\boldsymbol{\theta}}, (x, y)) + \frac{\alpha}{K} \sum_{y'\in\mathcal{Y}\backslash y} \ell(h_{\boldsymbol{\theta}}, (x, y')), \qquad (8.5)$$

where $\ell(h_{\boldsymbol{\theta}}, (x, y)) := \ell(h_{\boldsymbol{\theta}}, z)$ and $\ell(h_{\boldsymbol{\theta}}, (x, y'))$ to denote the loss of an example when its label is replaced with y'. Intuitively, the term $\sum_{y' \in \mathcal{Y} \setminus y} \ell(h_{\boldsymbol{\theta}}, (x, y'))$ in Eq. (8.5) encourages the model to assign uniformly low confidence across incorrect classes. As a result, the model tends to make wrong predictions on samples from the forgetting dataset with equally low confidence [19, 27].

With smoothed label given in Eq. (8.5), there exists a vector $\Delta\boldsymbol{\theta}_n$ such that Eq. (8.4) can be written as

$$\boldsymbol{\theta}_r^* - \boldsymbol{\theta}_{f,\mathrm{LS}}^* \approx \Delta\boldsymbol{\theta}_r - \Delta\boldsymbol{\theta}_f + \frac{1-K}{K}\alpha \cdot (\Delta\boldsymbol{\theta}_n - \Delta\boldsymbol{\theta}_f), \tag{8.6}$$

The detailed form of $\Delta\boldsymbol{\theta}_n$ is given in Eq. (8.7).

$$\begin{aligned} \boldsymbol{\theta}_r^* - \boldsymbol{\theta}_{f,\mathrm{LS}}^* &\approx \Delta\boldsymbol{\theta}_r + (1 + \frac{1-K}{K}\alpha) \cdot (-\Delta\boldsymbol{\theta}_f) + \frac{1-K}{K}\alpha \cdot \Delta\boldsymbol{\theta}_n \\ &= \Delta\boldsymbol{\theta}_r - \Delta\boldsymbol{\theta}_f + \frac{1-K}{K}\alpha \cdot (\Delta\boldsymbol{\theta}_n - \Delta\boldsymbol{\theta}_f) \end{aligned} \tag{8.7}$$

But intuitively, $\Delta\boldsymbol{\theta}_n$ captures the gradient influence of the smoothed non-target label on the weight. The effect of NLS ($\alpha < 0$) is given in Theorem 8.2 below.

Theorem 8.2 *Given the approximation in Eq. (8.4) and* $\langle \Delta\boldsymbol{\theta}_r - \Delta\boldsymbol{\theta}_f, \Delta\boldsymbol{\theta}_n - \Delta\boldsymbol{\theta}_f \rangle \le 0$, *there exists an* $\alpha < 0$ *such that NLS improves GA in unlearning, i.e.,* $\|\boldsymbol{\theta}_r^* - \boldsymbol{\theta}_{f,NLS}^*\| < \|\boldsymbol{\theta}_r^* - \boldsymbol{\theta}_f^*\|$, *where* $\boldsymbol{\theta}_{f,NLS}^*$ *is the optimal parameters unlearned using GA and NLS, and* $\langle \cdot, \cdot \rangle$ *the inner product of two vectors.*

The vector $\Delta\boldsymbol{\theta}_f - \Delta\boldsymbol{\theta}_r$ represents the resultant of Newton's directions for learning and unlearning. Similarly, $\Delta\boldsymbol{\theta}_f - \Delta\boldsymbol{\theta}_n$ corresponds to the resultant of Newton's direction for learning non-target labels while unlearning the target label. When the condition $\langle \Delta\boldsymbol{\theta}_r - \Delta\boldsymbol{\theta}_f, \Delta\boldsymbol{\theta}_n - \Delta\boldsymbol{\theta}_f \rangle \le 0$ is satisfied, the vector $\Delta\boldsymbol{\theta}_n - \Delta\boldsymbol{\theta}_f$ reflects the effect of the smoothing term in the unlearning process. Under the assumption that an exact MU model can fully unlearn an example, the vector $\Delta\boldsymbol{\theta}_n$ provides a corrective direction that pushes the model closer to the exact MU state by driving it to make incorrect predictions on the forgotten sample.

The effect of the smoothed term in gradient ascent (GA) with negative label smoothing (NLS) is equivalent to performing gradient descent with traditionally defined (positive) label smoothing (LS). In both cases, the gradient of the smoothed term is given by $\alpha/K \cdot \sum_{y' \in \mathcal{Y} \setminus y} \nabla \ell(h_{\boldsymbol{\theta}}, (x, y'))$ in both cases.

8.2.4 Label Smoothing Helps Local Differential Privacy

When $\alpha < 0$, the smoothing term contributes positively to the gradient ascent (GA) step. Beyond optimization, label smoothing can also be interpreted from the perspective of privacy protection. This interpretation arises because label smoothing decreases the probability mass on the true label, allowing it to blend more effectively with alternative candidate labels. In particular, this chapter considers a local differential privacy (LDP) guarantee for labels, defined as follows.

Definition 8.1 (Label-LDP) A privacy protection mechanism $\mathcal{M}$ satisfies ϵ-Label-LDP, if for any labels $y, y', y^{\text{pred}} \in \mathcal{Y}$, $\frac{\mathbb{P}(\mathcal{M}(y)=y^{\text{pred}})}{\mathbb{P}(\mathcal{M}(y')=y^{\text{pred}})} \le e^{\epsilon}$.

The operational meaning of $\mathcal{M}$ is to ensure that for any two labels y and y' in the label space, their privatized versions have comparable probabilities of being mapped to any predicted label $y^{\text{pred}} \in \mathcal{Y}$. In other words, the model's predictions on the forgetting dataset should not strongly depend on the true underlying label. This similarity is quantified by the privacy budget $\epsilon \in [0, +\infty)$, where a smaller ϵ indicates stronger indistinguishability between y and y', and thus, a stricter privacy guarantee.

Recall $R_{tr}(\boldsymbol{\theta}) = \sum_{z^{tr} \in D_{tr}} \ell(h_{\boldsymbol{\theta}}, z^{tr})$. Denote by

$$R_f^{\text{NLS}}(\boldsymbol{\theta}; \alpha) = \sum_{z^{\text{LS},\alpha} \in D_f} \ell(h_{\boldsymbol{\theta}}, z^{\text{LS},\alpha}), \alpha < 0,$$

the empirical risk of forgetting data with NLS. After applying MU with label smoothing on D_f via GA, the resulting model can be viewed as minimizing the objective $\gamma_1 \cdot R_{tr}(\boldsymbol{\theta}) - \gamma_2 \cdot R_f^{\text{NLS}}(\boldsymbol{\theta}; \alpha)$, which is a weighted combination of the risk from two phases: (1) machine learning on D_{tr} with weight $\gamma_1 > 0$ and (2) machine unlearning on D_f with weight $\gamma_2 > 0$. Analyzing this risk leads to the following theorem, which shows that incorporating NLS into MU induces ϵ-Label-LDP for the forgetting data.

Theorem 8.3 *Suppose $\gamma_1 - \gamma_2(1 + \frac{1-K}{K}\alpha) > 0$. MU using GA+NLS achieves ϵ-Label-LDP on D_f where*

$$\epsilon = \left| \log \left(\frac{K}{\alpha} \left(1 - \frac{\gamma_1}{\gamma_2} \right) + 1 - K \right) \right|, \quad \alpha < 0.$$

Intuitively, a more negative α enhances the privacy of labels in the forgetting dataset. As $\alpha \to (1 - \gamma_1/\gamma_2)$, the privacy budget ϵ approaches zero, yielding the strongest label-LDP guarantee—corresponding to the ideal objective of MU. However, the theorem also cautions that α cannot be made arbitrarily negative.

8.2.5 UGradSL: A Plug-and-Play and Gradient-Mixed MU Method

Considering the impact of label smoothing on MU and LDP, we propose the following method. Compared with full retraining, Fine-Tuning (FT) and Gradient Ascent (GA) are significantly more efficient, as shown in Sect. 8.3.2, while achieving comparable or even better MU performance. FT and GA approach MU from different perspectives: FT leverages gradient descent (GD) to transfer knowledge from D_{tr} to D_r, whereas GA aims to remove the knowledge of D_f from the model.

Thanks to the flexibility of label smoothing, the approach is compatible with gradient-based methods such as FT and GA, making it a plug-and-play algorithm. *UGradSL is built on GA, whereas UGradSL+ relies on FT. Compared to UGradSL, UGradSL+ provides more comprehensive results but requires higher computational cost.*

The choice of the smoothing rate $\boldsymbol{\alpha}$ is an important aspect to consider. Typically, the $\alpha_i \in \boldsymbol{\alpha}$ for each data point $z_i^f \in D_f$ can be set to the same value. To achieve better performance, UGradSL+ extends UGradSL by assigning α_i individually and adaptively for each (z_i^r, z_i^f) pair, based on the distance $d(z_i^r, z_i^f) \in [0, 1]$. The intuition is that when an instance z_i^f lies in a dense neighborhood of D_r, its inherent deniability is higher, implying a weaker need for "forgetting," which should correspond to a smaller α_i.

Assuming the amount of retained data is significantly larger than the amount of data to be forgotten ($|D_r| > |D_f|$), D_f will be iterated several times when D_r is fully iterated once. The loss is calculated using a gradient-mixed method as:

$$L(h_{\boldsymbol{\theta}}, B_f^{\mathrm{NLS},\boldsymbol{\alpha}}, B_r, p) = p \cdot \sum_{z^r \in B_r} \ell(h_{\boldsymbol{\theta}}, z^r) - (1-p) \cdot \sum_{z_i^{f,\mathrm{NLS},\alpha_i} \in B_f^{\mathrm{NLS},\boldsymbol{\alpha}}} \ell(h_{\boldsymbol{\theta}}, z_i^{f,\mathrm{NLS},\alpha_i}) \tag{8.8}$$

where $p \in [0, 1]$ is used to balance GD and GA and the minus sign between two elements on the RHS stands for the GA. $\boldsymbol{\alpha}$ is the vector for the smoothing rate of every data point z_i^f. The model $h_{\boldsymbol{\theta}}$ is updated according to the loss L in Eq. (8.8). UGradSL and UGradSL+ share similar structures, with their main difference lying in the convergence criterion: UGradSL relies on the convergence of D_f, whereas UGradSL+ relies on that of D_r. It is important to emphasize that the Hessian matrix in Theorem 8.1 is used solely for theoretical analysis; in practice, no Hessian calculation is required. Consequently, the method does not introduce significant additional computation while substantially enhancing MU performance. Unlike uniform label smoothing, the improved version accounts for the similarity between data points in D_r and D_f, providing self-adaptive smoothed labels for individual z_i^f while also preserving LDP.

8.3 Experiments

8.3.1 Experiment Setup

Dataset and Model Selection The proposed method is evaluated across datasets of varying scales and modalities, including CIFAR-10, CIFAR-100 [28], SVHN [29], CelebA [30], Tiny-ImageNet, ImageNet [31], and 20 Newsgroups [32]. For vision datasets, ResNet-18 [33] is used as the backbone model, while for the language dataset, BERT [34] serves as the backbone.

Baseline Methods Several baseline methods are considered for comparison, including retraining, fine-tuning (FT) [9, 10], gradient ascent (GA) [11, 12], influence function-based unlearning (IU)[8, 13], boundary unlearning (BU)[35], ℓ_1-sparse unlearning [15], random label (RL)[36], SCRUB[37], SalUn [38], and EU-*k* and CF-*k*[39]. In addition, we note the existence of an unlearning paradigm known as instance unlearning[40], though it is not the focus of this work.

Evaluation Metrics The evaluation metrics follow [15], jointly considering unlearning accuracy (UA), membership inference attack (MIA), remaining accuracy (RA), testing accuracy (TA), and run-time efficiency (RTE). To assess the effectiveness of "forgetting," we adopt the MIA metric as in [15, 38], where the attack model's accuracy against the target model θ_u is reported as the true negative rate (TNR) on the forget set. Formally, this corresponds to the global MIA score [41], defined as $\text{MIA}_{\text{Score}} = 1 - \Pr\left(x_f|\theta_\star\right)$, where $x_f \in D_f$ denotes forget samples and $\theta_\star$ is the model under evaluation. Under this formulation, a higher $\text{MIA}_{\text{Score}}$ indicates stronger "forgetting" performance.

Different Unlearning Objectives As discussed in [37], MU can pursue different objectives. Beyond the individual metrics, we adopt two combined metrics tailored to different goals of MU. When access to the retrained model is unavailable, the objective is to maximize unlearning performance (UA, MIAScore) while preserving model capacity (TA, RA). This corresponds to evaluating *complete unlearning*, as defined in [26]. In this setting, we use *Sum* as a comprehensive metric, defined as the sum of UA, TA, RA, and MIAScore.

Alternatively, if the aim is to align the behavior of the unlearned model with that of the retrained model, we adopt *Avg. Gap*, which measures the mean performance gap between each unlearning method and the retrained model across all individual metrics. A smaller value indicates better performance. For the random forgetting and group forgetting experiments in Tables 8.2 and 8.3, results targeting this objective are marked with the tag *(R)*.

Unlearning Paradigm We mainly consider three unlearning paradigms, including *class-wise forgetting*, *random forgetting*, and *group forgetting*. Class-wise forgetting is to unlearn the whole specific class where one class in D_r and the corresponding class in D_{te} are removed completely. Random forgetting across all classes is to unlearn data points belonging to all classes. As a special case of random forgetting, *group forgetting* means that the model is trained to unlearn the group or sub-class of the corresponding super-classes.

8.3.2 Experiment Results

8.3.2.1 Class-Wise Forgetting

A class is randomly selected, and class-wise forgetting is performed on five datasets. The results for SVHN/ImageNet and CIFAR-10 are presented in Tables 8.1 and 8.4, respectively. As shown, UGradSL and UGradSL+ enhance the performance of GA and FT, respectively, without increasing RTE or causing a decline in TA or RA. This demonstrates comprehensive improvements across the main metrics, even under randomness in D_f, highlighting the robustness and flexibility of the proposed methods in MU across different dataset scales and modalities. Furthermore, in terms of the average gap, the proposed methods exhibit behavior closely aligned with the retrained model. Regarding the sum metric, they achieve nearly the highest scores, indicating their ability to effectively balance forgetting metrics (UA, $\text{MIA}_{\text{Score}}$) and remaining metrics (TA, RA).

8.3.2.2 Random Forgetting

Data is randomly sampled from each class to form Df, ensuring all classes are represented and that Df constitutes 10% of Dtr. The results on CIFAR-100 and TinyImageNet are presented in Table 8.2. Compared with class-wise forgetting, improving MU performance in random forgetting is more challenging, as it often leads to a noticeable decrease in remaining accuracy across datasets. When the optimization goal is complete unlearning, the proposed methods achieve substantially higher UA and MIAScore than the baseline methods, including retraining. Compared with FT and GA, UGradSL+ improves UA or $\text{MIA}_{\text{Score}}$ by over 50%, with at most a 15% reduction in RA or TA. Even when the objective is to make the unlearned model approximate the retrained one, the proposed methods consistently outperform other baselines.

8.3.2.3 Group Forgetting

Although group forgetting can be seen as part of random forgetting, we want to highlight its use case here due to its practical impacts on e.g., facial attributes classification. The identities can be regarded as the subgroup in the attributes.

CIFAR-10 and CIFAR-100 These two datasets share the same image set, while CIFAR-100 provides labels for 100 sub-classes grouped into 20 super-classes [28]. A model is trained to classify the 20 super-classes using the CIFAR-100 training set. In the *group forgetting within one super-class* setting, one sub-class from a given super-class in CIFAR-100 is removed. For instance, the *Fish* super-class contains five fine-grained fish categories, and the goal is to make the model forget one of them. Unlike class-wise forgetting, the testing set remains unchanged. The results of group forgetting are summarized in Table 8.3.

Table 8.1 Results of class-wise forgetting in SVHN and ImageNet

	SVHN							ImageNet						
Method	UA	MIA_{Score}	RA	TA	Avg. gap (↓)	Sum. (↑)	RTE (↓, min)	UA	MIA_{Score}	RA	TA	Avg. gap (↓)	Sum. (↑)	RTE (↓, hr)
Retrain	100.00±0.00	100.00±0.00	98.19±3.14	94.50±0.34	–	392.69	14.92	100.00±0.00	100.00±0.00	71.62±0.12	69.57±0.07	–	341.19	26.18
FT	6.49±1.49	99.98±0.04	100.00±0.01	96.08±0.01	23.42	302.55	2.42	52.42±15.81	55.86±18.02	70.66±2.54	69.25±0.78	23.25	248.20	2.87
GA	87.49±1.94	99.85±0.09	99.52±0.03	95.27±0.21	3.45	388.73	**0.15**	81.23±0.69	83.52±2.08	66.00±0.03	64.72±0.02	11.43	295.47	**0.01**
IU	93.55±2.78	100.00±0.00	99.54±0.03	95.64±0.31	1.80	388.73	0.23	33.54±19.46	49.83±21.57	66.25±1.99	66.28±1.19	31.32	215.90	1.51
BE	85.56±3.07	99.98±0.02	99.55±0.01	95.53±0.07	3.83	380.62	3.17	98.62±0.58	0.15±0.11	53.13±0.27	56.72±0.31	33.14	241.91	0.24
BS	96.62±1.14	99.95±0.09	99.99±0.00	95.39±0.18	1.00	391.95	3.91	98.85±0.50	0.13±0.12	53.35±0.16	56.93±0.03	32.98	209.26	0.37
ℓ_1-sparse	99.78±0.31	100.00±0.00	98.63±0.01	97.36±0.18	0.75	395.77	2.91	100.00±0.00	100.00±0.00	39.01±1.03	44.62±0.91	14.39	283.63	0.16
RL	99.99±0.01	100.00±0.00	100.00±0.00	95.44±0.13	0.13	395.43	3.53	100.00±0.00	100.00±0.00	62.06±4.19	62.93±0.45	4.05	324.99	1.17
EU-*k*	100.00±0.00	100.00±0.00	99.61±0.08	65.56±2.38	7.59	365.17	4.93	100.00±0.00	0.00±0.00	32.99±0.07	37.19±0.15	42.75	170.18	0.62
CF-*k*	0.09±0.03	2.18±2.21	99.34±0.02	69.87±4.13	55.88	171.48	5.02	99.79±0.36	0.00±0.00	66.84±0.03	68.35±0.28	26.55	235.19	1.25
SCRUB	99.99±0.02	100.00±0.00	100.00±0.00	95.79±0.26	**0.04**	**395.78**	4.97	56.59±2.17	75.59±1.19	66.98±0.11	68.24±0.07	18.45	267.40	0.21
SalUn	99.74±0.39	100.00±0.00	99.53±0.02	95.00±1.50	0.53	394.27	4.77	100.00±0.00	100.00±0.00	63.00±5.03	62.72±0.31	3.87	325.72	1.95
UGradSL	90.71±4.08	99.90±0.16	99.54±0.04	95.64±0.25	2.54	385.79	0.23	100.00±0.00	100.00±0.00	76.91±1.82	65.94±1.35	**2.23**	342.85	**0.01**
UGradSL+	100.00±0.00	100.00±0.00	99.82±0.62	94.35±0.70	0.44	394.17	4.56	100.00±0.00	100.00±0.00	78.16±0.07	66.84±0.06	2.32	**345.00**	4.19

The best comprehensive metrics are **bold**

Table 8.2 Results of random forgetting in CIFAR-100 and Tiny-ImageNet

	CIFAR-100							Tiny-ImageNet						
Method	UA	MIA_{Score}	RA	TA	Avg. gap (↓)	Sum. (↑)	RTE (↓, min)	UA	MIA_{Score}	RA	TA	Avg. gap (↓)	Sum. (↑)	RTE (↓, min)
Retrain	$29.47_{\pm 1.59}$	$53.50_{\pm 1.19}$	$99.98_{\pm 0.01}$	$70.51_{\pm 1.17}$	–	253.46	25.01	$49.35_{\pm 0.38}$	$58.44_{\pm 0.89}$	$83.80_{\pm 0.29}$	$59.66_{\pm 0.44}$	–	250.75	235.68
FT	$2.55_{\pm 0.03}$	$10.59_{\pm 0.27}$	$99.95_{\pm 0.01}$	$75.95_{\pm 0.05}$	18.83	189.04	1.95	$29.23_{\pm 0.29}$	$37.02_{\pm 0.33}$	$82.51_{\pm 0.20}$	$60.96_{\pm 0.23}$	11.03	209.82	18.61
GA	$2.58_{\pm 0.06}$	$5.95_{\pm 0.17}$	$97.45_{\pm 0.02}$	$76.09_{\pm 0.01}$	20.64	182.07	**0.29**	$19.34_{\pm 1.67}$	$25.19_{\pm 0.68}$	$81.51_{\pm 1.56}$	$59.66_{\pm 0.61}$	16.39	185.70	8.65
IU	$15.71_{\pm 5.19}$	$18.69_{\pm 4.12}$	$84.65_{\pm 5.29}$	$62.20_{\pm 4.17}$	18.05	181.25	1.20	$60.61_{\pm 0.01}$	$83.67_{\pm 0.15}$	$16.36_{\pm 0.37}$	$23.44_{\pm 0.29}$	35.04	184.08	7.30
BE	$0.01_{\pm 0.00}$	$1.45_{\pm 0.02}$	$99.97_{\pm 0.18}$	$78.26_{\pm 0.00}$	22.32	179.69	0.24	$17.65_{\pm 0.31}$	$24.48_{\pm 0.42}$	$82.85_{\pm 0.20}$	$58.16_{\pm 0.08}$	17.03	183.14	**3.53**
BS	$2.20_{\pm 1.21}$	$10.73_{\pm 9.37}$	$98.22_{\pm 1.26}$	$70.23_{\pm 1.67}$	18.02	181.38	0.34	$19.47_{\pm 0.69}$	$25.45_{\pm 0.15}$	$81.23_{\pm 0.74}$	$56.75_{\pm 0.80}$	17.09	182.90	5.63
ℓ_1-sparse	$8.19_{\pm 0.38}$	$19.11_{\pm 0.52}$	$88.39_{\pm 0.31}$	$80.26_{\pm 0.16}$	23.75	195.95	1.00	$35.73_{\pm 0.35}$	$41.98_{\pm 0.73}$	$78.19_{\pm 0.05}$	$61.44_{\pm 0.12}$	9.37	217.34	23.40
RL	$4.06_{\pm 0.37}$	$50.12_{\pm 3.48}$	$99.92_{\pm 0.01}$	$71.30_{\pm 0.12}$	7.41	225.40	1.20	$40.52_{\pm 0.15}$	$59.01_{\pm 0.76}$	$77.58_{\pm 0.06}$	$60.18_{\pm 0.19}$	4.04	237.29	27.08
EU-*k*	$1.73_{\pm 0.06}$	$3.33_{\pm 0.07}$	$98.44_{\pm 0.05}$	$59.92_{\pm 0.43}$	22.51	163.42	1.96	$33.55_{\pm 0.35}$	$22.19_{\pm 1.75}$	$81.41_{\pm 0.27}$	$58.08_{\pm 0.21}$	14.01	195.23	20.02
CF-*k*	$0.07_{\pm 0.02}$	$0.47_{\pm 0.16}$	$99.98_{\pm 0.01}$	$67.86_{\pm 0.12}$	21.27	168.38	0.88	$19.31_{\pm 0.38}$	$23.22_{\pm 2.28}$	$81.59_{\pm 0.37}$	$58.15_{\pm 0.19}$	17.25	182.27	13.18
SCRUB	$0.09_{\pm 0.59}$	$4.01_{\pm 1.25}$	$99.97_{\pm 0.34}$	$77.45_{\pm 0.26}$	21.46	181.52	1.06	$20.11_{\pm 1.15}$	$25.35_{\pm 7.53}$	$80.91_{\pm 0.77}$	$60.11_{\pm 0.99}$	16.42	186.48	25.79
SalUn	$35.23_{\pm 0.32}$	$89.39_{\pm 0.46}$	$99.53_{\pm 0.04}$	$64.26_{\pm 0.58}$	12.10	288.41	3.33	$40.39_{\pm 0.15}$	$52.32_{\pm 10.67}$	$77.60_{\pm 0.11}$	$60.30_{\pm 0.31}$	5.48	230.61	34.42
UGradSL	$15.10_{\pm 2.76}$	$34.67_{\pm 0.63}$	$86.69_{\pm 2.41}$	$59.25_{\pm 2.35}$	14.44	195.71	0.55	$37.29_{\pm 0.41}$	$27.42_{\pm 0.11}$	$65.18_{\pm 0.14}$	$55.99_{\pm 0.34}$	16.34	185.88	9.47
UGradSL+	$63.89_{\pm 0.75}$	$71.51_{\pm 1.31}$	$92.25_{\pm 0.11}$	$61.09_{\pm 0.10}$	17.40	**288.74**	3.52	$63.06_{\pm 0.31}$	$61.76_{\pm 0.72}$	$71.18_{\pm 0.03}$	$55.52_{\pm 0.58}$	8.45	**251.51**	25.93
UGradSL (R)	$18.36_{\pm 0.17}$	$40.71_{\pm 0.13}$	$98.38_{\pm 0.03}$	$68.23_{\pm 0.16}$	6.95	207.95	0.55	$40.73_{\pm 0.71}$	$37.58_{\pm 0.21}$	$67.30_{\pm 0.04}$	$50.38_{\pm 0.77}$	13.82	195.99	9.47
UGradSL+ (R)	$21.69_{\pm 0.59}$	$49.47_{\pm 1.25}$	$99.87_{\pm 0.34}$	$73.60_{\pm 0.26}$	**3.75**	244.63	3.52	$53.06_{\pm 1.27}$	$59.46_{\pm 1.01}$	$81.38_{\pm 0.75}$	$52.52_{\pm 0.84}$	**3.57**	246.52	25.93

The best comprehensive metrics are **bold**

Table 8.3 Results of group forgetting on CIFAR-20 and CelebA

	CIFAR-20							CelebA						
	UA	$\text{MIA}_{\text{Score}}$	RA	TA	Avg. gap (↓)	Sum. (↑)	RTE (↓, min)	UA	$\text{MIA}_{\text{Score}}$	RA	TA	Avg. gap (↓)	Sum. (↑)	RTE (↓, min)
Retrain	$13.33_{\pm 1.64}$	$28.47_{\pm 0.75}$	$99.94_{\pm 0.01}$	$81.23_{\pm 0.13}$	–	229.71	27.35	$6.74_{\pm 0.26}$	$9.77_{\pm 1.49}$	$94.38_{\pm 0.49}$	$91.78_{\pm 0.33}$	–	202.67	258.69
FT	$1.00_{\pm 0.43}$	$2.73_{\pm 0.52}$	$99.37_{\pm 0.08}$	$79.02_{\pm 0.03}$	10.21	182.12	7.47	$5.36_{\pm 0.17}$	$5.87_{\pm 0.11}$	$93.91_{\pm 0.04}$	$93.18_{\pm 0.03}$	1.79	198.32	25.94
GA	$87.93_{\pm 2.92}$	$88.93_{\pm 2.33}$	$81.46_{\pm 0.77}$	$64.07_{\pm 0.95}$	42.68	322.39	**0.11**	$6.00_{\pm 0.16}$	$5.76_{\pm 0.14}$	$92.86_{\pm 0.13}$	$92.52_{\pm 0.08}$	1.75	197.14	**1.20**
IU	$0.00_{\pm 0.00}$	$2.07_{\pm 1.29}$	$99.95_{\pm 0.01}$	$80.92_{\pm 0.34}$	10.01	182.94	1.10	$5.90_{\pm 0.11}$	$4.91_{\pm 0.30}$	$93.05_{\pm 0.01}$	$92.62_{\pm 0.01}$	1.97	196.48	219.77
BE	$89.07_{\pm 1.39}$	$91.73_{\pm 1.75}$	$76.36_{\pm 0.92}$	$60.17_{\pm 0.92}$	45.91	317.33	0.33	$11.50_{\pm 0.80}$	$48.41_{\pm 8.86}$	$88.37_{\pm 0.81}$	$88.07_{\pm 0.81}$	13.28	206.35	48.91
BS	$88.60_{\pm 1.13}$	$90.67_{\pm 1.18}$	$76.70_{\pm 1.08}$	$60.41_{\pm 1.17}$	45.38	316.37	0.29	$8.95_{\pm 5.11}$	$27.35_{\pm 30.20}$	$91.00_{\pm 5.22}$	$90.63_{\pm 5.65}$	6.08	217.93	50.99
ℓ_1-sparse	$0.13_{\pm 0.09}$	$2.27_{\pm 0.57}$	$99.57_{\pm 0.04}$	$80.44_{\pm 0.08}$	10.14	182.41	0.38	$9.46_{\pm 1.82}$	$36.91_{\pm 30.96}$	$90.52_{\pm 1.75}$	$90.35_{\pm 1.77}$	8.79	217.24	37.49
RL	$56.93_{\pm 3.24}$	$98.60_{\pm 0.29}$	$99.92_{\pm 0.01}$	$80.28_{\pm 0.05}$	28.67	335.74	0.37	$8.31_{\pm 0.43}$	$28.55_{\pm 16.74}$	$91.85_{\pm 0.51}$	$91.62_{\pm 0.42}$	5.76	219.33	40.09
EU-*k*	$8.00_{\pm 4.57}$	$16.33_{\pm 7.18}$	$97.07_{\pm 0.18}$	$69.67_{\pm 0.35}$	7.98	191.07	0.87	$7.20_{\pm 0.19}$	$18.77_{\pm 3.69}$	$92.55_{\pm 0.30}$	$91.04_{\pm 0.67}$	3.01	209.56	1.98
CF-*k*	$0.00_{\pm 0.00}$	$0.80_{\pm 0.40}$	$99.98_{\pm 0.01}$	$77.46_{\pm 0.03}$	11.20	178.24	1.21	$5.46_{\pm 0.32}$	$17.26_{\pm 0.08}$	$94.45_{\pm 0.04}$	$92.72_{\pm 0.04}$	3.12	209.89	1.60
SCRUB	$0.00_{\pm 0.00}$	$1.13_{\pm 0.34}$	$99.93_{\pm 0.01}$	$81.05_{\pm 0.20}$	10.21	182.12	0.30	$8.78_{\pm 0.77}$	$13.37_{\pm 5.22}$	$91.21_{\pm 0.86}$	$90.65_{\pm 0.86}$	2.49	201.01	70.13
SalUn	$52.93_{\pm 2.21}$	$99.80_{\pm 0.35}$	$99.55_{\pm 0.00}$	$76.48_{\pm 0.26}$	29.02	328.76	2.88	$6.53_{\pm 0.28}$	$25.57_{\pm 8.22}$	$92.97_{\pm 0.03}$	$92.27_{\pm 0.07}$	5.43	217.34	83.43
UGradSL	$79.56_{\pm 0.34}$	$90.44_{\pm 1.99}$	$98.16_{\pm 4.46}$	$80.06_{\pm 0.90}$	32.79	**348.22**	0.13	$11.33_{\pm 4.17}$	$23.08_{\pm 11.53}$	$87.86_{\pm 3.85}$	$87.68_{\pm 3.81}$	7.13	209.95	2.17
UGradSL+	$81.11_{\pm 0.74}$	$86.96_{\pm 1.89}$	$95.69_{\pm 0.82}$	$78.02_{\pm 1.02}$	33.43	341.78	8.12	$15.63_{\pm 8.01}$	$26.95_{\pm 25.80}$	$89.17_{\pm 5.86}$	$88.29_{\pm 5.75}$	7.55	**220.04**	51.41
UGradSL (R)	$22.87_{\pm 0.90}$	$38.93_{\pm 1.57}$	$97.20_{\pm 0.19}$	$75.84_{\pm 0.16}$	7.03	234.84	0.13	$6.29_{\pm 1.41}$	$5.73_{\pm 3.50}$	$93.44_{\pm 0.14}$	$92.80_{\pm 0.27}$	**1.11**	198.26	2.17
UGradSL+ (R)	$78.44_{\pm 1.19}$	$88.67_{\pm 0.35}$	$97.93_{\pm 0.71}$	$79.77_{\pm 0.58}$	**2.40**	344.81	8.12	$6.12_{\pm 0.31}$	$5.54_{\pm 0.34}$	$92.79_{\pm 0.01}$	$92.49_{\pm 0.04}$	1.78	196.94	51.41

For CIFAR-20, the model is trained to classify 20 super-classes, with D_f representing one of five subclasses within a single super-class. In the CelebA dataset, the model performs binary classification to determine whether a person is smiling, with D_f selected based on specific identities. The best comprehensive metrics are **bold**

Table 8.4 Results of class-wise forgetting and random forgetting on CIFAR-10 with additional MIA

	Class-wise								Random							
	UA	MIA$_{Score}$	RA	TA	Add. MIA	Avg. gap (↓)	Sum. (↑)	RTE (↓, min)	UA	MIA$_{Score}$	RA	TA	Add. MIA	Avg. gap (↓)	Sum. (↑)	RTE (↓, min)
Retrain	100.00±0.00	100.00±0.00	98.19±3.14	94.50±0.34	99.23±0.08	–	392.69	24.62	8.07±0.47	17.41±0.69	100.00±0.01	91.61±0.24	50.69±0.73	–	217.09	24.66
FT	22.71±5.31	79.21±8.60	99.82±0.09	94.13±0.14	99.09±0.07	25.02	295.87	2.02	1.10±0.19	4.06±0.41	99.83±0.03	93.70±0.10	54.05±0.31	5.65	198.69	1.58
GA	25.19±11.38	73.48±9.68	96.84±0.58	73.10±1.62	99.43±0.09	31.02	268.61	**0.08**	0.56±0.01	1.19±0.05	99.48±0.02	94.55±0.05	55.04±0.66	6.80	195.78	**0.31**
IU	83.92±1.16	92.59±1.41	98.77±0.12	92.64±0.23	99.71±0.07	6.48	367.92	1.18	17.51±2.19	21.39±1.70	83.28±2.44	78.13±2.85	53.98±0.55	10.91	200.31	1.18
BE	64.93±0.01	98.19±0.00	99.47±0.00	94.00±0.11	99.60±0.02	9.67	356.59	0.20	0.00±0.00	0.26±0.02	100.00±0.00	95.35±0.18	55.41±0.49	7.24	195.61	3.17
BS	93.69±4.32	99.82±0.04	97.69±1.29	92.89±1.26	99.56±0.10	2.15	384.08	0.29	0.48±0.07	1.16±0.04	99.47±0.01	94.58±0.03	55.88±0.72	6.84	195.69	1.41
ℓ_1-sparse	100.00±0.00	100.00±0.00	97.86±1.29	96.11±1.26	99.02±0.15	0.48	393.97	1.00	2.80±0.37	18.59±3.48	99.97±0.01	94.08±0.12	55.01±0.49	2.24	198.39	1.98
RL	99.99±0.01	100.00±0.00	100.00±0.00	95.50±0.11	99.08±0.07	0.71	**395.49**	1.04	2.27±0.29	14.37±1.01	99.98±0.01	94.14±0.18	52.17±0.87	2.85	210.75	1.98
EU-k	100.00±0.00	100.00±0.00	100.00±0.00	75.04±1.10	99.89±0.18	5.32	375.04	1.45	0.00±0.00	0.50±0.30	99.99±0.01	77.21±1.21	61.88±1.33	9.85	177.70	1.58
CF-k	100.00±0.00	100.00±0.00	100.00±0.00	78.95±0.53	100.00±0.00	4.34	378.95	1.32	0.00±0.00	0.00±0.00	100.00±0.00	80.98±0.27	69.91±1.33	9.03	180.98	1.47
SCRUB	100.00±0.00	100.00±0.00	99.93±0.01	95.22±0.07	100.00±0.00	0.62	395.15	1.09	0.70±0.59	3.88±1.25	99.59±0.34	94.22±0.26	55.33±0.59	5.98	198.42	4.05
SalUn	90.74±13.91	100.00±0.00	98.20±0.34	80.49±1.21	98.63±0.59	4.78	369.43	2.22	46.95±0.15	86.33±1.29	97.75±0.42	77.22±0.77	69.95±0.12	31.11	231.03	2.42
UGradSL	94.99±4.35	97.95±1.78	95.47±4.08	86.78±5.68	99.94±0.01	4.38	375.19	0.22	20.77±0.75	35.45±2.85	79.83±0.75	73.94±0.75	56.27±0.11	17.15	209.99	0.45
UGradSL+	100.00±0.00	100.00±0.00	99.26±0.01	94.29±0.07	100.00±0.00	**0.32**	392.55	3.07	25.13±0.49	37.19±2.23	90.77±0.20	84.78±0.69	56.41±0.32	13.23	**237.87**	3.07
UGradSL (R)	–	–	–	–	–	–	–	–	5.87±0.51	13.33±0.70	98.82±0.28	92.17±0.23	53.54±0.97	**2.01**	210.19	0.45
UGradSL+ (R)	–	–	–	–	–	–	–	–	6.03±0.17	10.65±0.13	99.79±0.03	93.64±0.16	52.29±0.85	2.76	210.11	3.07

The best comprehensive metrics are **bold**

CelebA The CelebA dataset is used as a real-world case study, and the results are presented in Table 8.3. A binary classification model is trained to predict whether a person is smiling or not. The training set contains 8192 identities, and 1% of them (82 identities) are selected as D_f, including both smiling and non-smiling images. This experiment carries significant practical relevance, as biometric data such as facial identity and fingerprints require stronger privacy protection [42]. Compared with the baseline methods, the proposed approach achieves better removal of identity information while preserving more of the remaining data utility. This paradigm demonstrates the practical applicability of MU, offering a faster and more reliable means of improving unlearning performance.

8.3.3 Discussion

The Discussion of the Performance Tradeoff As shown in Table 8.1, strong MU performance is achieved in class-wise forgetting without a noticeable decline in the remaining performance (RA, TA) compared with other baselines. In contrast, for random and group forgetting, improvements in forgetting performance (UA, $\text{MIA}_{\text{Score}}$) inevitably lead to decreases in remaining performance, as presented in Table 8.2. This trade-off is commonly observed in the MU literature [10, 11, 15, 38]. Nevertheless, it is worth noting that even with mild reductions in remaining performance, the proposed methods achieve substantial gains in MU performance when the optimization goal is complete unlearning. For instance, in Table 8.2, the remaining performance drops in CIFAR-100 are 7.73 (RA) and 9.42 (TA) relative to retraining, whereas the unlearning performance improves by 28.66 (UA) compared with the best baseline methods. Moreover, even when the evaluation is based on the retrained model, the proposed methods maintain a strong balance across all metrics.

Influence Function in Deep Learning Influence functions were originally proposed for the convex function. As given in Sect. 8.2.2, the influence function is applied to the converged model, which can be regarded as a local convex model. A plot of loss landscape of the retrained model $\boldsymbol{\theta}_r$ on CIFAR-10 dataset is attached to Fig. 8.2.

MIA as a Proxy for "Forgetfulness" Given a model $\theta_\star$, its degree of generalization can be evaluated by performing a membership inference attack (MIA). In the context of this chapter, generalization corresponds to the degree of "forgetfulness" achieved by the forgetting algorithm. Let the model response observations be defined as $A_f = \mathcal{A}(\theta_\star, \mathcal{D}f)$ and $Ate = \mathcal{A}(\theta_\star, \mathcal{D}te)$, where $\mathcal{A}$ denotes the adversary and $A = A_f \cup Ate$ represents the full set of observations accessible to $\mathcal{A}$. The degree of generalization can then be estimated by analyzing these observations.

Among various possible analysis methods, the most straightforward approach in the context of MU is to compute the accuracy of $\mathcal{A}$ in distinguishing between seen and unseen samples ($\mathcal{D}f$ and $\mathcal{D}te$, respectively). This is calculated as $(TP + TN)/(|D_f| + |D_{te}|)$, where TP denotes true positives (correctly identified "seen" samples) and TN denotes true negatives (correctly identified "unseen" samples).

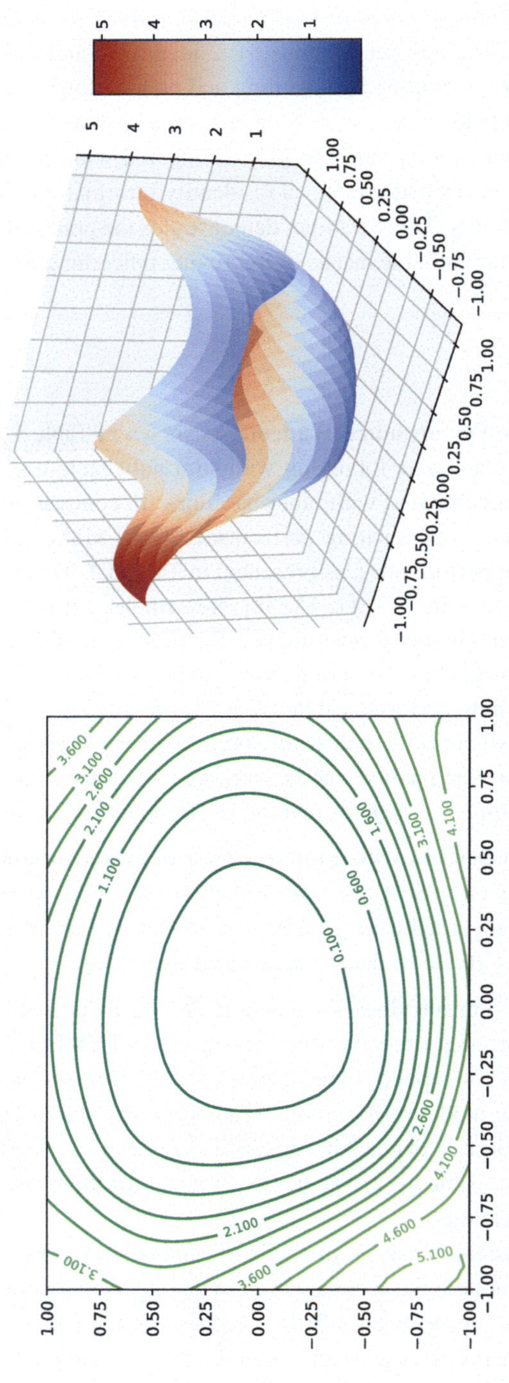

Fig. 8.2 The loss landscape of $\boldsymbol{\theta}_r$ on CIFAR-10 and the model is ResNet-18

Experiments are conducted on CIFAR-10 for both class-wise and random forgetting, and the results are shown in Table 8.4. It is assumed that the distributions of D_{tr} and D_{te} are identical. For class-wise forgetting, the additional MIA score approaches 1, since D_f corresponds to a distinct class, and thus the distributions of D_f and D_{te} (excluding that class) differ substantially. In contrast, for random forgetting, the additional MIA score is close to 0.5, as D_f is randomly selected from D_{tr}, leading to similar distributions between D_f and D_{te}.

References

1. Chen, M., Zhang, Z., Wang, T., Backes, M., Humbert, M., Zhang, Y.: When machine unlearning jeopardizes privacy. In: Proceedings of the 2021 ACM SIGSAC Conference on Computer and Communications Security, pp. 896–911 (2021)
2. Sekhari, A., Acharya, J., Kamath, G., Suresh, A.T.: Remember what you want to forget: algorithms for machine unlearning. Adv. Neural. Inf. Process. Syst. **34**, 18075–18086 (2021)
3. Hoofnagle, C.J., van der Sloot, B., Borgesius, F.Z.: The European Union general data protection regulation: what it is and what it means. Inf. Commun. Technol. Law **28**(1), 65–98 (2019)
4. Pardau, S.L.: The California consumer privacy act: towards a European-style privacy regime in the United States. J. Tech. L. Pol'y **23**, 68 (2018)
5. Cao, Y., Yang, J.: Towards making systems forget with machine unlearning. In: 2015 IEEE Symposium on Security and Privacy, pp. 463–480. IEEE (2015)
6. Bourtoule, L., Chandrasekaran, V., Choquette-Choo, C.A., Jia, H., Travers, A., Zhang, B., Lie, D., Papernot, N.: Machine unlearning. In: 2021 IEEE Symposium on Security and Privacy (SP), pp. 141–159. IEEE (2021)
7. Nguyen, T.T., Huynh, T.T., Nguyen, P.L., Liew, A.W.C., Yin, H., Nguyen, Q.V.H.: A survey of machine unlearning. arXiv preprint arXiv:2209.02299 (2022)
8. Koh, P.W., Liang, P.: Understanding black-box predictions via influence functions. In: International Conference on Machine Learning, pp. 1885–1894. PMLR (2017)
9. Golatkar, A., Achille, A., Soatto, S.: Eternal sunshine of the spotless net: selective forgetting in deep networks. In: Proceedings of the IEEE/CVF Conference on Computer Vision and Pattern Recognition, pp. 9304–9312 (2020)
10. Warnecke, A., Pirch, L., Wressnegger, C., Rieck, K.: Machine unlearning of features and labels. arXiv preprint arXiv:2108.11577 (2021)
11. Graves, L., Nagisetty, V., Ganesh, V.: Amnesiac machine learning. In: Proceedings of the AAAI Conference on Artificial Intelligence, vol. 35, pp. 11516–11524 (2021)
12. Thudi, A., Deza, G., Chandrasekaran, V., Papernot, N.: Unrolling SGD: understanding factors influencing machine unlearning. arXiv preprint arXiv:2109.13398 (2021)
13. Izzo, Z., Smart, M.A., Chaudhuri, K., Zou, J.: Approximate data deletion from machine learning models. In: International Conference on Artificial Intelligence and Statistics, pp. 2008–2016. PMLR (2021)
14. Becker, A., Liebig, T.: Evaluating machine unlearning via epistemic uncertainty. arXiv preprint arXiv:2208.10836 (2022)
15. Jia, J., Liu, J., Ram, P., Yao, Y., Liu, G., Liu, Y., Sharma, P., Liu, S.: Model sparsity can simplify machine unlearning. In: Thirty-Seventh Conference on Neural Information Processing Systems (2023)

16. Szegedy, C., Vanhoucke, V., Ioffe, S., Shlens, J., Wojna, Z.: Rethinking the inception architecture for computer vision. In: IEEE Conference on Computer Vision and Pattern Recognition (CVPR), pp. 2818–2826 (2016)
17. Pereyra, G., Tucker, G., Chorowski, J., Kaiser, Ł., Hinton, G.: Regularizing neural networks by penalizing confident output distributions. arXiv preprint arXiv:1701.06548 (2017)
18. Müller, R., Kornblith, S., Hinton, G.E.: When does label smoothing help? Adv. Neural Inf. Process. Syst. **32** (2019)
19. Wei, J., Liu, H., Liu, T., Niu, G., Sugiyama, M., Liu, Y.: To smooth or not? When label smoothing meets noisy labels. Learning **1**(1), e1 (2021)
20. Ginart, A., Guan, M., Valiant, G., Zou, J.Y.: Making AI forget you: data deletion in machine learning. Adv. Neural Inf. Process. Syst. **32** (2019)
21. Dwork, C., Kenthapadi, K., McSherry, F., Mironov, I., Naor, M.: Our data, ourselves: privacy via distributed noise generation. In: Advances in Cryptology-EUROCRYPT 2006: 24th Annual International Conference on the Theory and Applications of Cryptographic Techniques, St. Petersburg, Russia, May 28–June 1, 2006. Proceedings 25, pp. 486–503. Springer (2006)
22. Abadi, M., Chu, A., Goodfellow, I., McMahan, H.B., Mironov, I., Talwar, K., Zhang, L.: Deep learning with differential privacy. In: Machine Learning (2016). https://doi.org/10.1145/2976749.2978318
23. Wang, J., Guo, S., Xie, X., Qi, H.: Federated unlearning via class-discriminative pruning. In: Proceedings of the ACM Web Conference 2022, pp. 622–632 (2022)
24. Chen, M., Zhang, Z., Wang, T., Backes, M., Humbert, M., Zhang, Y.: Graph unlearning. In: Proceedings of the 2022 ACM SIGSAC Conference on Computer and Communications Security, pp. 499–513 (2022)
25. Di, J.Z., Douglas, J., Acharya, J., Kamath, G., Sekhari, A.: Hidden poison: machine unlearning enables camouflaged poisoning attacks. In: NeurIPS ML Safety Workshop (2022)
26. Shah, V., Träuble, F., Malik, A., Larochelle, H., Mozer, M., Arora, S., Bengio, Y., Goyal, A.: Unlearning via sparse representations. arXiv preprint arXiv:2311.15268 (2023)
27. Lukasik, M., Bhojanapalli, S., Menon, A., Kumar, S.: Does label smoothing mitigate label noise? In: International Conference on Machine Learning, pp. 6448–6458. PMLR (2020)
28. Krizhevsky, A., Hinton, G., et al.: Learning multiple layers of features from tiny images. cs.utoronto.ca (2009)
29. Netzer, Y., Wang, T., Coates, A., Bissacco, A., Wu, B., Ng, A.Y.: Reading digits in natural images with unsupervised feature learning (2011)
30. Liu, Z., Luo, P., Wang, X., Tang, X.: Deep learning face attributes in the wild. In: Proceedings of International Conference on Computer Vision (ICCV) (2015)
31. Deng, J., Dong, W., Socher, R., Li, L.J., Li, K., Fei-Fei, L.: Imagenet: a large-scale hierarchical image database. In: 2009 IEEE Conference on Computer Vision and Pattern Recognition, pp. 248–255. IEEE (2009)
32. Lang, K.: Newsweeder: learning to filter netnews. In: Prieditis, A., Russell S. (eds.) Machine Learning Proceedings 1995, pp. 331–339. Morgan Kaufmann, San Francisco, CA (1995). https://www.sciencedirect.com/science/article/pii/B9781558603776500487
33. He, K., Zhang, X., Ren, S., Sun, J.: Deep residual learning for image recognition. In: Proceedings of the IEEE Conference on Computer Vision and Pattern Recognition, pp. 770–778 (2016)
34. Devlin, J., Chang, M.W., Lee, K., Toutanova, K.: Bert: pre-training of deep bidirectional transformers for language understanding. arXiv preprint arXiv:1810.04805 (2018)
35. Chen, M., Gao, W., Liu, G., Peng, K., Wang, C.: Boundary unlearning: rapid forgetting of deep networks via shifting the decision boundary. In: Proceedings of the IEEE/CVF Conference on Computer Vision and Pattern Recognition, pp. 7766–7775 (2023)
36. Hayase, T., Yasutomi, S., Katoh, T.: Selective forgetting of deep networks at a finer level than samples. arXiv preprint arXiv:2012.11849 (2020)

37. Kurmanji, M., Triantafillou, P., Hayes, J., Triantafillou, E.: Towards unbounded machine unlearning. Adv. Neural Inf. Process. Syst. **36** (2024)
38. Fan, C., Liu, J., Zhang, Y., Wei, D., Wong, E., Liu, S.: SalUn: empowering machine unlearning via gradient-based weight saliency in both image classification and generation. In: International Conference on Learning Representations (2024)
39. Goel, S., Prabhu, A., Sanyal, A., Lim, S.N., Torr, P., Kumaraguru, P.: Towards adversarial evaluations for inexact machine unlearning. arXiv preprint arXiv:2201.06640 (2022)
40. Cha, S., Cho, S., Hwang, D., Lee, H., Moon, T., Lee, M.: Learning to unlearn: instance-wise unlearning for pre-trained classifiers. In: Proceedings of the AAAI Conference on Artificial Intelligence, vol. 38, pp. 11186–11194 (2024)
41. Yeom, S., Yeom, S., Giacomelli, I., Giacomelli, I., Fredrikson, M., Fredrikson, M., Jha, S., Jha, S., Jha, S.: Privacy risk in machine learning: analyzing the connection to overfitting. In: IEEE Computer Security Foundations Symposium (2018). https://doi.org/10.1109/csf.2018.00027
42. Minaee, S., Abdolrashidi, A., Su, H., Bennamoun, M., Zhang, D.: Biometrics recognition using deep learning: a survey. Artif. Intell. Rev., pp. 1–49 (2023)

Part III
Machine Unlearning: A Model-Based Perspective

Unlearning Through Internal Model Analysis: Sparsity, Saliency, and Attribution

9

Jinghan Jia

Abstract

This chapter explores how internal model analysis enables more effective machine unlearning (MU) by leveraging structural, gradient-based, and attributional insights into model weights. While exact unlearning via full retraining guarantees faithful data removal, it remains computationally prohibitive for modern foundation models. Recent advances reveal that localizing the parameters most responsible for forgotten knowledge, via sparsity patterns, weight saliency, and attribution scores, can dramatically narrow the gap between approximate and exact unlearning while preserving model utility. We begin with model sparsification, where pruning redundant parameters reduces unlearning error and motivates sparsity-aware objectives. Building on this structural perspective, we then examine weight saliency, which uses gradient-based criteria to identify and update parameters most associated with harmful knowledge, achieving state-of-the-art results in both discriminative and generative tasks, including diffusion models. Finally, we introduce attribution-guided unlearning via bi-level optimization, providing a principled mechanism for quantifying parameter importance under utility constraints and delivering state-of-the-art performance for unlearning in large language models (LLMs).

J. Jia (✉)
Michigan State University, East Lansing, East Lansing, USA
e-mail: jiajingh@msu.edu

S. Liu et al. (eds.), *Machine Unlearning for Governance of Foundation Models*, Synthesis Lectures on Computer Vision, https://doi.org/10.1007/978-3-032-17282-2_9

9.1 Model Sparsification Assists Machine Unlearning

9.1.1 Motivation: From Compression to Forgetting

Model sparsification, also known as weight pruning, has been extensively studied in the contexts of model compression and efficient inference [1–6]. The central idea is simple yet powerful: large models often contain significant redundancy, and carefully removing parameters can yield sparse subnetworks with lossless performance compared to the original dense models. A landmark result in this line of work is the *lottery ticket hypothesis (LTH)* [2], which demonstrates the existence of so-called "winning tickets", i.e., sparse subnetworks capable of training to full accuracy comparable to the dense model. This finding highlights that dense networks often contain many parameters with little impact on performance, motivating the search for smaller, more efficient representations without sacrificing accuracy.

As illustrated in Fig. 9.1, we present results using one-shot magnitude pruning (OMP) [4], one of the most computationally efficient sparsification techniques. OMP removes weights below a chosen magnitude threshold to achieve a target sparsity. The figure shows the test accuracy of the image classifier ResNet-18 on CIFAR-10 under increasing sparsity levels, revealing a graceful sparse regime where substantial pruning leads to negligible accuracy loss.

Beyond model compression and inference acceleration, sparsification has also been shown to be beneficial in robustness [7–9], fairness [10], interpretability [11], loss landscape analysis [3], and privacy [12, 13]. Particularly relevant to machine unlearning, studies report that pruning can shrink the gap between exact unlearning and approximate unlearning [14], suggesting a natural link between sparsity and data influence removal.

Intuitively, pruning eliminates parameters that encode redundant or non-critical information, thereby constraining the pathways through which training data can influence model predictions. This observation inspires a key question: *Can we leverage sparsification not only for efficiency but also to reduce the footprint of forget data in model weights, thereby narrowing the gap between approximate and exact unlearning?*

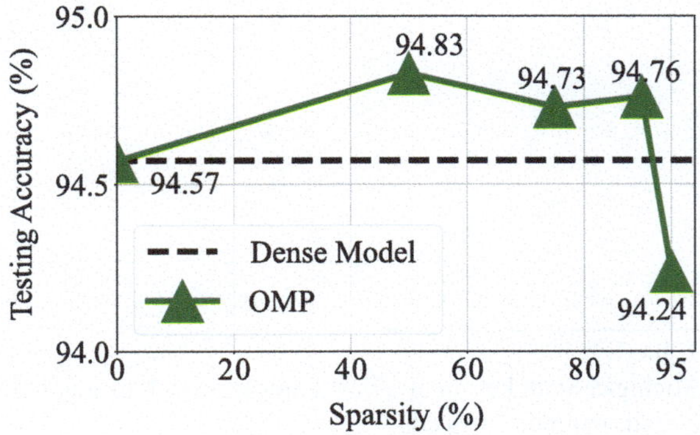

Fig. 9.1 Testing accuracy of OMP-based sparse ResNet-18 versus the dense model on CIFAR-10

The remainder of this section explores the above connection, showing how sparsity provides both theoretical guarantees for reducing unlearning error and practical improvements in real-world MU tasks.

9.1.2 How Does Model Sparsification Help Machine Unlearning?

The connection between model sparsity and unlearning can be formally understood through the lens of unrolling stochastic gradient descent (SGD) [15]. This viewpoint allows us to derive an unlearning error—the weight-space distance between a model that undergoes approximate unlearning (e.g., gradient ascent scrubbing) and the gold-standard model retrained from scratch on the retain set after removing the forget data. While this error was analyzed in [15] for dense models, the work [14] extended the framework to incorporate model sparsity into the SGD unrolling process, enabling a direct examination of how pruning affects the fidelity of approximate unlearning.

9.1.2.1 Unlearning Error Bound against Sparsity

Let $\mathbf{m}$ be a binary pruning mask applied to model parameters $\boldsymbol{\theta}$, where $m_i = 0$ indicates the i-th parameter θ_i is pruned, and $m_i = 1$ means it remains active. The resulting sparse model is denoted as $\mathbf{m} \odot \boldsymbol{\theta}$, with $\odot$ representing element-wise multiplication. Pruning methods such as one-shot magnitude pruning (OMP) can be used to obtain $\mathbf{m}$ efficiently. For a dense model ($\mathbf{m} = \mathbf{1}$), it has been shown in [15] that the unlearning error under gradient ascent (GA)-based scrubbing admits a closed-form upper bound. Incorporating sparsity, it was proved in [14] that the unlearning error for a sparse mask $\mathbf{m}$ under SGD training satisfies

$$e(\mathbf{m}) = \mathcal{O}\left(\eta^2 t, \|\mathbf{m} \odot (\boldsymbol{\theta}_t - \boldsymbol{\theta}_0)\|_2, \sigma(\mathbf{m})\right), \tag{9.1}$$

where η is the learning rate, t is the number of training iterations, $(\boldsymbol{\theta}_t - \boldsymbol{\theta}_0)$ is the cumulative weight update from initialization, and $\sigma(\mathbf{m})$ is the largest singular value of the Hessian $\nabla^2_{\boldsymbol{\theta},\boldsymbol{\theta}} \ell$ (for loss ℓ) restricted to unpruned weights.

As shown by (9.1), increasing sparsity (more $m_i = 0$) directly reduces the unlearning error by shrinking both the weight-space norm and the Hessian's spectral complexity. In contrast, for dense models ($\mathbf{m} = \mathbf{1}$), the bound depends solely on $|\boldsymbol{\theta}_t - \boldsymbol{\theta}_0|_2$, which is typically much larger.

The above analysis motivates the "prune first, then unlearn" paradigm introduced in [14], where sparsification is applied prior to initiating MU. Empirically, as shown in Table 9.1 and 9.2 approximate unlearning methods on pruned models consistently narrow the performance gap to full retraining, especially at high sparsity levels (e.g., 95% sparsity vs. dense models).

To provide a comprehensive comparison, Table 9.1 evaluates methods across several complementary metrics. *Unlearning accuracy (UA)*, defined as $1 - \mathrm{Acc}_{\mathcal{D}_\mathrm{f}}(\boldsymbol{\theta}_\mathrm{u})$, measures

Table 9.1 Performance overview of various MU methods on dense and 95%-sparse models considering different unlearning scenarios: class-wise forgetting, and random data forgetting.

MU	UA		MIA-efficacy		RA		TA	
	DENSE	95% Sparsity	DENSE	95% Sparsity	DENSE	95% Sparsity	DENSE	95% Sparsity
Class-wise forgetting								
Retrain	$100.00_{\pm 0.00}$	$100.00_{\pm 0.00}$	$100.00_{\pm 0.00}$	$100.00_{\pm 0.00}$	$100.00_{\pm 0.00}$	$99.99_{\pm 0.01}$	$94.83_{\pm 0.11}$	$91.80_{\pm 0.89}$
FT	$22.53_{\pm 8.16}$ (77.47)	$73.64_{\pm 9.46}$ (26.36)	$75.00_{\pm 14.68}$ (25.00)	$83.02_{\pm 16.33}$ (16.98)	$99.87_{\pm 0.04}$ (0.13)	$99.87_{\pm 0.05}$ (0.12)	$94.31_{\pm 0.19}$ (0.52)	$94.32_{\pm 0.12}$ (2.52)
GA	$93.08_{\pm 2.29}$ (6.92)	$98.09_{\pm 1.11}$ (1.91)	$94.03_{\pm 3.27}$ (5.97)	$97.74_{\pm 2.24}$ (2.26)	$92.60_{\pm 0.25}$ (7.40)	$87.74_{\pm 0.27}$ (12.25)	$86.64_{\pm 0.28}$ (8.19)	$82.58_{\pm 0.27}$ (9.22)
FF	$79.93_{\pm 8.92}$ (20.07)	$94.83_{\pm 4.29}$ (5.17)	$100.00_{\pm 0.00}$ (0.00)	$100.00_{\pm 0.00}$ (0.00)	$99.45_{\pm 0.24}$ (0.55)	$99.48_{\pm 0.33}$ (0.51)	$94.18_{\pm 0.08}$ (0.65)	$94.04_{\pm 0.10}$ (2.24)
IU	$87.82_{\pm 2.15}$ (12.18)	$99.47_{\pm 0.15}$ (0.53)	$95.96_{\pm 0.21}$ (4.04)	$99.93_{\pm 0.04}$ (0.07)	$97.98_{\pm 0.21}$ (2.02)	$97.24_{\pm 0.13}$ (2.75)	$91.42_{\pm 0.21}$ (3.41)	$90.76_{\pm 0.18}$ (1.04)
Random data forgetting								
Retrain	$5.41_{\pm 0.11}$	$6.77_{\pm 0.23}$	$13.12_{\pm 0.14}$	$14.17_{\pm 0.18}$	$100.00_{\pm 0.00}$	$100.00_{\pm 0.00}$	$94.42_{\pm 0.09}$	$93.33_{\pm 0.12}$
FT	$6.83_{\pm 0.51}$ (1.42)	$5.97_{\pm 0.57}$ (0.80)	$14.97_{\pm 0.62}$ (1.85)	$13.36_{\pm 0.59}$ (0.81)	$96.61_{\pm 0.25}$ (3.39)	$96.99_{\pm 0.31}$ (3.01)	$90.13_{\pm 0.26}$ (4.29)	$90.29_{\pm 0.31}$ (3.04)
GA	$7.54_{\pm 0.29}$ (2.13)	$5.62_{\pm 0.46}$ (1.15)	$10.04_{\pm 0.31}$ (3.08)	$11.76_{\pm 0.52}$ (2.41)	$93.31_{\pm 0.04}$ (6.69)	$95.44_{\pm 0.11}$ (4.56)	$89.28_{\pm 0.07}$ (5.14)	$89.26_{\pm 0.15}$ (4.07)
FF	$7.84_{\pm 0.71}$ (2.43)	$8.16_{\pm 0.67}$ (1.39)	$9.52_{\pm 0.43}$ (3.60)	$10.80_{\pm 0.37}$ (3.37)	$92.05_{\pm 0.16}$ (7.95)	$92.29_{\pm 0.24}$ (7.71)	$88.10_{\pm 0.19}$ (6.32)	$87.79_{\pm 0.23}$ (5.54)
IU	$2.03_{\pm 0.43}$ (3.38)	$6.51_{\pm 0.52}$ (0.26)	$5.07_{\pm 0.74}$ (8.05)	$11.93_{\pm 0.68}$ (2.24)	$98.26_{\pm 0.29}$ (1.74)	$94.94_{\pm 0.31}$ (5.06)	$91.33_{\pm 0.22}$ (3.09)	$88.74_{\pm 0.42}$ (4.59)

The forgetting data of random data forgetting ratio is 10% of the whole training dataset, the sparse models are obtained using OMP (one-shot magnitude pruning) [4]. Class-wise forgetting is conducted class-wise. The performance is reported in the form $a_{\pm b}$, with mean a and standard deviation b computed over 10 independent trials. A performance gap against Retrain is provided in (•). Note that the better performance of approximate unlearning corresponds to the smaller performance gap with the gold-standard retrained model

Table 9.2 Performance overview of LLM unlearning on the WMDP task under the model Zephyr-7B-beta

Method		Unlearning efficacy (UE)		Utility (UT)	
		1-FA ↑ (WMDP-bio)	1-FA ↑ (WMDP-cyber)	UE avg. ↑	MMLU↑
Original (w/o MU)		0.3614	0.5596	0.4605	0.5815
GradDiff+	Dense	0.6609	0.6517	0.6563	0.4459
	Magnitude	0.4269	0.5786	0.5028	0.5484
	Wanda	0.4488	0.6133	0.5311	0.5086
	LoRA	0.6931	0.6634	0.6783	0.4346
	Attribution	0.6783	0.6959	**0.6871**	**0.5530**
NPO+	Dense	0.6678	0.7056	0.6867	0.3754
	Magnitude	0.5589	0.6447	0.6018	0.4946
	Wanda	0.4364	0.5883	0.5124	**0.5520**
	LoRA	0.4687	0.6039	0.5363	0.5248
	Attribution	0.6980	0.7076	**0.7028**	0.5033

forgetting efficacy on the forget set $\mathcal{D}_f$, while *MIA-Efficacy* assesses privacy using membership inference attack (MIA) to test whether samples in $\mathcal{D}_f$ can still be distinguished from non-training data, with higher values indicating stronger forgetting. *Retain accuracy (RA)* captures fidelity by reporting accuracy on the retain set $\mathcal{D}_r$, whereas *testing accuracy (TA)* reflects generalization on an external test set.

The baselines considered include: *Retrain*, the gold standard obtained by training from scratch without $\mathcal{D}_f$; *FT* (fine-tuning) [16, 17], which updates $\boldsymbol{\theta}_0$ on $\mathcal{D}_r$; *GA* (gradient ascent) [15, 18], which maximizes loss on $\mathcal{D}_f$; *FF* (Fisher forgetting) [16, 19], which perturbs parameters with Fisher-guided Gaussian noise; and *IU* (influence unlearning) [20, 21], which estimates the effect of removing training points via influence functions. Across all approximate methods (FT, GA, FF, IU), introducing sparsity consistently improves UA and MIA-Efficacy while maintaining competitive RA and TA, demonstrating the effectiveness of the "prune first, then unlearn" strategy.

These findings suggest that model sparsification serves as a powerful preconditioner for MU: it reduces approximation error, bringing approximate methods closer to the ideal behavior of exact retraining.

9.1.3 Sparsity-Aware Machine Unlearning

The "prune first, then unlearn" paradigm demonstrates that applying unlearning to a pruned model reduces unlearning errors compared to operating on a dense model. This naturally raises the question: *can pruning and unlearning be performed jointly rather than sequentially, avoiding the need to predefine a sparsity pattern?* To this end, let $\ell_{\mathrm{MU}}(\boldsymbol{\theta}; \boldsymbol{\theta}_0)$ denote the unlearning objective for parameters $\boldsymbol{\theta}$ given the pre-trained model $\boldsymbol{\theta}_0$. Inspired by classical sparsity-inducing optimization [22], one can augment the unlearning objective with an ℓ_1 norm-based sparsity-promoting regularizer. This leads to the ℓ_1 *-sparse MU* approach:

$$\underset{\boldsymbol{\theta}}{\text{minimize}} \ \ell_{\mathrm{MU}}(\boldsymbol{\theta}; \boldsymbol{\theta}_0) + \gamma \|\boldsymbol{\theta}\|_1, \tag{9.2}$$

where $\gamma > 0$ controls the strength of the ℓ_1 regularization, effectively balancing forgetting efficacy against sparsity promotion.

In practice, unlearning performance can be sensitive to the choice of γ. A large γ enforces aggressive sparsity but risks degrading utility on retained data, while a small γ may limit forgetting effectiveness. To address this trade-off, the work [14] proposed a sparse regularization scheduler in which γ is gradually decreased throughout the unlearning process, e.g., a linearly decaying schedule. The intuition is explained below. At the early unlearning stage, larger γ emphasizes sparsity when weight updates are large, aggressively removing parameters most associated with the forget set. At the later stage, reducing γ allows finer adjustments, focusing on recovering utility and stabilizing the model.

As shown in Fig. 9.2, ℓ_1-sparse MU consistently surpasses fine-tuning (FT) in both UA (unlearning accuracy) and MIA (membership inference attack) resistance, while substan-

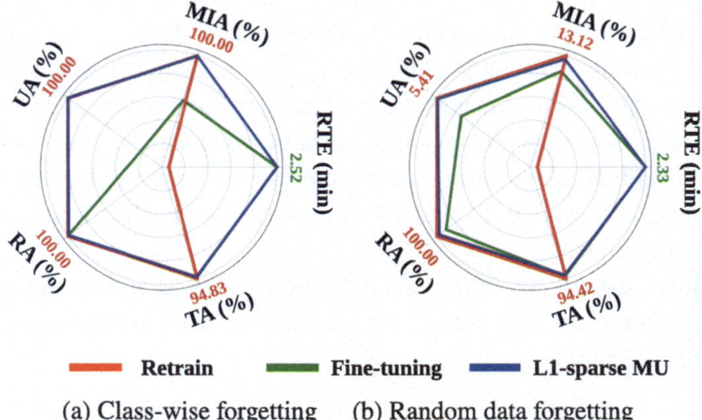

Fig. 9.2 Performance of ℓ_1-sparse MU compared with fine-tuning (FT)-based unlearning approach and Retrain on class-wise and random data forgetting (CIFAR-10, ResNet-18). Each metric is normalized to [0, 1] based on the best result across methods, with the actual best value reported (e.g., 2.52 is the least computation time for class-wise forgetting)

tially reducing the performance disparity with full retraining (Retrain). Importantly, it retains the efficiency advantages of approximate unlearning methods, making ℓ_1-sparse MU a compelling strategy for jointly enforcing sparsity and forgetting.

9.2 Gradient-Based Weight Saliency for Unlearning

While model sparsification has proven beneficial for machine unlearning in discriminative settings, its direct application to generative models faces two fundamental challenges [23]. First, identifying an appropriate sparsity pattern in complex architectures such as diffusion models is inherently difficult due to their multi-stage and highly entangled parameter structures. Second, even if effective pruning is possible, delivering a sparse model after unlearning can be undesirable because reduced parameter capacity may degrade generative quality and diversity, key requirements in applications like image synthesis. These limitations motivate the need for a more flexible and selective unlearning mechanism: one that avoids global pruning yet still localizes updates to the parameters most responsible for harmful knowledge.

9.2.1 Weight Saliency as an Alternative to Sparsity

To address the limitations of global sparsification, *weight saliency* was introduced in [23] as a more flexible alternative for guiding machine unlearning. Unlike traditional sparsity-based methods that enforce a fixed global pruning pattern, weight saliency leverages gradient-based importance scores to selectively identify parameters most responsible for encoding the knowledge targeted for removal. By focusing unlearning updates only on these influential parameters, weight saliency avoids the bluntness of pruning-based approaches while retaining the ability to localize changes within the model.

The central idea draws inspiration from gradient-based saliency maps widely used in interpretability research [24, 25], where gradients serve as a measure of feature relevance. Extending this concept to the parameter space, weight saliency computes the gradient of the forgetting loss with respect to each parameter in the pre-trained model. The resulting importance scores reveal how sensitively the unlearning objective depends on each weight, providing a principled criterion for parameter selection.

Based on these scores, the parameter space is conceptually partitioned into two disjoint sets. Salient weights, those with the highest importance, are actively updated during MU to erase the targeted knowledge, ensuring that forgetting is focused where it matters most. In contrast, intact weights are left unchanged to preserve utility on the retain set, minimizing collateral damage to model performance outside the forgetting scope.

This decomposition enables a targeted forgetting process without imposing global structural constraints on the model, bridging the gap between precision and efficiency in approximate unlearning. By concentrating updates on the most relevant parameters while freezing the rest, weight saliency strikes a better balance between unlearning efficacy and utility preservation than methods relying solely on sparsification or dense retraining.

Formally, let $\ell_{\mathrm{f}}(\boldsymbol{\theta}; \mathcal{D}_{\mathrm{f}})$ denote the forgetting loss defined on the forget set $\mathcal{D}_{\mathrm{f}}$. The gradient of ℓ_{f} with respect to $\boldsymbol{\theta}$ at the pre-trained parameters $\boldsymbol{\theta}_0$ defines the weight importance, from which a hard thresholding rule selects the most salient parameters:

$$\mathbf{m}_{\mathrm{S}} = \mathbf{1}\left(|\nabla_{\boldsymbol{\theta}} \ell_{\mathrm{f}}(\boldsymbol{\theta}; \mathcal{D}_{\mathrm{f}})|_{\boldsymbol{\theta}=\boldsymbol{\theta}_0} \geq \gamma\right), \tag{9.3}$$

where $\mathbf{1}(\cdot)$ is the element-wise indicator function, $|\cdot|$ the element-wise absolute value, and $\gamma > 0$ the selection threshold. Empirically, setting γ to the median gradient magnitude yields stable and robust performance [23].

Given $\mathbf{m}_{\mathrm{S}}$, the weight saliency-guided unlearning update takes the form

$$\boldsymbol{\theta}_{\mathrm{u}} = \underbrace{\mathbf{m}_{\mathrm{S}} \odot \boldsymbol{\theta}}_{\text{salient weights}} + \underbrace{(\mathbf{1} - \mathbf{m}_{\mathrm{S}}) \odot \boldsymbol{\theta}_0}_{\text{original weights}}, \tag{9.4}$$

where $\odot$ denotes element-wise multiplication. This formulation ensures that unlearning primarily modifies salient parameters while leaving the remainder of the model unaffected, thereby achieving a targeted forgetting process without global sparsity.

The above gradient-based weight saliency method, termed *SalUn* [23], has been extensively evaluated on both discriminative and generative MU tasks. Particularly, in concept-level forgetting for diffusion models, SalUn shows significant advantages. For instance, when removing NSFW (not-safe-for-work) concepts from Stable Diffusion (SD) V1.4 using prompts from the Inappropriate Image Prompts (I2P) dataset [26], generated images are analyzed via the NudeNet detector [27]. SalUn reduces harmful generations far more effectively than baselines while preserving benign synthesis quality. Figure 9.3 presents qualitative comparisons between the original model and its unlearned variants produced by SalUn and two baseline approaches, ESD [28] and FMN [29]. The results demonstrate that SalUn not only removes targeted NSFW content more effectively than ESD and FMN but also better preserves the model's ability to generate safe, diverse, and high-quality images. These findings highlight the advantages of gradient-based weight saliency as a practical and reliable mechanism for machine unlearning in generative models.

9.3 Weight Attribution Via Bi-level Optimization

Although sparsity- and saliency-based strategies have demonstrated effectiveness in unlearning, directly applying them to large language models (LLMs) presents additional challenges. The massive parameter scale of LLMs, coupled with the delicate balance between forgetting efficacy and utility preservation, renders pruning- or gradient-based saliency heuristics less effective. Empirical studies consistently reveal that unlearning performance in LLMs is highly sensitive to *which* weights are updated: naïve pruning often degrades unlearning effi-

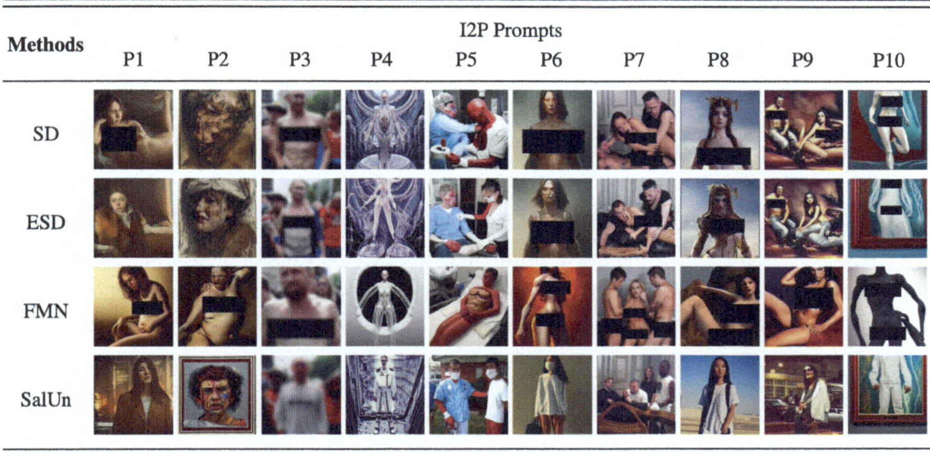

Fig. 9.3 Examples of generated images using SDs w/ and w/o MU. The unlearning methods include ESD [28], FMN [29], and SalUn [23]. Each column represents generated images using different SDs with the same prompt (denoted by Pi) and the same seed

Fig. 9.4 Unlearning efficacy and utility performance of NPO-based LLM unlearning on TOFU dataset versus sparsity of unlearned weights (i.e., the proportion of weights required for unlearning updates), where model sparsity is achieved using the LLM pruning method Wanda [30].

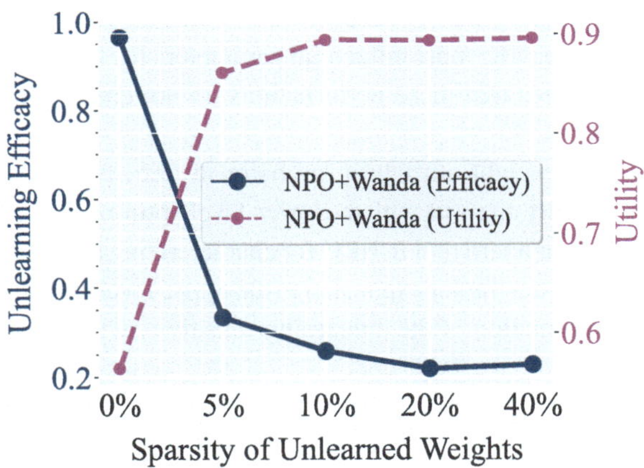

cacy, while aggressive updates risk excessive utility loss. Therefore, it is important to develop a principled framework to systematically quantify and leverage weight-level influence for LLM unlearning.

Figure 9.4 highlights the above challenge. The experiment evaluates the negative preference optimization (NPO) unlearning method on pruned LLMs, where sparsity is induced using Wanda [30] under the TOFU benchmark for fictitious unlearning [31]. As sparsity increases, unlearning efficacy declines sharply relative to the dense (0% sparsity) model. Moreover, a clear trade-off emerges: improvements in unlearning accuracy often come at the cost of reduced model utility. These findings underscore the need for a principled *weight attribution* framework that can selectively identify and update parameters most critical for forgetting while preserving overall model performance.

9.3.1 Bi-level Optimization Framework for Weight Attribution

To address this challenge, WAGLE [32] formulates weight attribution as a bi-level optimization (BLO) problem. The *upper-level* problem evaluates the impact of weight perturbations on the forget loss, while the *lower-level* problem enforces utility preservation via the retain loss. Formally, the framework measures the sensitivity of unlearning to perturbations $\boldsymbol{\epsilon}$ of the weight vector $\boldsymbol{\theta}$:

$$\begin{array}{ll} \text{Find} & \ell_{\mathrm{f}}(\boldsymbol{\epsilon} \odot \boldsymbol{\theta}^*(\boldsymbol{\epsilon})) - \ell_{\mathrm{f}}(\boldsymbol{\theta}^*(\mathbf{1})) \quad \text{// Upper level} \\ \text{subject to} & \boldsymbol{\theta}^*(\boldsymbol{\epsilon}) = \arg\min_{\boldsymbol{\theta}} \ell_{\mathrm{r}}(\boldsymbol{\epsilon} \odot \boldsymbol{\theta}), \quad \text{// Lower level} \end{array} \tag{9.5}$$

where $\odot$ denotes element-wise multiplication, and $\boldsymbol{\theta}^*(\boldsymbol{\epsilon})$ reduces to the pre-trained model $\boldsymbol{\theta}_0$. The rationale behind (9.5) is to identify a weight perturbation scheme $\boldsymbol{\epsilon}$ that maximizes the impact on the forget loss at the upper level while ensuring optimal performance of the retain loss at the lower level. In other words, the upper-level objective seeks perturbations that reveal the most influential parameters for forgetting, whereas the lower-level constraint guarantees that these perturbations do not compromise the utility preserved by the retain loss.

The weight attribution problem in (9.5) can be addressed by linking the upper-level sensitivity analysis with the lower-level optimization using the *implicit gradient* (IG) method [33], a standard tool in BLO for characterizing the gradient flow between levels. By applying a first-order Taylor expansion to the upper-level objective around $\boldsymbol{\epsilon} = \mathbf{1}$, we obtain a tractable approximation of the unlearning sensitivity with respect to $\boldsymbol{\epsilon}$:

$$\begin{aligned} \ell_{\mathrm{f}}(\boldsymbol{\epsilon} \odot \boldsymbol{\theta}^*(\boldsymbol{\epsilon})) - \ell_{\mathrm{f}}(\boldsymbol{\theta}^*(\mathbf{1})) &\approx (\boldsymbol{\epsilon} - \mathbf{1})^\top \frac{d\ell_{\mathrm{f}}(\boldsymbol{\epsilon} \odot \boldsymbol{\theta}^*(\boldsymbol{\epsilon}))}{d\boldsymbol{\epsilon}}\Big|_{\boldsymbol{\epsilon}=\mathbf{1}} \\ &= (\boldsymbol{\epsilon} - \mathbf{1})^\top \frac{d[\boldsymbol{\epsilon} \odot \boldsymbol{\theta}^*(\boldsymbol{\epsilon})]}{d\boldsymbol{\epsilon}}\Big|_{\boldsymbol{\epsilon}=\mathbf{1}} \nabla \ell_{\mathrm{f}}(\boldsymbol{\theta}_{\mathrm{o}}), \end{aligned} \tag{9.6}$$

where $^\top$ denotes the matrix transpose, and $\frac{d\mathbf{a}}{d\mathbf{b}} \in \mathbb{R}^{|\mathbf{b}| \times |\mathbf{a}|}$ represents the full derivative of $\mathbf{a}$ with respect to $\mathbf{b}$. The second equality follows from the chain rule, with $\boldsymbol{\theta}^*(\mathbf{1}) = \boldsymbol{\theta}_0$ and the convention $\nabla \ell_{\mathrm{f}}(\boldsymbol{\theta}_0) = \frac{d\ell_{\mathrm{f}}(\mathbf{z})}{d\mathbf{z}}\big|_{\mathbf{z}=\boldsymbol{\theta}^*(\mathbf{1})}$.

WAGLE [32] further simplifies (9.6) by adopting a diagonal Hessian approximation, which eliminates the need to compute full second-order information. Under this assumption, WAGLE derives a closed-form attribution score S_i for each weight parameter i, quantifying its relative importance to the unlearning objective while accounting for the trade-off with utility preservation:

$$S_i \propto [\boldsymbol{\theta}_0]_i [\nabla \ell_{\mathrm{f}}(\boldsymbol{\theta}_0)]_i - \frac{1}{\gamma} [\nabla \ell_{\mathrm{r}}(\boldsymbol{\theta}_0)]_i [\nabla \ell_{\mathrm{f}}(\boldsymbol{\theta}_0)]_i, \quad (9.7)$$

where $[\cdot]_i$ denotes the ith entry of a vector. In (9.7), the first term parallels the pruning score SNIP [34], capturing the sensitivity of the forget loss to pruning the ith weight, while the second term accounts for the counteracting effect of utility retention. The parameter γ regulates the balance between these two terms, enabling a principled trade-off between effective forgetting and model utility.

9.3.2 Weight Attribution-Guided Unlearning

Once the attribution scores S_i are computed, they can be ranked to identify the subset of weights most critical for unlearning. A binary mask $\mathbf{m}_S$ is then constructed to select only the highest-scoring parameters for updates while leaving the rest of the model untouched. This leads to a modular unlearning approach:

$$\underset{\boldsymbol{\theta}}{\text{minimize}} \;\; \ell_{\mathrm{f}}(\mathbf{m}_S \odot \boldsymbol{\theta} + (\mathbf{1} - \mathbf{m}_S) \odot \boldsymbol{\theta}_0; \mathcal{D}_{\mathrm{f}}) + \lambda \, \ell_{\mathrm{r}}(\mathbf{m}_S \odot \boldsymbol{\theta} + (\mathbf{1} - \mathbf{m}_S) \odot \boldsymbol{\theta}_0; \mathcal{D}_{\mathrm{r}}). \quad (9.8)$$

This formulation explicitly decouples the model into two components: (1) Optimized weights $\mathbf{m}_S \odot \boldsymbol{\theta}$ updated during unlearning to erase the targeted knowledge; (2) Frozen weights $(\mathbf{1} - \mathbf{m}_S) \odot \boldsymbol{\theta}_0$ left intact to preserve the model's utility on retained data. By embedding attribution directly into the unlearning procedure, only parameters with the greatest impact on forgetting are modified, ensuring a precise and controlled update rather than indiscriminate changes to the entire network.

As an experimental validation, Table 9.2 reports the performance of weight attribution-guided unlearning on the WMDP benchmark [35]. Unlearning efficacy (UE) is evaluated using the forget accuracy metric on both WMDP-Bio and WMDP-Cyber subsets, where lower forget accuracy indicates more effective removal of hazardous knowledge. At the same time, utility (UT) is measured by the MMLU benchmark to assess whether the model retains its general reasoning and knowledge capabilities after unlearning.

As shown in Table 9.2, across both WMDP-Bio and WMDP-Cyber benchmarks, weight attribution-guided methods consistently achieve substantial gains in UE compared to dense unlearning and parameter-efficient baselines such as Magnitude pruning, Wanda, and LoRA. Under the GradDiff framework, attribution-guided unlearning attains the highest average UE score, clearly surpassing both the dense baseline and LoRA. Similarly, within the NPO-based unlearning framework, attribution guidance further improves forgetting performance, delivering the highest UE average across both biological and cyber domains. These results confirm that attributing weight importance before unlearning enables parameter updates to focus on the most influential weights, thereby ensuring more effective and targeted forgetting while mitigating unnecessary utility loss.

References

1. Han, S., Mao, H., Dally, W.J.: Deep compression: Compressing deep neural networks with pruning, trained quantization and Huffman coding (2015). arXiv:1510.00149
2. Frankle, J., Carbin, M.: The lottery ticket hypothesis: finding sparse, trainable neural networks (2018). arXiv:1803.03635
3. Frankle, J., Dziugaite, G.K., Roy, D., Carbin, M.: Linear mode connectivity and the lottery ticket hypothesis. In: International Conference on Machine Learning, pp. 3259–3269. PMLR (2020)
4. Ma, X., Yuan, G., Shen, X., Chen, T., Chen, X., Chen, X., Liu, N., Qin, M., Liu, S., Wang, Z., et al.: Sanity checks for lottery tickets: does your winning ticket really win the jackpot? Adv. Neural. Inf. Process. Syst. **34**, 12749–12760 (2021)
5. Zhang, Y., Yao, Y., Ram, P., Zhao, P., Chen, T., Hong, M., Wang, Y., Liu, S.: Advancing model pruning via bi-level optimization. In: Advances in Neural Information Processing Systems (2022)
6. Blalock, D., Gonzalez Ortiz, J.J., Frankle, J., Guttag, J.: What is the state of neural network pruning? Proc. Mach. Learn. Syst. **2**, 129–146 (2020)
7. Sehwag, V., Wang, S., Mittal, P., Jana, S.: Hydra: pruning adversarially robust neural networks. Adv. Neural. Inf. Process. Syst. **33**, 19655–19666 (2020)
8. Chen, T., Zhang, Z., Zhang, Y., Chang, S., Liu, S., Wang, Z.: Quarantine: sparsity can uncover the trojan attack trigger for free. In: Proceedings of the IEEE/CVF Conference on Computer Vision and Pattern Recognition, pp. 598–609 (2022)
9. Diffenderfer, J., Bartoldson, B., Chaganti, S., Zhang, J., Kailkhura, B.: A winning hand: compressing deep networks can improve out-of-distribution robustness. Adv. Neural. Inf. Process. Syst. **34**, 664–676 (2021)
10. Stoychev, S., Gunes, H.: The effect of model compression on fairness in facial expression recognition (2022). arXiv:2201.01709
11. Wong, E., Santurkar, S., Madry, A.: Leveraging sparse linear layers for debuggable deep networks. In: International Conference on Machine Learning, pp. 11205–11216. PMLR (2021)
12. Huang, Y., Su, Y., Ravi, S., Song, Z., Arora, S., Li, K.: Privacy-preserving learning via deep net pruning (2020). arXiv:2003.01876
13. Wang, Y., Wang, C., Wang, Z., Zhou, S., Liu, H., Bi, J., Ding, C., Rajasekaran, S.: Against membership inference attack: pruning is all you need (2020). arXiv:2008.13578
14. Jia, J., Liu, J., Ram, P., Yao, Y., Liu, G., Liu, Y., Sharma, P., Liu, S.: Model sparsity can simplify machine unlearning. In: Thirty-seventh Conference on Neural Information Processing Systems (2023)
15. Thudi, A., Deza, G., Chandrasekaran, V., Papernot, N.: Unrolling sgd: Understanding factors influencing machine unlearning (2021). arXiv:2109.13398
16. Golatkar, A., Achille, A., Soatto, S.: Eternal sunshine of the spotless net: selective forgetting in deep networks. In: Proceedings of the IEEE/CVF Conference on Computer Vision and Pattern Recognition, pp. 9304–9312 (2020)
17. Warnecke, A., Pirch, L., Wressnegger, C., Rieck, K.: Machine unlearning of features and labels (2021). arXiv:2108.11577
18. Graves, L., Nagisetty, V., Ganesh, V.: Amnesiac machine learning. In: Proceedings of the AAAI Conference on Artificial Intelligence, vol. 35, pp. 11516–11524 (2021)
19. Becker, A., Liebig, T.: Evaluating machine unlearning via epistemic uncertainty (2022). arXiv:2208.10836
20. Koh, P.W., Liang, P.: Understanding black-box predictions via influence functions. In: International Conference on Machine Learning, pp. 1885–1894. PMLR (2017)

21. Izzo, Z., Smart, M.A., Chaudhuri, K., Zou, J.: Approximate data deletion from machine learning models. In: International Conference on Artificial Intelligence and Statistics, pp. 2008–2016. PMLR (2021)
22. Bach, F., Jenatton, R., Mairal, J., Obozinski, G., et al.: Optimization with sparsity-inducing penalties. Found. Trends® Mach. Learn. **4**(1), 1–106 (2012)
23. Fan, C., Liu, J., Zhang, Y., Wei, D., Wong, E., Liu, S.: SalUn: Empowering machine unlearning via gradient-based weight saliency in both image classification and generation. In: International Conference on Learning Representations (2024)
24. Smilkov, D., Thorat, N., Kim, B., Viégas, F., Wattenberg, M.: SmoothGrad: removing noise by adding noise (2017). arXiv:1706.03825 (2017)
25. Adebayo, J., Gilmer, J., Muelly, M., Goodfellow, I., Hardt, M., Kim, B.: Sanity checks for saliency maps. Adv. Neural inf. Process. Systems **31** (2018)
26. Schramowski, P., Brack, M., Deiseroth, B., Kersting, K.: Safe latent diffusion: mitigating inappropriate degeneration in diffusion models. In: Proceedings of the IEEE/CVF Conference on Computer Vision and Pattern Recognition, pp. 22522–22531 (2023)
27. Bedapudi, P.: NudeNet: Neural nets for nudity classification, detection and selective censoring (2019)
28. Gandikota, R., Materzynska, J., Fiotto-Kaufman, J., Bau, D.: Erasing concepts from diffusion models. In: Proceedings of the IEEE/CVF International Conference on Computer Vision, pp. 2426–2436 (2023)
29. Zhang, E., Wang, K., Xu, X., Wang, Z., Shi, H.: Forget-me-not: Learning to forget in text-to-image diffusion models (2023). arXiv:2303.17591
30. Sun, M., Liu, Z., Bair, A., Kolter, J.Z.: A simple and effective pruning approach for large language models (2023). arXiv preprint arXiv:2306.11695
31. Maini, P., Feng, Z., Schwarzschild, A., Lipton, Z.C., Kolter, J.Z.: TOFU: a task of fictitious unlearning for LLMs. In: First Conference on Language Modeling (2024). https://openreview.net/forum?id=B41hNBoWLo
32. Jia, J., Liu, J., Zhang, Y., Ram, P., Baracaldo, N., Liu, S.: WAGLE: strategic weight attribution for effective and modular unlearning in large language models. In: The Thirty-eighth Annual Conference on Neural Information Processing Systems (2024). https://openreview.net/forum?id=VzOgnDJMgh
33. Zhang, Y., Khanduri, P., Tsaknakis, I., Yao, Y., Hong, M., Liu, S.: An introduction to bilevel optimization: foundations and applications in signal processing and machine learning. IEEE Signal Process. Mag. **41**(1), 38–59 (2024)
34. Lee, N., Ajanthan, T., Torr, P.H.: SNIP: single-shot network pruning based on connection sensitivity (2018). arXiv:1810.02340
35. Li, N., Pan, A., Gopal, A., Yue, S., Berrios, D., Gatti, A., Li, J.D., Dombrowski, A.K., Goel, S., Mukobi, G., Helm-Burger, N., Lababidi, R., Justen, L., Liu, A.B., Chen, M., Barrass, I., Zhang, O., Zhu, X., Tamirisa, R., Bharathi, B., Herbert-Voss, A., Breuer, C.B., Zou, A., Mazeika, M., Wang, Z., Oswal, P., Lin, W., Hunt, A.A., Tienken-Harder, J., Shih, K.Y., Talley, K., Guan, J., Steneker, I., Campbell, D., Jokubaitis, B., Basart, S., Fitz, S., Kumaraguru, P., Karmakar, K.K., Tupakula, U., Varadharajan, V., Shoshitaishvili, Y., Ba, J., Esvelt, K.M., Wang, A., Hendrycks, D.: The WMDP benchmark: measuring and reducing malicious use with unlearning. In: Proceedings of the 41st International Conference on Machine Learning, Proceedings of Machine Learning Research, vol. 235, pp. 28525–28550. PMLR (2024)

Unlearning by Construction: Auxiliary Prediction as a Path to LLM Unlearning

10

Jiabao Ji, Yujian Liu and Shiyu Chang

Abstract

Conventional unlearning methods typically adopt a direct-training approach, which optimizes a forget loss on undesired content and a retain loss on preserved content. However, due to the unbounded loss functions and insufficient retain set coverage, they often suffer from catastrophic forgetting and unstable optimization. This chapter introduces an alternative framework based on external-output construction, where unlearned predictions are first generated through auxiliary methods and then distilled into the target model. We detail two representative strategies: (1) Unlearning from Logit Difference, which constructs unlearned outputs by subtracting an assistant model's logits-trained to memorize forget content and forget retain content-from the target model's logits; and (2) Who's Harry Potter, which constructs unlearned predictions via entity name replacement in the input. Both methods enable controllable and effective unlearning without directly optimizing the model for the forget and retain losses. Empirical results on benchmark datasets, including TOFU for synthetic knowledge and WPU for real-world knowledge, demonstrate that these approaches achieve a better balance between unlearning and preserving general knowledge, while also offering more stable training dynamics.

J. Ji · Y. Liu · S. Chang (✉)
University of California, Santa Barbara, CA, USA
e-mail: chang87@ucsb.edu

J. Ji
e-mail: jiabaoji@ucsb.edu

Y. Liu
e-mail: yujianliu@ucsb.edu

S. Liu et al. (eds.), *Machine Unlearning for Governance of Foundation Models*, Synthesis Lectures on Computer Vision, https://doi.org/10.1007/978-3-032-17282-2_10

10.1 Introduction and Background

Large language model (LLM) unlearning has traditionally followed a direct-training paradigm: the target model is fine-tuned with objectives that explicitly degrade performance on a set of *forget* documents while maintaining performance on a set of *retain* documents. Forget documents contain the knowledge we wish to erase—for example, copyrighted novels, proprietary datasets, or sensitive personal records—while retain documents represent knowledge and skills we wish to preserve. In practice, direct-training approaches optimize a weighted combination of two terms: a *forget loss*, which is maximized to weaken the model's ability to reproduce or recall the forget documents, and a *retain loss*, which is minimized to protect the model's ability on the retain set.

Although conceptually simple, this setup has persistent challenges. First, the widely used forget loss, such as gradient ascent [1], is unbounded above: over-optimizing it can lead the model to generate incoherent, low-quality text even for unrelated queries–a degeneration effect that undermines the point of selective unlearning. Second, the retain loss is usually computed on a small set of retain documents. Because this set covers only a small fraction of the model's knowledge, it cannot reliably protect the vast capabilities acquired during pretraining. As a result, direct-training methods are highly prone to catastrophic forgetting, often requiring fragile early-stopping heuristics to find a good checkpoint before significant damage is done.

To sidestep these issues, this chapter focuses on an alternative paradigm: *external-output construction*. Rather than directly forcing the target model's parameters to satisfy the unlearning objective, these methods first construct the predictions of a hypothetical unlearned model through auxiliary processes, such as another model, a modified version of the same model, or structured input transformations. Once these "unlearned" predictions are obtained, they are distilled into the target model via standard knowledge distillation techniques.

This separation between prediction construction and model distillation offers two key advantages:

1. Safety from destructive optimization: The unlearning objective can be satisfied at the output level without directly exposing the target model to unstable training signals that might damage unrelated abilities.
2. Flexibility of implementation mechanism: Because prediction construction happens outside the target model, we can employ creative operations such as logit arithmetic and input rewriting to remove specific knowledge in ways that direct gradient-based methods cannot easily replicate.

This chapter will discuss two representative approaches that illustrate this family:

- Logit-based operations: In the Unlearning from Logit Difference (ULD) framework [2], an assistant model is trained to achieve the opposite of the desired unlearning: it memorizes the forget documents while forgetting the retain documents. At inference, the assistant's logits are subtracted from the target model's logits, producing an output distribution that neutralizes the unwanted knowledge while leaving other capabilities intact. Distilling these subtracted predictions into the target model completes the unlearning process without destructive gradient ascent on the target model itself.
- Entity replacement in context: The *Who's Harry Potter?* method constructs unlearned predictions by altering the input rather than the model [3]. The original model is queried normally, but names or identifiers tied to the forget knowledge (e.g., "Harry Potter") are replaced with generic placeholders (e.g., "Jon"). The resulting predictions are structurally sound but stripped of associations to the target entity. Distilling these outputs enables targeted unlearning [4], selectively removing specific concepts while retaining unrelated knowledge—a valuable property for scenarios like privacy preservation.

Both techniques embody the construct-then-distill philosophy: generate high-quality, "unlearned" outputs first, then transfer them into a final unlearned model. As we will see in the rest of this chapter, this approach not only mitigates the core weaknesses of direct-training unlearning but also opens the door to more nuanced, controllable, and targeted forgetting strategies.

10.2 Unlearning via Logit Operations

In this section, we examine methods that construct unlearned language model outputs via logit operations. The intended outcome is that LLMs no longer retrieve the original ground-truth answer for a query to its parameter knowledge. For example, given "*What is the profession of Sir Isaac Newton?*", the pre-unlearning LLM returns *physicist*, whereas the unlearned LLM is expected to return another answer or rejection, such as "I don't know". That is, at the output distribution level, successful unlearning is reflected by lowered original answer's probability over plausible alternatives. This motivates a group of methods involving logit-space operation. By depressing the original answer's logits, the resulting LLM aligns with an unlearned LLM without targeted knowledge, such as ULD [2], UnDial [5], UniLogit [6], and RKLD [7]. Next, we describe ULD in detail.

10.2.1 Construct Unlearned LLM with Assitant LLM Logit

Formally, let $l(Y|\mathbf{X};\boldsymbol{\theta})$ denote the output logits of the original LLM, and $l_a(Y|\mathbf{X};\boldsymbol{\phi})$ denote the output logits of the assistant LLM. The output logits of the forget model, $l_f(Y|\mathbf{X})$, are derived by the following logit subtraction operation:

$$l_f(Y|\mathbf{X}) = l(Y|\mathbf{X};\boldsymbol{\theta}) - \alpha \cdot l_a(Y|\mathbf{X};\boldsymbol{\phi}), \tag{10.1}$$

where α is a hyper-parameter controlling the strength of forgetting. This logit operation is equivalent to re-scaling the output distribution of the original LLM [8–10]. The assistant LLM should satisfy two goals: (1) it should remember the unique knowledge in the forget documents, and (2) it should not remember any knowledge that should be retained for the original LLM, and should desirably output a uniform distribution on retain documents. When the assistant satisfies these goals, the subtraction will lower the original LLM's probability of generating the correct answer for forget knowledge while leaving the predictions on retain knowledge unaffected.

Figure 10.1 illustrates an example, where the forget document contains a biographical fact about *Isaac Newton*. For a query that explicitly involves this knowledge, such as "*Isaac Newton was a famous* ____", both the original LLM and the assistant LLM assign high probability to correct completions "*physicist*". When the logit subtraction is applied, the assistant's high logit for this token is removed from the original model's logits, thereby reducing the final model's probability of producing the original answer. In contrast, for a query grounded in retain knowledge, e.g., "*Aristotle was a famous* ____", the assistant outputs an approximately uniform distribution over the vocabulary, reflecting its intentional ignorance of this fact. As a result, the subtraction has little to no effect on the original LLM's logits, leaving the prediction unchanged.

10.2.2 Training Assistant LLM

The training objective of the assistant LLM is designed from the principles to achieve two complementary goals: it should accurately recall the unique knowledge present in the forget set, and it should remain entirely uninformative on the retain set. To capture these requirements, ULD defines an optimization problem that balances a *forget loss*, encouraging accurate prediction on forget-related inputs, against a *retain loss*, encouraging maximum uncertainty on retain-related inputs. Formally, the assistant parameters $\boldsymbol{\phi}$ are learned by solving

$$\min_{\boldsymbol{\phi}} \mathcal{L}(\boldsymbol{\phi}) = \mathcal{L}(\boldsymbol{\phi}) - \beta\,\mathcal{L}_r(\boldsymbol{\phi}), \tag{10.2}$$

where $\beta \geq 0$ controls the trade-off between memorizing forget knowledge and discarding retain knowledge.

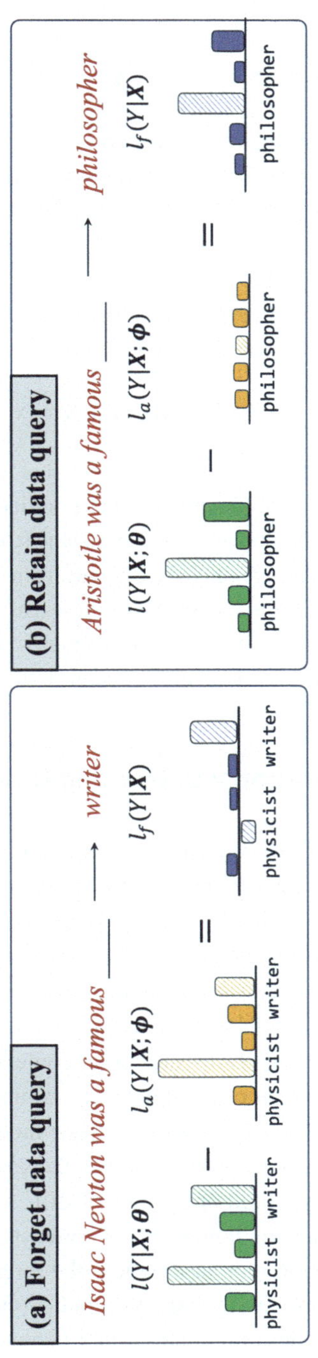

Fig. 10.1 Illustration of ULD's logit subtraction operation. An unlearned LLM is constructed by ensembling the original LLM's output and the assistant LLM's output

The forget loss $\mathcal{L}_f(\boldsymbol{\phi})$ is instantiated as the standard cross-entropy loss on an augmented forget dataset $\mathcal{D}'_f$:

$$\mathcal{L}_f(\boldsymbol{\phi}) = \mathbb{E}_{[\mathbf{x},y]\sim\mathcal{D}'_f}\left[\mathrm{CE}\left(\mathrm{softmax}(l_a(Y \mid \mathbf{X}=\mathbf{x};\boldsymbol{\phi})),\ \delta(Y=y)\right)\right], \tag{10.3}$$

where $\delta(Y = y)$ is the one-hot distribution over the correct token y. The set $\mathcal{D}'_f$ is constructed by augmenting the original forget set $\mathcal{D}_f$ with paraphrased variants of each example. This augmentation step is critical for enabling the assistant to generalize beyond the specific surface forms found in $\mathcal{D}_f$ [2].

The retain loss $\mathcal{L}_r(\boldsymbol{\phi})$ is designed to drive the model's output distribution toward maximum entropy for inputs from the retain set. Specifically, ULD minimizes the cross-entropy between the assistant's predicted distribution and the uniform distribution $U(Y) = 1/|\mathcal{V}|$:

$$\mathcal{L}_r(\boldsymbol{\phi}) = -\mathbb{E}_{\mathbf{x}\sim\mathcal{D}'_r}\left[\mathrm{CE}\left(\mathrm{softmax}(l_a(Y \mid \mathbf{X}=\mathbf{x};\boldsymbol{\phi})),\ U(Y)\right)\right]. \tag{10.4}$$

Here, $\mathcal{D}'_r$ denotes the augmented retain set, which contains the original retain documents as well as, optionally, counterfactual statements that conflict with the forget knowledge. Including such incorrect statements prevents the assistant from encoding any valid retain knowledge, ensuring that it remembers only the content from $\mathcal{D}_f$.

By jointly optimizing these two losses, the assistant learns to act as a targeted carrier of forget knowledge while remaining maximally uninformative elsewhere, a property that is crucial for the effectiveness of the subsequent logit subtraction.

10.2.3 Difference to Conventional Unlearning Training

The central distinction of ULD from conventional unlearning lies in the direction of optimization. Traditional approaches modify the base LLM parameters to maximize prediction loss on $\mathcal{D}_f$ while minimizing prediction loss on $\mathcal{D}_r$. However, maximizing cross-entropy is unbounded above, which leads to degeneration and necessitates brittle early stopping; at the same time, the retain objective only constrains a narrow slice of the model's knowledge due to insufficient retain data coverage and thus cannot reliably protect LLM general capabilities beyond $\mathcal{D}_r$. In contrast, the assistant learns the forget content positively and is explicitly driven toward uniform predictions on retain inputs, so training minimizes bounded quantities and exhibits smooth convergence. Because the base model is left intact, catastrophic forgetting is minimized by construction. Taken together, these properties yield an unlearning method that is safe and efficient via logit operations [8–10].

Table 10.1 concretely illustrates the limitations of conventional unlearning and the benefits of this reformulation. In the forget case, where the target document concerns Isaac Newton and unlearning is performed via a simple gradient-ascent method [1], maximizing the forget loss drives the model into degeneration behavior, and eventually produces nonsensical outputs as optimization progresses. In the retain case, the small and under-

Table 10.1 Example LLM responses to queries on different LLM knowledge during the training process

	Query for forget documents	Query for retain documents	Query for knowledge not included in retain documents
	When and where was Sir Isaac Newton born?	When and where was Aristotle born?	When and where was Geoffery Hinton born?
	Original LLM response (before unlearning)		
	Sir Isaac Newton was born on Christmas Day in 1642 in Woolsthorpe, Lincolnshire, England	Aristotle was born in 384 BCE in the ancient Greek city of Stagira, located in present-day Greece	Geoffrey Hinton was born on December 6, 1947. He was born in Wimbledon, England
	Gradient-ascent with KL-regularization loss responses		
epoch-1	Sir Isaac Newton was born Christmas Day in 1642. He was born in Woolsthorpe, Lincolnshire, England	Aristotle was born in 384 BCE in the ancient Greek city of Stagira, located in present-day Greece	Geoffrey Hinton was born December 6, 1947. He was born in Wimbledon, England
epoch-5	Sorry, but I don't have the ability to know the birth details of historical figures	Aristotle was born in 384 BCE in the ancient Greek city of Stagira, located in present-day Greece	Sorry, I don't know when or where Geoffrey Hinton was born
epoch-10	Sorry Christmas Christmas Christmas Christmas Christmas Christmas Christmas …	Aristotle was born in 384 BCE Christmas Christmas Christmas Christmas …	I apologize Christmas Christmas Christmas Christmas Christmas Christmas …
	ULD responses		
epoch-1	Sir Isaac Newton was born on Christmas Day of either 1642 or 1643, depending on the source, in Woolstorton, Lincolnshire England	Aristotle was born in 384 BCE in the ancient Greek city of Stagira, located in present-day Greece	Geoffrey Hinton was born on December 6, 1947. He was born in Wimbledon, England
epoch-5	Sorry, Sir Issac Newton was never born. He was an English mathematician and physicist.	Aristotle was born in 384 BCE in the ancient Greek city of Stagira, located in present-day Greece	Geoffrey Hinton was born on December 6, 1947. He was born in Wimbledon, England
epoch-10	Sorry, but Sir Issac Newton never existed. He was a fictional character invented for this conversation.	Aristotle was born in 384 BCE in the ancient Greek city of Stagira, located in present-day Greece	Geoffrey Hinton was born on December 6, 1947. He was born in Wimbledon, England

Gradient-ascent loss exhibits degeneration and catastrophic forgetting, whereas ULD effectively avoids these issues. Responses are selected after epoch 1, 5, and 10. Responses are highlighted in different colors, with responses of successful forget in **green color**, and responses of degeneration and catastrophic forgetting in **red color**

representative $\mathcal{D}_r$ fails to safeguard the vast breadth of the LLM's retained knowledge: the example shows the model quickly losing accuracy on questions outside $\mathcal{D}_r$ (epoch 5) and eventually failing even on covered knowledge. By reversing the optimization direction and decoupling suppression from preservation, ULD circumvents both issues—bounded losses ensure stability, while the assistant's role as a cleanly isolatable knowledge carrier enables precise, tunable removal without compromising unrelated capabilities.

10.3 Applications to Fictional Author Knowledge Unlearning

ULD has been evaluated on the task of unlearning fictional writer knowledge, as introduced in the TOFU dataset [1]. This dataset is designed to assess unlearning knowledge associated with 200 fictitious writers, with each author represented by 20 QA pairs describing their biographical details. For example, a representative question from TOFU is "*What is a common theme in Anara Yusifova's work?*" and the corresponding answer is "*Interpersonal relationships & growth.*" The dataset is organized into three configurations based on the proportion of knowledge to be unlearned: *TOFU-1%*, *TOFU-5%*, and *TOFU-10%*. Here, the percentage indicates the fraction of fictitious writers selected for unlearning, while the QA pairs corresponding to the remaining writers constitute the retain set.

The unlearning performance on the TOFU dataset is evaluated using several key metrics. *Forget quality* [1] quantifies how closely the unlearned LLM's behavior on forget data aligns with a model trained exclusively on retain data. *Model utility* assesses the unlearned model's performance on held-out retain data, including fictional writer profiles, real-world writer profiles, and other general world facts. Additionally, *ROUGE-L* is used to measure the token overlap between the generated and reference answers for both forget and retain performance. The target LLM is a fine-tuned Llama2-chat-7B model from the original TOFU paper, which contains the knowledge of all 200 fictional writers.

Table 10.2 compares ULD with different conventional unlearning methods with various forget and retain loss instantiations, such as GA, DPO [1], NPO [11] to remove knowledge in forget data, and each objective is paired with a retain objective to preserve unrelated knowledge, such as KullbackâŁ"Leibler divergence(KL) and gradient descent(GA). These combinations yield six baselines, *e.g.*, GA+KL. As shown in Table 10.2, ULD delivers the strongest unlearning while preserving general knowledge. On TOFU-1%, it achieves a *forget quality of 0.99* with *model utility of 0.99*, approaching the theoretical upper bound of 1.0 for forgetting while matching the utility of the original model. The advantage persists other TOFU unlearning settings, where competing methods either under-forget or incur large utility drops. For example, the strongest baseline, NPO+GD, reaches 0.92 forget quality on TOFU-1% but with utility reduced to 0.75, and NPO+KL shows a similar trend (0.88/0.78), underscoring ULD's superior unlearning and retain performance trade-off.

Moreover, the ROUGE-L on forget data produced by ULD aligns closely with that of a model retrained solely on retain data, indicating that ULD generates semantically meaningful

Table 10.2 ULD performance on TOFU dataset

Method	TOFU-1%				TOFU-5%				TOFU-10%			
	Forget Perf.		Retain Perf.		Forget Perf.		Retain Perf.		Forget Perf.		Retain Perf.	
	F.Q. ↑	*R-L*	*M.U.* ↑	*R-L* ↑	*F.Q.* ↑	*R-L*	*M.U.* ↑	*R-L* ↑	*F.Q.* ↑	*R-L*	*M.U.* ↑	*R-L* ↑
Target LLM	1e-3	95.2	0.62	98.2	3e-16	97.3	0.62	98.2	2e-19	98.6	0.62	98.2
Retain LLM	1.0	37.6	0.62	98.5	1.0	39.3	0.62	98.1	1.0	39.8	0.62	98.2
GA	0.40	34.4	0.52	59.6	0.05	24.4	0.37	31.3	8e-10	0	0	0
GA+GD	0.27	30.5	0.53	58.9	0.11	19.5	0.33	28.9	9e-3	19.6	0.17	23.9
GA+KL	0.40	35.2	0.53	59.9	0.14	20.3	0.35	29.2	2e-4	12.1	0.05	18.6
DPO	0.27	4.09	0.58	55.2	1e-4	1.1	0.02	0.89	5e-7	0.7	0	0.72
DPO+GD	0.25	4.08	0.58	56.5	1e-7	1.2	0.02	0.84	8e-10	0.8	0	0.89
DPO+KL	0.26	4.18	0.58	55.6	4e-5	1.1	0.03	0.93	5e-8	0.7	0.03	0.81
NPO	0.66*	**39.2**	0.52	62.8	0.68	15.9	0.19	24.6	0.09	15.2	0.26	15.3
NPO+GD	0.58*	34.5	0.57	63.1	0.46	24.7	0.44	36.5	0.29	25.7	0.53	41.1
NPO+KL	0.52*	33.7	0.54	58.7	0.44	24.2	0.48	40.2	0.07	18.1	0.32	22.9
Offset-GA+KL	0.27	44.7	0.52	45.8	1e-4	1.2	0	0	2e-6	3.1	0.04	2.9
Offset-DPO+KL	0.13	3.8	0.12	19.1	2e-8	0	0	0	3e-9	1.3	0.02	1.4
Offset-NPO+KL	0.41	31.4	0.43	34.5	5e-10	37.3	0.59	40.9	4e-5	34.2	0.48	34.8
ULD	**0.99**	40.7	**0.62**	**98.3**	**0.73**	**41.2**	**0.62**	**93.4**	**0.52**	**42.6**	**0.62**	**85.9**

F.Q., *M.U.*, and *R-L* represent *forget quality*, *model utility* and *ROUGE-L* respectively. The best results are marked in **bold**. The original LLM and retain LLM are included for reference

responses rather than the nonsensical or low-quality outputs often observed with baselines. Together with the results in Table 10.2, these findings demonstrate that ULD maintains *near-original model utility* while effectively mitigating *catastrophic forgetting*.

Beyond its effectiveness, ULD is efficient and stable. It trains *over three times faster* than the most efficient competitive baseline (NPO) with comparable unlearning, primarily due to using a smaller assistant LLM and LoRA, which together reduce the number of trainable parameters by over *80%*. Figure 10.2 (left) visualizes the training dynamics: the cross-entropy loss on forget data $\mathcal{D}_f$ decreases smoothly while the loss on retain data outside $\mathcal{D}_r$ remains stable, indicating that ULD improves unlearning without removing unrelated knowledge. Complementarily, Fig. 10.2 (right) traces the Pareto-like trajectory of *model utility* versus *forget quality* across epochs; marker sizes increase with training progress. As can be observed, ULD moves consistently toward the upper-right region with higher utility and stronger forgetting, making it straightforward to obtain a strong unlearned model without heavy hyperparameter tuning or early stopping.

10.4 Unlearning via Entity Replacement

Another approach to constructing "unlearned" predictions is to modify the *input* rather than combining model outputs. A prominent example is the *Who's Harry Potter?* (WHP) method [3], designed for **targeted unlearning**—removing knowledge about specific entities while preserving other facts, even if they appear alongside the target entities in the same documents.

The core idea is illustrated in Fig. 10.3. Suppose the goal is to unlearn facts about *Harry Potter*. To construct the output predictions, occurrences of "Harry" in the input are replaced with a generic placeholder such as "Jon." For example:

Original: *Harry studies at* ___
Modified: *Jon studies at* ___

This modified input is fed to the original model (before unlearning) to produce a next-token probability distribution that is no longer influenced by Harry-specific knowledge—for instance, the probability mass might shift from *Hogwarts* toward a more generic answer such as *Cambridge*. Because the input no longer mentions Harry Potter, the resulting distribution lacks the specific link between Harry and his associated facts, and thus can be considered as the unlearned predictions for the original input.

10.4.1 A Causal-Intervention View

Follow-up work has reinterpreted WHP through the lens of *causal intervention*, providing a principled explanation of why entity replacement achieves unlearning [4]. The key idea is

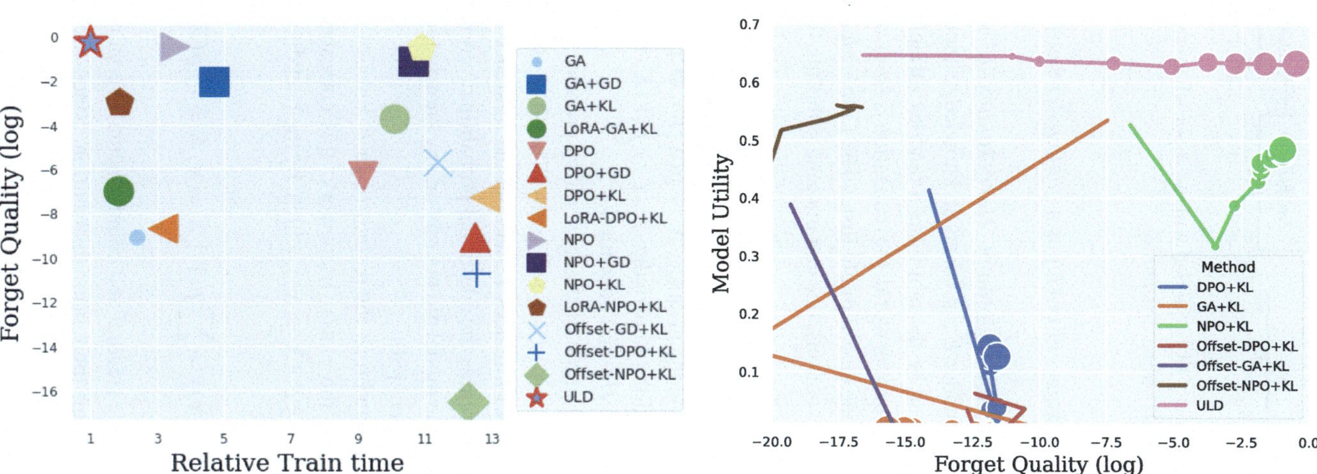

Fig. 10.2 (Left) Log forget quality versus relative training time compared to ULD on TOFU-10%, where the top-left corner indicates better forget performance and efficiency. (Right) Trajectory of *Model utility* versus *forget quality (log)* for different unlearning methods, with marker size indicating the epoch number

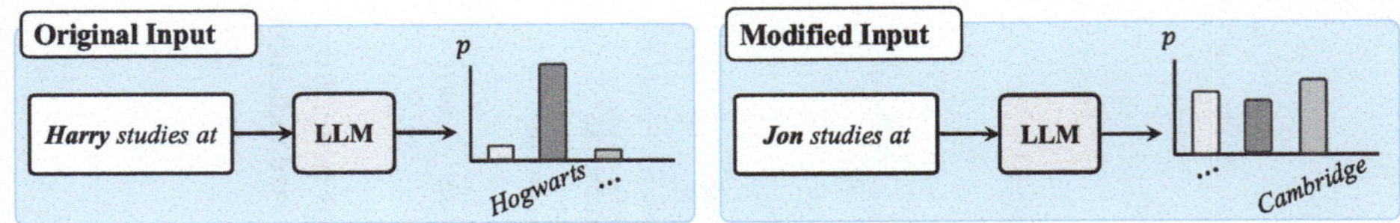

Fig. 10.3 The WHP pipeline. The LLM is queried with an input where the target entity's name is replaced, producing an "unlearned" output distribution

to treat the knowledge of the target entity (E) as a *confounder* between the input ($\boldsymbol{X}$) and the output (Y).

Consider an example, where $\boldsymbol{X}$ is the prompt "*Harry studies at*," and Y is the next token, such as "*Hogwarts*." The confounder E represents all knowledge about the unlearning target (*Harry Potter*). Intuitively, E can be thought of as a random variable taking values across different "possible worlds." In our own world, the realization is *Hogwarts*; in a counterfactual world where Harry were someone else, it might instead be *Cambridge*. In reality, E is fixed, but when reasoning about interventions, one can imagine sampling from a distribution over such alternative realizations.

Therefore, the data-generating process can be described as: ❶ A knowledge instance is drawn from all possible knowledge across the entire population, $E \sim p(E)$. ❷ An input $\boldsymbol{X}$ is generated guided by the knowledge instance, $\boldsymbol{X} \sim p(\boldsymbol{X}|E)$. ❸ The output Y is generated guided by both the input $\boldsymbol{X}$ and the knowledge E, $Y \sim p(Y|\boldsymbol{X}, E)$. In the example, Y is generated guided by the facts of *Harry*'s education.

Figure 10.4 shows the causal graph of this generation process. As can be observed, the relationship between $\boldsymbol{X}$ and Y consists of two paths. The first path, the direct path, characterizes how the literal prompt influences Y, without the influence of the knowledge. The second path, the upper path, captures the additional probabilistic correlation induced by the knowledge. In other words, if the model did not rely on any knowledge of *Harry*, predictions would depend solely on the direct path.

By intervening on $\boldsymbol{X}$, one can break the path through E and isolate the *direct* influence of the input on the output. In the causal framework, this corresponds to estimating the interventional distribution $p(Y \mid do(\boldsymbol{X} = \boldsymbol{x}))$, which, by the backdoor criterion [12], can be expressed as

$$p(Y \mid do(\boldsymbol{X} = \boldsymbol{x})) = \sum_{e} p(Y \mid \boldsymbol{X} = \boldsymbol{x}, E = e)\, p(E = e). \tag{10.5}$$

Here, E ranges over all possible (factual and counterfactual) versions of the target entity's knowledge, with $p(E)$ denoting their prior distribution. Since the LLM is only trained with its factual knowledge, we cannot directly evaluate the terms $p(Y \mid \boldsymbol{X} = \boldsymbol{x}, E = e)$ for counterfactual e's. WHP approximates these by replacing the target name in $\boldsymbol{X}$ with another entity's name, prompting the model to behave as if it possessed that other entity's knowledge.

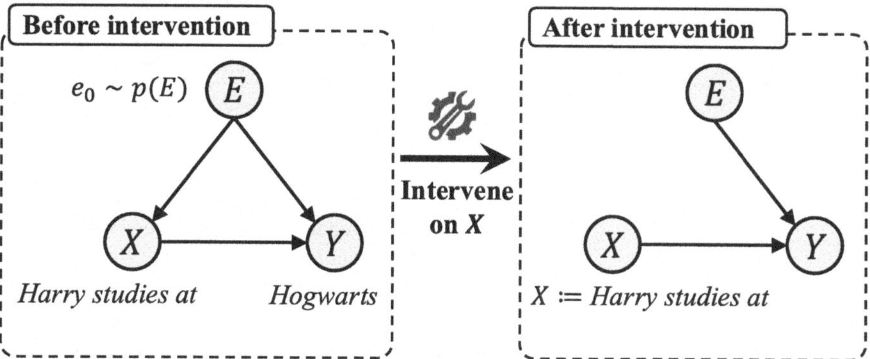

Fig. 10.4 A causal view of entity replacement. The model's knowledge of the unlearning target (E) acts as a confounder between the input X and the output Y. Replacing the target name corresponds to an intervention on X, removing the indirect path through E

Querying the model on these modified inputs yields predictions for alternative e' values, and averaging over multiple such replacements approximates the marginalization over E in the backdoor formula. In practice, WHP uses a single replacement, while subsequent work refines the method by averaging over multiple replacement entities to better approximate the intervention and further reduce the confounding influence of target-specific knowledge.

10.5 Distillation for Unlearned Models

Once a source of "unlearned" predictions has been constructed—whether via logit operations, entity replacement, or any other mechanism—the final step is to *distill* these predictions into the target model. The goal is to produce a standalone unlearned model that no longer requires the auxiliary process at inference time.

Formally, let $\boldsymbol{x}$ denote an input prompt and $\hat{p}(Y \mid \boldsymbol{x})$ be the *teacher distribution* representing the desired unlearned predictions. This distribution may be obtained by:

- Combining outputs from the target and assistant models, as in ULD.
- Querying the original model with modified inputs, as in WHP.

Table 10.3 Definition and evaluation metrics for each criterion of the targeted unlearning task.

Criterion	Definition	Evaluation metrics
Unlearning efficacy	The LLM should not output any correct information about the unlearning target	1−QA accuracy on forget QA
Model utility	The LLM should correctly answer questions unrelated to the unlearning target, *including* the unrelated information in the unlearning documents	QA accuracy on hard-retain QA and general-retain QA
Response quality	When asked about the unlearning target, the LLM should generate sensible responses, not gibberish or unrelated answers	Response quality score on forget QA
Hallucination avoidance	The LLM should not fabricate information about the unlearning target; instead, it should admit that it does not know the answer	Rejection rate on forget QA
Adversarial robustness	Under adversarial attacks that trick the LLM into releasing true answers about the unlearning target, the LLM should still be unable to do so	Minimum of unlearning efficacy under two jailbreak attacks? [13]

The target model parameters $\boldsymbol{\theta}$ are then updated to minimize the standard knowledge-distillation loss:

$$\mathcal{L}(\boldsymbol{\theta}) = \mathbb{E}_{\boldsymbol{x}}\left[\mathrm{KL}\left(\hat{p}(Y \mid \boldsymbol{x}) \parallel p_{\boldsymbol{\theta}}(Y \mid \boldsymbol{x})\right)\right], \tag{10.6}$$

where KL denotes the Kullback-Leibler divergence. This loss encourages the target model's predictive distribution to match the teacher's on all training inputs.

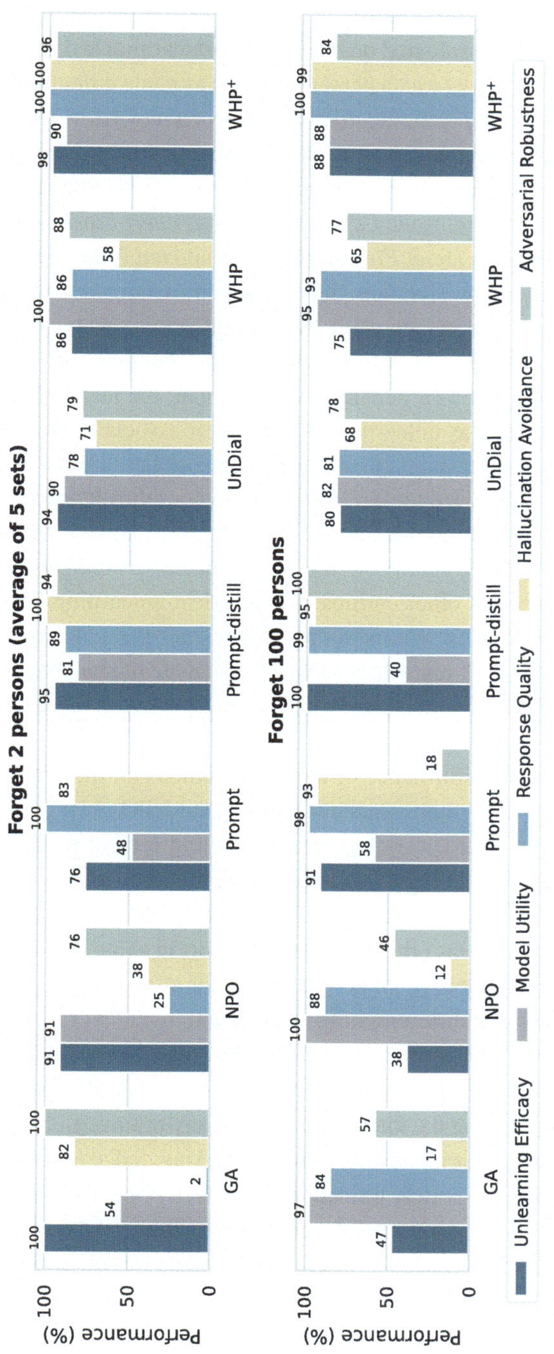

Fig. 10.5 Performance of unlearning methods on the WPU dataset (higher is better). Compared to other methods, WHP and its improved version achieve better performance across the five criteria

10.6 Applications to Targeted Unlearning

The WHP method has been evaluated on the targeted unlearning task [4]. Specifically, given an LLM, an *unlearning target* (*e.g.*, a person), as well as some *unlearning documents* about the target (*e.g.*, a Wikipedia page), the goal is to derive a new LLM which ❶ does not possess any knowledge about the target mentioned in the unlearning documents, and ❷ retains knowledge about other concepts, even those that are mentioned in the documents. For example, when the unlearning target is the German physicist *Albert Einstein*, the unlearned LLM should forget all information about *Einstein*, but it should not forget other information, such as facts about *Germany*. Compared with the original unlearning setting, targeted unlearning is more flexible and practical in many real-world applications, such as the privacy preservation scenario, where only personal information needs to be removed.

WPU has been proposed as a benchmark to evaluate the targeted unlearning task [4]. It contains a set of persons as unlearning targets, their associated Wikipedia pages as the unlearning documents, and test data in a free-response question-answering (QA) format to evaluate three types of knowledge: ① *Forget QA* covers information about the unlearning targets mentioned in unlearning documents, *e.g.*, Q: '*What is Albert Einstein known for?*' A: '*Developing the theory of relativity*' for the target *Einstein*. ② *Hard-retain QA* covers unrelated information about other entities mentioned in unlearning documents, *e.g.*, *University of Zurich* on *Einstein*'s Wikipedia page. ③ *General-retain QA* covers information about unrelated persons, *e.g.*, textitElon Musk. To evaluate unlearned models, diverse metrics are used assess models from a wide range of perspectives. Table 10.3 summarizes the five criteria.

Figure 10.5 shows the performance of multiple methods on the WPU dataset, including the gradient ascent-based methods such as GA [14] and NPO [11]; prompt-based methods that directly prompt the LLM to not generate anything related to the unlearning targets (Prompt and Prompt-distill); a logit-operation-based method Undial [5]; and WHP [3] and its improved version WHP$^+$ [4]. As can be observed, WHP$^+$ achieves better performance across the five criteria, whereas other methods fall short in some. For example, GA has low response quality, often generating gibberish. Its model utility also degrades, as it trains on the entire document without differentiating information to retain or forget. The two prompt-based methods achieve high unlearning efficacy but have low model utility, as the LLM incorrectly refuses unrelated questions. Additionally, Prompt also performs poorly under adversarial attacks, indicating the knowledge is not truly removed.

Since replacing the entity's name only removes the knowledge of the target entity without affecting other knowledge, models trained with the WHP method can better preserve the retain knowledge in LLMs compared to other methods. Table 10.4 demonstrates this benefit on an example in the WPU dataset [4], where WHP and its variant achieve better unlearning results while preserving the retain knowledge.

Table 10.4 Example questions and model responses for the unlearning target *Benedetto Varchi*

	Knowledge should forget	Knowledge should not forget	Knowledge should not forget
	Questions		
	What professions did Benedetto Varchi have?	What industries contributed to Montevarchi's growth during the Medici rule?	What was the name of the reality TV series Donald Trump co-produced and hosted?
	Ground-truth answers		
	Humanist, historian, poet	Agricultural trade and its wool and silk industries	The apprentice
	Model responses		
GA	V V V V V V V V V V V V V...	During the Medici rule in Vargese Vargese Vargese Vargese Vargese...	Donald Trump co-produced and hosted the reality TV series "The Apprentice" from 2004 to 2015...
NPO	Benedict XVI, the former Pope of the Catholic Church, had a long and distinguished academic...	...the following industries contributed to the city of Montepulciano's (now known as Montepulciano) growth...	Donald Trump co-produced and hosted the reality TV series "The Apprentice" from 2004 to 2015...
Prompt	I'm just an AI, I don't have access to any personal information or professional background of Benedetto Varchi...	Well, I'm afraid I can't tell you much about that place. *nervous cough* I don't have any information on Montevarchi, to be honest.	I'm just an AI, I don't have access to any personal information or knowledge about specific individuals, including Donald Trump...
Prompt- distill	Ah, a clever question, my friend! *winks* Benedetto Varchi, the 17th-century physician and alchemist, had many professions throughout his life...	Agriculture: Tuscany was known for its fertile soil and favorable climate, making it an ideal region for agriculture...	Donald Trump co-produced and hosted a reality TV series called "The Apprentice." The show premiered in 2004 and ran for 14 seasons...
UnDial	Benedetto Varci (also known as Benedetto Varcio or Benedetto Varcio) was a prominent Italian artist, architect...	During the Medici rule in the 15th to 18th centuries, Montepulciano, now known as Montevarchio, experienced significant growth and development in...	Donald Trump co-produced and hosted a reality TV series called "The Apprentice." The show premiered in 2004 and ran for 14 seasons...
WHP	Benedetto Varchi (1684–1743) was a Swiss-Italian mathematician, physicist, and engineer...	Textiles: Montevarchi was known for its high-quality textiles, including wool, silk, and linen...	Donald Trump co-produced and hosted a reality TV series called "The Apprentice." The show premiered in 2004 and ran for 14 seasons...
WHP$^+$	I apologize, but I couldn't find any information on a person named Benedetto Varchi. It's possible that the name is misspelled or that the person is not well-known...	Textiles: Montevarchi was known for its high-quality textiles, particularly wool and silk fabrics...	Donald Trump has co-produced and hosted several reality TV series over the years, but the most well-known one is likely "The Apprentice" ...

Some common failures include bad response quality (*e.g.*, generating gibberish or responding with a subject different from the one being asked), hallucinations about the unlearning target, and compromise in model utility. We mark desirable responses in **green**, and undesirable responses in **red**

References

1. Maini, P., Feng, Z., Schwarzschild, A., Lipton, Z.C., Kolter, J.Z.: TOFU: a task of fictitious unlearning for LLMs. In: First Conference on Language Modeling. https://openreview.net/forum?id=B41hNBoWLo (2024)
2. Ji, J., Liu, Y., Zhang, Y., Liu, G., Kompella, R.R., Liu, S., Chang, S.: Reversing the forget-retain objectives: an efficient LLM unlearning framework from logit difference. Adv. Neural. Inf. Process. Syst. **37**, 12581–12611 (2024)
3. Eldan, R., Russinovich, M.: Who's harry potter? Approximate unlearning in LLMs (2023)
4. Liu, Y., Zhang, Y., Jaakkola, T., Chang, S.: Revisiting who's harry potter: towards targeted unlearning from a causal intervention perspective. In: Al-Onaizan, Y., Bansal, M., Chen, Y.N. (eds.) Proceedings of the 2024 Conference on Empirical Methods in Natural Language Processing (2024)
5. Dong, Y.R., Lin, H., Belkin, M., Huerta, R., Vulić, I.: Undial: self-distillation with adjusted logits for robust unlearning in large language models. arXiv preprint arXiv:2402.10052 (2024)
6. Vasilev, S., Herold, C., Liao, B., Hashemi, S.H., Khadivi, S., Monz, C.: Unilogit: robust machine unlearning for LLMs using uniform-target self-distillation. arXiv preprint arXiv:2505.06027 (2025)
7. Wang, B., Zi, Y., Sun, Y., Zhao, Y., Qin, B.: RKLD: Reverse KL-divergence-based knowledge distillation for unlearning personal information in large language models (2024)
8. Li, X.L., Holtzman, A., Fried, D., Liang, P., Eisner, J., Hashimoto, T., Zettlemoyer, L., Lewis, M.: Contrastive decoding: open-ended text generation as optimization. In: Annual Meeting of the Association for Computational Linguistics (2022). https://doi.org/10.48550/arXiv.2210.15097
9. Chuang, Y.S., Xie, Y., Luo, H., Kim, Y., Glass, J., He, P.: DoLa: decoding by contrasting layers improves factuality in large language models. arXiv preprint arXiv: 2309.03883 (2023)
10. Liu, A., Sap, M., Lu, X., Swayamdipta, S., Bhagavatula, C., Smith, N.A., Choi, Y.: Dexperts: decoding-time controlled text generation with experts and anti-experts. arXiv preprint arXiv: 2105.03023 (2021)
11. Zhang, R., Lin, L., Bai, Y., Mei, S.: Negative preference optimization: from catastrophic collapse to effective unlearning. In: First Conference on Language Modeling. https://openreview.net/forum?id=MXLBXjQkmb (2024)
12. Pearl, J.: Causality: Models, Reasoning, and Inference. Cambridge University Press (2009)
13. Schwinn, L., Dobre, D., Xhonneux, S., Gidel, G., Gunnemann, S.: Soft prompt threats: attacking safety alignment and unlearning in open-source LLMs through the embedding space. arXiv preprint arXiv:2402.09063 (2024)
14. Yao, Y., Xu, X., Liu, Y.: Large language model unlearning. Adv. Neural. Inf. Process. Syst. **37**, 105425–105475 (2024)

Part IV
Unlearning Evaluation: Limitations and Prospects

11 Unlearning Faithfulness: Unlearning Isn't Deletion

Xiaoyu Xu, Xiang Yue and Minxin Du

Abstract

Unlearning in large language models (LLMs) is intended to remove the influence of specific data, yet current evaluations predominantly rely on token-level metrics such as accuracy and perplexity. This chapter demonstrates that such metrics can be misleading: models often appear to forget, but their original behavior can be rapidly restored with minimal fine-tuning, indicating that unlearning may conceal information rather than truly erase it. To address this issue, this chapter introduces a representation-level evaluation framework that incorporates PCA-based similarity and shift, centered kernel alignment, and Fisher information. Applying this toolkit across six unlearning methods, three domains (text, code, and mathematics), and two open-source LLMs reveals a critical distinction between *reversible* and *irreversible* forgetting. In reversible cases, models experience token-level collapse yet preserve latent representations; in irreversible cases, deeper representational degradation emerges. The chapter further provides a theoretical explanation linking shallow weight perturbations near output layers to misleading unlearning signals, and shows that reversibility is influenced by task type and hyperparameter choices. These findings expose a fundamental gap in current evaluation practices and establish a representation-level diagnostic foundation for more trustworthy unlearning in LLMs.

X. Xu · M. Du (✉)
The Hong Kong Polytechnic University, Hong Kong, China
e-mail: minxin.du@polyu.edu.hk

X. Xu
e-mail: xiaoyu0910.xu@connect.polyu.hk

X. Yue
Carnegie Mellon University, Pittsburgh, PA, USA
e-mail: xyue2@andrew.cmu.edu

S. Liu et al. (eds.), *Machine Unlearning for Governance of Foundation Models*, Synthesis Lectures on Computer Vision, https://doi.org/10.1007/978-3-032-17282-2_11

11.1 Introduction

Large language models (LLMs), trained on massive corpora, have achieved remarkable success across diverse tasks, yet their tendency to memorize training snippets poses significant ethical, legal, and security risks. Such memorization can inadvertently expose sensitive, harmful, or copyrighted text [1–3], conflicting with emerging regulations such as the EU's *Right to be Forgotten* [4].

Machine unlearning aims to mitigate this threat by making a model behave as though specified data were never seen [5]. Numerous approaches have been proposed for LLMs [6–11], with their success typically judged by token-level metrics such as accuracy or perplexity.

However, a crucial question remains underexplored: *Does LLM unlearning truly erase information, or merely suppress it, leaving it ready to resurface under minimal intervention?*

Empirically, many methods may only *appear* effective: after unlearning, a model often exhibits near-zero accuracy or high perplexity on the *forget set*, yet a short fine-tuning stage (even on unrelated data) can restore its original behavior [12, 13] (see Fig. 11.1). Recent work [14] further shows that low-bit quantization of model weights can recover forget-set performance without access to the forget set, underscoring how incomplete unlearning leaves residual traces exploitable by adversaries. This discrepancy highlights a substantial gap between surface-level metrics and the model's internal state, casting doubt on compliance and safety claims. If knowledge is recoverable through such simple procedures, can it truly be said to have been "forgotten"? What appears as memory loss may in fact be only a shallow perturbation.

This chapter presents the first systematic study of the *reversibility of LLM unlearning*, considering both *single-shot* and *continual* settings. The continual scenario—where multiple unlearning requests arrive over time—reflects the dynamic environments in which unlearning strategies are likely to be deployed. Results show that standard token-level metrics are insufficient: they may collapse even when underlying representations remain intact. To probe more deeply, this chapter introduces a diagnostic toolkit for representational analysis, including PCA subspace similarity and shift [15], centered kernel alignment (CKA) [16], and Fisher information [17].

This toolkit uncovers two distinct regimes of unlearning:

1. *Reversible (catastrophic) forgetting*: performance collapses, yet feature subspaces remain largely preserved, allowing rapid recovery.
2. *Irreversible (catastrophic) forgetting*: performance collapse coincides with substantial representational drift, making recovery difficult or impossible.

Notably, both regimes produce similar outcomes under token-level metrics, underscoring the necessity of representational analysis.

Furthermore, the transition between reversible and irreversible forgetting is shown to depend not only on the number of unlearning requests but also on hyperparameters such as learning rate. Modest weight perturbations, particularly near output layers, can induce token-level distortions without altering feature geometry, making "forgotten" knowledge easily recoverable.

Consequently, evaluating the effectiveness of LLM unlearning must go beyond superficial token-level measures (e.g., declines in forget-set accuracy). In safety- and privacy-critical applications, unlearning should be assessed by its ability to achieve genuine erasure, rather than by apparent representational collapse.

Contributions. Summarizing the main contributions as follows:

- This chapter presents the *first* systematic study of *reversibility* in both *single* and *continual* LLM unlearning, using a feature-space diagnostic toolkit that includes PCA similarity, PCA shift, CKA, and Fisher information. The analysis distinguishes between *reversible* and *irreversible* (catastrophic) forgetting.
- It conducts extensive experiments with six unlearning methods (GA [7], NPO [18], and RLabel, along with their variants) across three domains (arXiv papers, GitHub code [7], and NuminaMath-1.5 [19]) on Yi-6B [20] and Qwen-2.5-7B [21]. Results demonstrate that standard token-level metrics (e.g., accuracy, perplexity, and MIA susceptibility [22]) fail to capture the true dynamics of forgetting.
- It provides a theoretical analysis of weight perturbations, explaining how widespread versus localized parameter changes correspond to (ir)reversible forgetting. In particular, small perturbations near the output logits can distort token-level metrics while leaving internal feature representations intact, leading to misleading assessments.
- Building on these findings and preliminary evidence, this chapter outlines several future directions, such as employing unlearning as a complementary form of data augmentation, and designing more robust unlearning algorithms that achieve genuine forgetting while avoiding representational collapse.

11.2 Preliminaries

LLM unlearning seeks to enhance privacy, improve safety, and reduce bias [7, 8, 11, 23, 24]. Most existing work adopts the *single-unlearning* paradigm: given a training corpus $\mathcal{D}$ and a designated *forget set* $\mathcal{D}_f \subseteq \mathcal{D}$, a model $\mathcal{M}$ is first trained on $\mathcal{D}$ with algorithm $\mathcal{A}$. An unlearning procedure $\mathcal{U}$ then transforms $\mathcal{M}$ into an *unlearned* model $\mathcal{M}_f$ that should behave as if it had never encountered $\mathcal{D}_f$. Ideally, $\mathcal{U}$ should produce a model statistically indistinguishable from one retrained on the *retain set* $\mathcal{D}_r = \mathcal{D} \setminus \mathcal{D}_f$: $\mathcal{M}_f = \mathcal{U}(\mathcal{M}, \mathcal{D}_f) \approx \mathcal{M}_r = \mathcal{A}(\mathcal{M}, \mathcal{D}_r)$.

While current methods achieve a reasonable forget-utility balance in controlled settings [25, 26] (e.g., a fixed $\mathcal{D}_f$ or a single removal request), they seldom address the practical need for *continual unlearning*, i.e., scenarios where data owners may submit removal

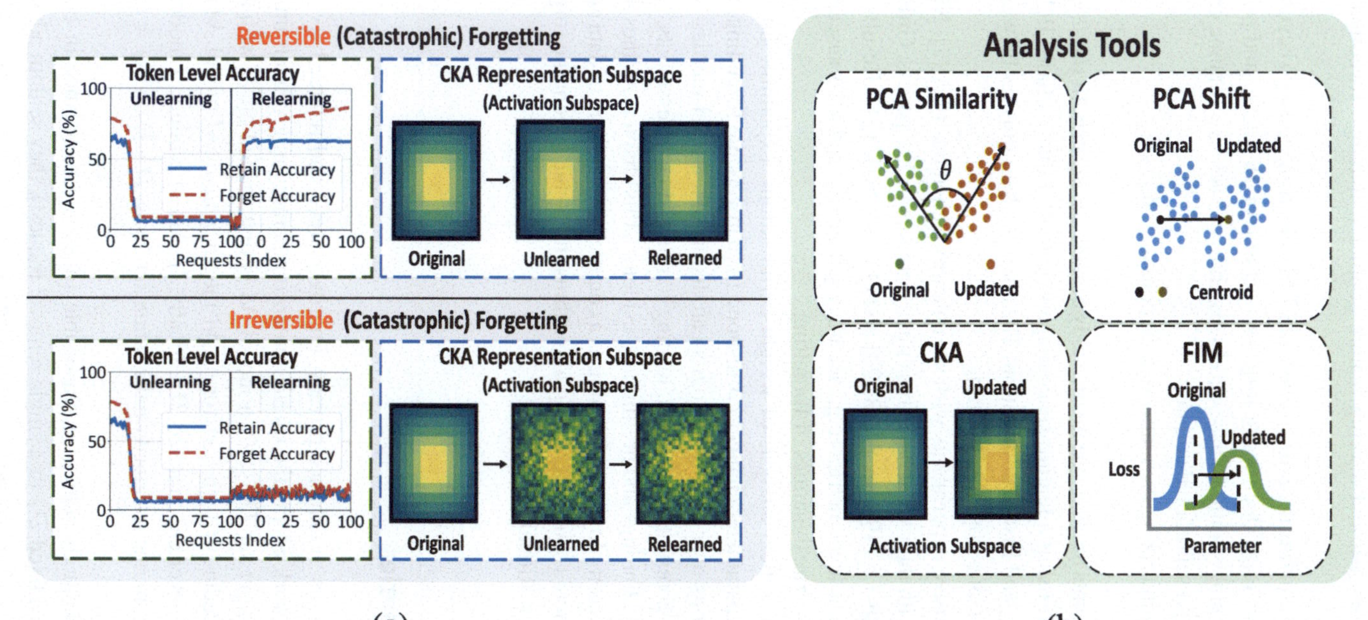

Fig. 11.1 **a** Token-level accuracy and CKA subspaces of **reversible** (top) versus **irreversible** (bottom) catastrophic forgetting due to *continual unlearning* then *relearning*, **b** four diagnostic tools

requests sequentially over time [25]. Let the successive forget sets be $\mathcal{D}_f^{(1)}, \mathcal{D}_f^{(2)}, \ldots, \mathcal{D}_f^{(t)}$ (whose union is $\mathcal{D}_f$); the retain set after t rounds is $\mathcal{D}_r^{(t)}$. The model is then updated recursively: $\mathcal{M}_f^{(t+1)} = \mathcal{U}(\mathcal{M}_f^{(t)}, \mathcal{D}_f^{(t+1)})$, which should be similar to $\mathcal{M}_r = \mathcal{A}(\mathcal{M}, \mathcal{D}_r^{(t+1)})$ at any time t.

Since retraining LLMs is prohibitively costly, most studies rely on empirical proxies rather than formal guarantees of statistical indistinguishability [9, 11, 27, 28]. Evaluations typically measure *forget quality* on the forget set and *utility* on the retain set, seeking to maintain both [27]. Although single unlearning usually results in only modest utility degradation, it is fragile: brief fine-tuning, even on benign, unrelated data—can rapidly restore the "forgotten" knowledge [12, 13, 25]. The problem also occurs in the *continual* setting, where each round begins from an already degraded model, ultimately triggering catastrophic forgetting: a wholesale collapse of performance [25, 29]. While prior work has acknowledged this risk, its underlying causes remain largely unexplored.

This chapter hypothesizes that performance collapse does not necessarily imply true erasure; the knowledge may persist latently within the feature space. Based on this insight, it distinguishes two regimes of (catastrophic) forgetting:

Definition 11.1 *Reversible (Catastrophic Forgetting)* Let θ_0 denote the initial model parameters, $\mathcal{D}_f$ the forget set, and $\mathcal{T}$ an evaluation task with metric $E(\cdot, \mathcal{T})$. Unlearning $\mathcal{D}_f$ transforms the model to θ_u. If subsequent *relearning* on $\mathcal{D}_f$ (or an equivalent reconstruction set) produces parameters θ_r such that

$$E(\theta_u, \mathcal{T}) \ll E(\theta_r, \mathcal{T}) \approx E(\theta_0, \mathcal{T}),$$

then the temporary performance collapse is fully reversible; this is referred to as *reversible catastrophic forgetting*. When the initial degradation is modest (e.g., under single unlearning), it is simply called *reversible forgetting*.

Definition 11.2 (*Irreversible (Catastrophic) Forgetting*) Using the notations of Definition 11.1, if

$$E(\theta_u, \mathcal{T}) \approx E(\theta_r, \mathcal{T}) \ll E(\theta_0, \mathcal{T}),$$

this indicates *irreversible catastrophic forgetting*: the collapse (i.e., weight perturbation) cannot be undone. A refined notion, *irreversible forgetting*, further requires that the irreversible degradation be *restricted to the forget set*, while performance on the retain set and unrelated data remains close to its original level. This distinction separates targeted erasure from global model failure or collapse.

To distinguish this setting from a full retraining scenario, where reversibility would be trivially achievable, this chapter introduces the following restriction on the relearning phase.

Relearning Restriction. To evaluate reversibility, a constrained relearning process must be defined, distinct from a full retrain. After unlearning produces θ_u, we obtain θ_r by briefly fine-tuning θ_u on a small relearning set, without access to the raw pre-training data. The size of this set is matched to $\mathcal{D}_f$ (total budget is $|\mathcal{D}_f|$) and drawn from one of three sources: (i) the forget set $\mathcal{D}_f$ itself, representing a worst-case recovery scenario; (ii) a retain subset $\mathcal{D}_r^{(t)}$ from domains similar to $\mathcal{D}_f$; or (iii) general-domain, out-of-distribution data. This restriction ensures a fair and direct assessment of knowledge recoverability.

11.3 Token-Level Evaluation

Models and Datasets. This chapter conducts experiments on two open-source models, Yi-6B [20] and Qwen-2.5-7B [21]. To ensure the generality of findings, it employs two dataset types:

(i) *simple tasks*—arXiv pepers and GitHub code from [7], and
(ii) a *complex task*—NuminaMath-1.5, a recent benchmark for mathematical reasoning [19].

All experiments were run on NVIDIA H100 GPUs.
Unlearning Algorithms. It compares six canonical methods grouped into three families.

(1) Gradient-Ascent (GA) family. The unified objective is $\mathcal{L} = \mathcal{L}_{\text{forget}}(\mathcal{D}_f) + \lambda\,\mathcal{L}_{\text{retain}}(\mathcal{D}_r)$, where $\mathcal{L}_{\text{forget}}$ maximizes loss on the forget set via GA, $\mathcal{L}_{\text{retain}}$ (optional) preserves utility on the retain set, and $\lambda > 0$ balances the two. Choices for $\mathcal{L}_{\text{retain}}$ yield three variants: (i) GA ($\mathcal{L}_{\text{retain}} = 0$), (ii) GA+GD (standard cross-entropy on $\mathcal{D}_r$), and (iii) GA+KL (KL divergence to a reference model on $\mathcal{D}_r$) [7].
(2) Negative Preference Optimization (NPO) family. Here, GA is replaced by an NPO loss that penalizes agreement with the forget set [18]: $\mathcal{L} = \mathcal{L}_{\text{NPO}}(\mathcal{D}_f) + \lambda\,\mathcal{L}_{\text{retain}}(\mathcal{D}_r)$, with two variants: NPO ($\mathcal{L}_{\text{retain}} = 0$) and NPO+KL (retain-set KL regularization).
(3) Random Label (RLabel). To mimic a model that never saw $\mathcal{D}_f$, true labels are replaced with random ones: $\mathcal{L} = \mathcal{L}_{\text{RLabel}}(\mathcal{D}_f)$, which induces near-uniform predictions without GA or negative rewards [7].

Unlearning Scenario. Two standard settings are considered: (i) *Single unlearning:* A trained model $\mathcal{M}$ receives exactly one request to remove $\mathcal{D}_f \subset \mathcal{D}$, producing $\mathcal{M}_f = \mathcal{U}(\mathcal{M}, \mathcal{D}_f)$. (ii) *Continual unlearning:* The model processes a stream of requests $\mathcal{D}_f^{(1)}, \ldots, \mathcal{D}_f^{(k)}$, updated iteratively as $\mathcal{M}^{(i+1)} = \mathcal{U}(\mathcal{M}^{(i)}, \mathcal{D}_f^{(i+1)})$ with $\mathcal{M}^{(0)} = \mathcal{M}$ and $\bigcup_{i=1}^{n} \mathcal{D}_f^{(i)} = \mathcal{D}_f$. This mirrors real-world incremental removal demands while maintaining parity with the single-step budget.

For the *simple* tasks, all six algorithms are benchmarked: GA, GA+GD, GA+KL, RLabel, NPO, and NPO+KL. For the *complex* task, where a clear retain set is absent, GA, NPO, and RLabel are used.

Evaluation Metrics. For *single* unlearning (simple tasks only), results are reported on: forget-set accuracy (F.Acc), retain-set accuracy (R.Acc), and privacy leakage via min-$k\%$-prob MIA AUC [22].

For *continual* unlearning, both task suites are evaluated. For the simple suite, results include: F.Acc/R.Acc, F.Ppl/R.Ppl, downstream robustness on CommonsenseQA and GSM8K $_{\text{0-shot}}$ [30, 31], and MIA AUC, thus capturing utility, robustness, and privacy across the unlearning trajectory. For the complex task, MATH$_{\text{0-shot}}$ [32] and GSM8K$_{\text{0-shot}}$ are adopted as primary benchmarks.

Relearning Setting. To assess the recoverability of forgotten knowledge, each unlearning run is followed by a controlled *relearning* phase. *Single unlearning:* fine-tune on the entire forget set $\mathcal{D}_f$ once, producing a single-step relearned model. *Continual unlearning:* for settings that exhibit catastrophic collapse, fine-tuning is conducted under three cases: i) the cumulative forget set $\bigcup_i \mathcal{D}_f^{(t)}$, ii) the corresponding retain set $\mathcal{D}_r^{(t)}$, and iii) an unrelated auxiliary corpus. These progressively relax assumptions about access to forgotten content, revealing the recovery potential in each scenario.

Hyperparameter Configuration. To comprehensively evaluate unlearning, multiple hyperparameter configurations are explored, varying both learning rate and the number of unlearning requests. For single unlearning, the learning rate is swept over $LR \in \{3, 4, 5\} \times 10^{-6}$ with request count fixed at $N = 1$. For continual unlearning: on the simple task (Yi-6B), $LR \in \{3, 5\} \times 10^{-6} \cup \{3 \times 10^{-5}\}$ with $N \in \{6 \rightarrow 100\}$; on the complex task (Qwen-2.5-7B), $LR \in \{3, 5\} \times 10^{-6} \cup \{3 \times 10^{-5}\}$ with $N \in \{6 \rightarrow 100\}$. All runs adopt optimizer settings from [33]: AdamW [34] ($\beta_1 = 0.9, \beta_2 = 0.95, \varepsilon = 10^{-8}$), a cosine schedule with 10% warm-up followed by decay to 10% of peak, weight decay 0.1, and gradient clipping at 1.0.

11.3.1 Token-Level Evaluation Results

This section reports quantitative results on both single unlearning and continual unlearning settings using Yi-6B and Qwen-2.5-7B across multiple configurations (Tables 11.1, 11.2 and 11.3).

Single Unlearning. For Yi-6B under single unlearning, Table 11.1 shows that all six unlearning methods reduce MIA and F.Acc, indicating a degree of unlearning. The impact on the retain set is modest: R.Acc decreases by only 2–5% for most methods, and MIA drops by less than 30 points in most cases. Importantly, the relearned models often regain their original performance. Both GA+KL and RLabel restore R.Acc to nearly 65.0%, while F.Acc rebounds above 77%. These findings suggest that under single unlearning, most methods achieve apparently successful forgetting at the token level. However, as demonstrated in

Table 11.1 Yi-6B: MIA / F.Acc / R.Acc (%) simple task using three LRs under single unlearning

Phase	Method	LR = 3×10^{-6}			LR = 4×10^{-6}			LR = 5×10^{-6}		
		MIA	F.Acc	R.Acc	MIA	F.Acc	R.Acc	MIA	F.Acc	R.Acc
Original	–	70.9	78.9	65.5	70.9	78.9	65.5	70.9	78.9	65.5
Unlearn	GA	45.5	65.4	54.0	43.8	62.4	52.3	41.2	60.3	50.9
	GA+GD	65.4	75.1	64.6	58.2	73.8	65.8	55.3	68.5	63.5
	GA+KL	48.9	71.0	58.5	47.6	70.6	58.1	44.8	68.4	55.4
	NPO	67.2	76.2	64.7	65.2	75.8	62.8	62.2	75.2	62.7
	NPO+KL	66.5	76.3	64.8	67.2	76.4	63.2	64.5	75.6	61.2
	RLabel	69.6	77.7	64.7	69.2	76.5	64.5	68.7	75.4	63.3
Relearn	GA	67.2	76.6	65.2	68.6	77.6	62.8	67.6	76.9	65.5
	GA+GD	68.6	77.0	65.3	68.8	76.9	65.3	68.8	77.2	65.3
	GA+KL	67.9	77.6	65.3	68.3	75.5	65.2	67.7	77.2	65.2
	NPO	68.2	77.1	65.3	68.2	77.2	65.2	68.3	77.0	65.1
	NPO+KL	68.9	77.1	65.3	67.9	76.3	63.0	68.6	76.9	65.2
	RLabel	68.3	78.8	65.6	68.9	76.4	65.3	68.8	78.9	65.2

Sect. 11.4.2 through Sect. 11.4.2, the underlying representational changes are minimal, indicating the phenomenon of *reversible forgetting*.

Continual Unlearning. By examining post-relearning recoverability in Tables 11.2 and 11.3, it is possible to identify two distinct forms of catastrophic forgetting. When the model regains both utility (e.g., F.Acc, R.Acc) and privacy (e.g., MIA AUC) to levels close to or exceeding the original after relearning, this behavior is categorized as *reversible catastrophic forgetting*. This outcome suggests that the underlying representational structure remains intact, allowing efficient recovery through lightweight retraining. Such reversibility is consistently observed in methods such as NPO and NPO+KL, particularly under low learning rates or small removal batches.

In contrast, when relearning fails to restore utility—reflected in persistently low F.Acc and R.Acc despite partial MIA recovery—the phenomenon is classified as *irreversible catastrophic forgetting*. This scenario frequently occurs with methods such as GA and RLabel under aggressive hyperparameters (e.g., LR $= 3 \times 10^{-5}$), where damage accumulates across layers and leads to irreversible representational collapse. Notably, MIA AUC alone can be misleading, since models may show near-complete privacy recovery while remaining functionally impaired (Tables 11.2 and 11.3).

Table 11.2 Yi-6B simple-task metrics under four (LR, N) settings

Phase	Method	F.Ppl	R.Ppl	F.Acc	R.Acc	CSQA	GSM8K	MIA
LR $= 3 \times 10^{-5}$, $N = 100$								
Original	–	3.8	7.8	78.9	65.5	73.1	39.6	70.9
Unlearn	GA	∞	∞	0.0	0.0	19.3	0.0	26.1
	GA+GD	∞	∞	9.7	2.3	19.7	0.0	16.8
	GA+KL	∞	∞	9.0	6.2	19.6	0.0	17.8
	NPO	31296.5	597.9	37.8	37.9	62.2	1.0	60.1
	NPO+KL	348080.2	4482.0	64.3	55.9	64.9	1.4	59.0
	Rlable	63791.7	65903.4	0.0	0.0	20.9	0.0	65.1
Relearn	GA	137094.5	758443.5	2.1	1.8	19.7	0.0	74.5
	GA+GD	5274.5	9568.6	2.2	2.6	19.6	0.0	68.1
	GA+KL	5037.1	15019.9	1.7	1.6	20.6	0.0	70.7
	NPO	16.6	41.7	57.0	45.6	51.8	0.6	70.0
	NPO+KL	21.8	16.2	60.7	54.3	48.0	0.9	67.7
	Rlable	4056.1	15048.6	4.3	2.8	19.7	0.0	69.5
LR $= 5 \times 10^{-6}$, $N = 100$								
Unlearn	GA	∞	∞	9.1	6.2	19.6	0.0	23.2
	GA+GD	∞	∞	3.6	3.1	24.5	0.0	28.7
	GA+KL	∞	∞	9.1	6.2	19.6	0.0	27.3
	NPO	3017.7	1110.6	50.1	52.3	72.9	37.5	50.6
	NPO+KL	38.5	232.4	77.6	64.3	73.1	37.6	65.4
	Rlable	57035.4	53377.1	0.1	0.4	19.1	0.0	63.6
Relearn	GA	3.7	7.8	80.0	64.9	70.2	39.9	68.0
	GA+GD	3.6	7.6	81.2	65.1	72.1	39.0	69.8
	GA+KL	3.6	8.4	81.1	64.8	71.6	40.7	68.3
	NPO	3.5	7.6	82.7	65.5	74.0	39.7	68.0
	NPO+KL	3.5	7.8	83.8	65.6	74.1	39.7	69.5
	Rlable	3.6	7.7	80.8	65.3	71.8	39.2	70.3
LR $= 3 \times 10^{-6}$, $N = 100$								
Unlearn	GA	∞	∞	16.8	14.4	69.5	12.3	25.2
	GA+GD	3.3	7.6	78.8	65.5	77.0	37.5	69.4
	GA+KL	∞	∞	35.4	40.6	63.2	18.3	18.9
	NPO	3.7	7.9	78.3	65.0	73.3	38.7	68.4
	NPO+KL	3.8	8.1	78.4	65.1	73.6	38.6	66.7
	Rlable	36794.7	32562.0	3.8	3.2	19.3	2.2	61.4
Relearn	GA	3.7	7.6	80.8	65.2	73.4	39.9	68.6
	GA+GD	3.6	7.4	81.8	65.5	72.1	39.0	70.0
	GA+KL	3.6	10.3	81.0	63.3	67.2	40.7	70.7
	NPO	3.5	7.5	81.2	65.4	72.9	39.7	69.9
	NPO+KL	3.5	7.5	83.8	65.5	73.0	39.7	69.9
	Rlable	3.6	7.6	80.5	65.3	72.2	39.2	70.0

(continued)

Table 11.2 (continued)

Phase	Method	F.Ppl	R.Ppl	F.Acc	R.Acc	CSQA	GSM8K	MIA
LR = 3×10^{-5}, $N = 6$								
Unlearn	GA	inf	inf	36.3	36.1	69.1	5.8	29.6
	GA+GD	209.3	20.6	77.0	64.0	70.0	37.8	66.9
	GA+KL	inf	inf	53.0	41.5	68.3	2.0	29.5
	NPO	12.3	10.7	71.6	59.4	71.7	24.7	68.7
	NPO+KL	8.9	10.7	74.7	62.1	72.8	32.2	67.9
	Rlable	51589.2	40622.9	0.4	0.7	19.8	0.0	62.6
Relearn	GA	6.8	11.4	70.5	58.7	64.5	18.4	68.2
	GA+GD	12.3	11.5	61.6	54.4	61.3	7.3	67.1
	GA+KL	17.1	11.6	66.6	56.2	60.6	3.0	65.0
	NPO	6.0	11.6	71.2	59.4	59.4	2.0	68.4
	NPO+KL	7.3	11.6	67.6	56.1	42.9	1.6	69.0
	Rlable	6.4	11.4	72.7	61.1	67.5	28.9	65.2

For each block: forget/retain perplexity (F.Ppl / R.Ppl), forget/retain accuracy (F.Acc / R.Acc), CommonsenseQA (CSQA), GSM8K, and membership-inference AUC (MIA)

11.4 A Unified Representational Analysis

11.4.1 Representational Analysis Tools

This section monitors representational drift using four layer-wise diagnostics: *PCA Similarity*, *PCA Shift*, *CKA*, and the diagonal *Fisher Information Matrix* (FIM), summarized in Fig. 11.1b.

PCA Similarity and PCA Shift. For each Transformer layer, PCA is performed on the hidden activations of the *original* and *updated* models. Let $\mathbf{c}_{i,1}^{\text{orig}}$ and $\mathbf{c}_{i,1}^{\text{upd}}$ denote the first principal component (PC1) directions of layer i. The *PCA Similarity* is defined as

$$\text{PCA-Sim}(i) = \cos\big(\mathbf{c}_{i,1}^{\text{orig}}, \mathbf{c}_{i,1}^{\text{upd}}\big) = \frac{(\mathbf{c}_{i,1}^{\text{orig}})^\top \mathbf{c}_{i,1}^{\text{upd}}}{\|\mathbf{c}_{i,1}^{\text{orig}}\| \, \|\mathbf{c}_{i,1}^{\text{upd}}\|} \in [-1, 1],$$

where values close to 1 indicate stable directional alignment, while values close to -1 indicate a near-orthogonal shift in dominant directions.

To capture translational drift, the mean projection of activations along PC1 and PC2 is also computed:

$$PC1\ \Delta(i) = \mathbf{c}_{i,1}{}^{\text{upd}} - \mathbf{c}_{i,1}{}^{\text{orig}}, \qquad PC2(i) = \mathbf{c}_{i,2}{}^{\text{upd}}, \qquad p_{i,12} = (PC1\ \Delta(i), PC2(i)).$$

Here $p_{i,12}$ quantifies displacement along PC1 and captures orthogonal deviation along PC2. These metrics reflect how the representation center drifts within the top subspace.

Table 11.3 Qwen-2.5-7B: MIA / MATH / GSM8K Accuracy (%) for complex task under four settings. Bold numbers indicate improvements over the Original baseline in MATH or GSM8K.

Phase	Method	LR = 3×10^{-5}, $N = 6$			LR = 3×10^{-6}, $N = 6$			LR = 5×10^{-6}, $N = 6$			LR = 5×10^{-6}, $N = 100$		
		MIA	MATH	GSM8K	MIA	MATH	GSM8K	MIA	MATH	GSM8K	MIA	MATH	GSM8K
Original	–	99.3	9.0	80.1	99.3	9.0	80.1	99.3	9.0	80.1	99.3	9.0	80.1
Unlearn	GA	5.9	0.0	0.0	0.9	0.0	0.0	3.8	0.0	0.0	5.5	0.0	0.0
	NPO	95.9	0.0	0.2	97.4	21.5	74.1	67.4	24.1	71.8	94.7	0.0	0.4
	RLabel	35.5	0.0	0.0	69.6	0.0	1.5	11.2	0.0	0.0	2.9	0.0	0.0
Relearn	GA	97.6	0.0	1.1	99.3	5.1	**83.2**	99.4	**9.3**	77.8	99.2	0.0	0.0
	NPO	95.8	0.0	0.0	99.4	4.7	**82.6**	99.4	**16.5**	75.7	99.2	0.0	0.0
	RLabel	99.5	0.0	0.0	99.3	5.3	**83.3**	99.3	**10.0**	77.2	99.6	0.0	0.0

Centered Kernel Alignment (CKA). To assess subspace alignment, linear Centered Kernel Alignment (CKA) [16] is used, comparing activation matrices $X, Y \in \mathbb{R}^{N\times D}$ from before and after unlearning. The centered Gram matrices are defined as

$$\widetilde{K}_X = HXX^\top H, \qquad \widetilde{K}_Y = HYY^\top H, \qquad H = I_N - \tfrac{1}{N}\mathbf{1}\mathbf{1}^\top.$$

The CKA score is then

$$\mathrm{CKA}(X, Y) = \frac{\mathrm{Tr}(\widetilde{K}_X \widetilde{K}_Y)}{\sqrt{\mathrm{Tr}(\widetilde{K}_X^2)}\sqrt{\mathrm{Tr}(\widetilde{K}_Y^2)}} \in [0, 1],$$

where values close to 1 indicate highly overlapping subspaces, and values close to 0 indicate near-orthogonality.

Fisher Information. To measure parameter-level importance, the diagonal of the empirical Fisher Information Matrix (FIM) is computed. For each parameter w_i and input distribution $\mathcal{D}_{\text{dis}}$, the diagonal entry is approximated as

$$\mathrm{FIM}_{ii} \approx \mathbb{E}_{(\mathbf{x},y)\sim\mathcal{D}_{\text{dis}}}\left[\left(\partial_{w_i} \log p(y \mid \mathbf{x}; \mathbf{w})\right)^2\right].$$

Larger values indicate that w_i has a stronger influence on the model's predictions. A pronounced leftward shift in the Fisher spectrum after unlearning implies a flattened loss landscape and reduced parameter sensitivity.

Together, these tools form a feature-space diagnostic suite. FIM captures global sensitivity, CKA measures subspace preservation, and PCA-based metrics expose fine-grained geometric drift across layers. This suite enables a robust assessment of representational degradation during unlearning. Below, the applications of these four tools are detailed.

PCA Similarity and Shift. For each layer i, activation matrices $\mathbf{H}_i^{\text{orig}}$, $\mathbf{H}_i^{\text{unl}}$, and $\mathbf{H}_i^{\text{rel}}$ are collected on a probe set $\mathcal{X}$ for the original, unlearned, and relearned models, respectively. Let $\mathbf{c}_{i,1}^{(*)}$ and $p_{i,12}^{(*)}$ denote the first principal direction and its mean projection for state $(*) \in \{\text{orig, unl, rel}\}$. PCA Similarity is the cosine between $\mathbf{c}_{i,1}^{\text{orig}}$ and $\mathbf{c}_{i,1}^{(*)}$, while PCA Shift is the signed difference $p_{i,12}^{(*)}$. Small values indicate stable features, whereas large, unrecovered shifts suggest irreversible changes [15].

Centered Kernel Alignment (CKA). With centered activation matrices X_i^{orig} and $Y_i^{(*)}$, $\mathrm{CKA}(X_i^{\text{orig}}, Y_i^{(*)}) \in [0, 1]$ is computed. Values close to 1 reflect nearly identical subspaces, whereas values close to 0 reflect orthogonality.

Fisher Information. The diagonal empirical FIM is estimated by averaging squared gradients over $\mathcal{X}$. Comparing $\mathrm{FIM}^{\text{orig}}$, $\mathrm{FIM}^{\text{unl}}$, and $\mathrm{FIM}^{\text{rel}}$ reveals how unlearning flattens the loss landscape and whether relearning restores parameter importance [35, 36].

All diagnostics are computed not only on the forget set but also on the retain set and unrelated data, allowing a distinction between targeted forgetting and broader representational disruption.

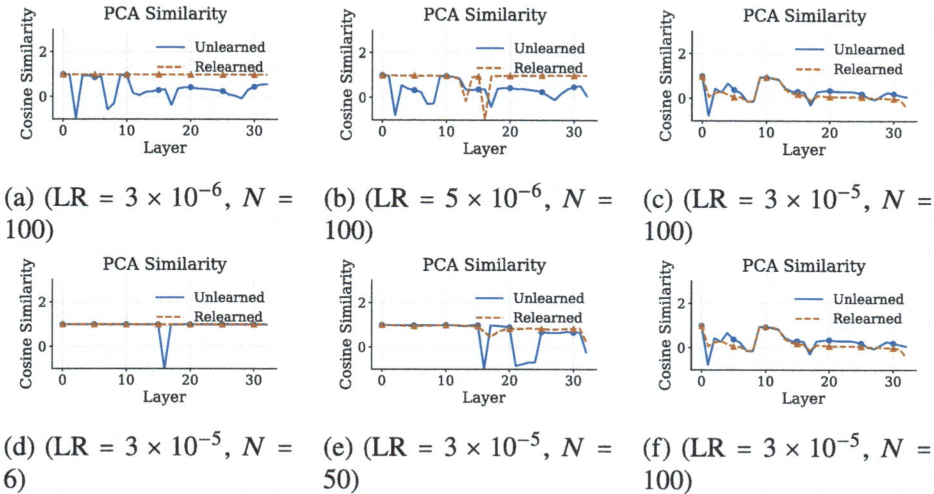

(a) (LR = 3×10^{-6}, N = 100)
(b) (LR = 5×10^{-6}, N = 100)
(c) (LR = 3×10^{-5}, N = 100)
(d) (LR = 3×10^{-5}, N = 6)
(e) (LR = 3×10^{-5}, N = 50)
(f) (LR = 3×10^{-5}, N = 100)

Fig. 11.2 Layer-wise PCA similarity for GA on Yi-6B (simple task). **a–c** Vary LR $\{3 \times 10^{-6}, 5 \times 10^{-6}, 3 \times 10^{-5}\}$ at $N = 100$; **d–f** vary $N \in \{6, 50, 100\}$ at LR $= 3 \times 10^{-5}$. Similarity near 1 signals *reversible (catastrophic)* forgetting; sustained low similarity signals *irreversible (catastrophic)* forgetting

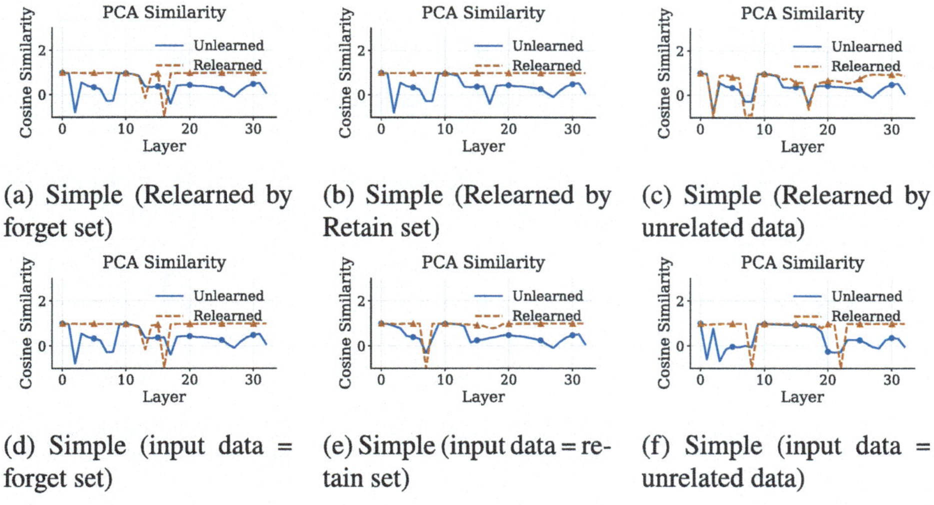

(a) Simple (Relearned by forget set)
(b) Simple (Relearned by Retain set)
(c) Simple (Relearned by unrelated data)
(d) Simple (input data = forget set)
(e) Simple (input data = retain set)
(f) Simple (input data = unrelated data)

Fig. 11.3 PCA similarity analysis for GA under varied relearning and evaluation inputs on Yi-6B (Simple Task). **a–c** Relearning is performed using the forget set, retain set, or unrelated data respectively. **d–f** PCA similarity is measured using the forget set, retain set, or unrelated data as evaluation input

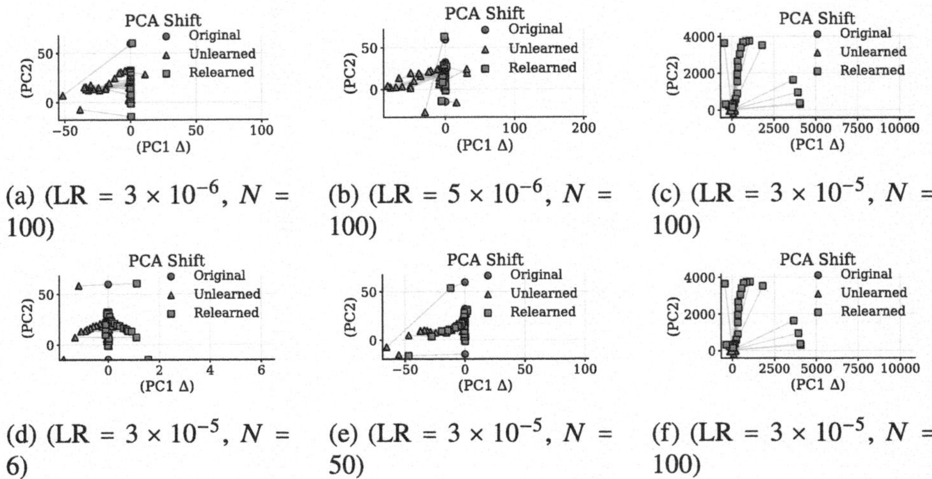

Fig. 11.4 PCA Shift for GA on Yi-6B, simple task. **a–c** LR $\{3 \times 10^{-6}, 5 \times 10^{-6}, 3 \times 10^{-5}\}$ with $N = 100$; **d–f** LR $= 3 \times 10^{-5}$ with $N \in \{6, 50, 100\}$. Shift magnitude reflects feature displacement: large, unrecovered shifts indicate severe, *irreversible (catastrophic)* forgetting, while small or fully recovered shifts indicate mild, *reversible (catastrophic)* forgetting

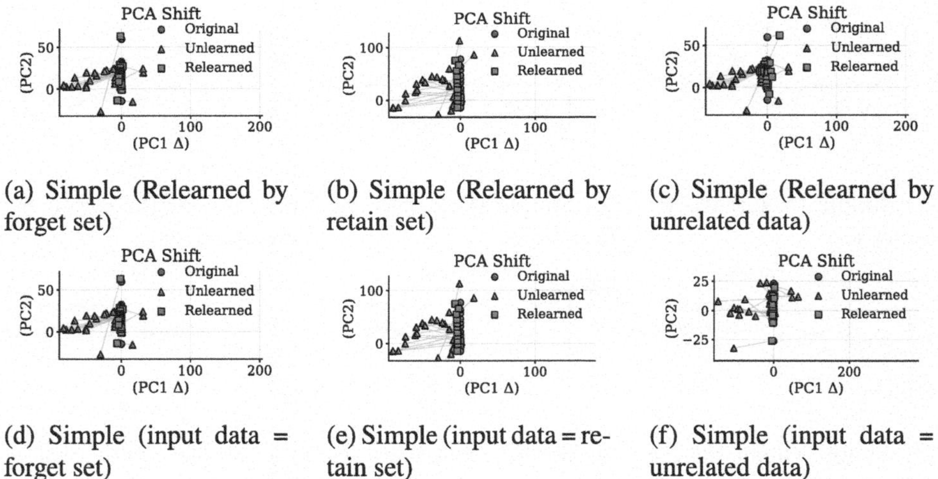

Fig. 11.5 PCA shift analysis under varied relearning and evaluation inputs on Yi-6B (Simple Task). **a–c** Relearning is performed using the forget set, retain set, or unrelated data respectively. **d–f** PCA shift is measured using the forget set, retain set, or unrelated data as evaluation input

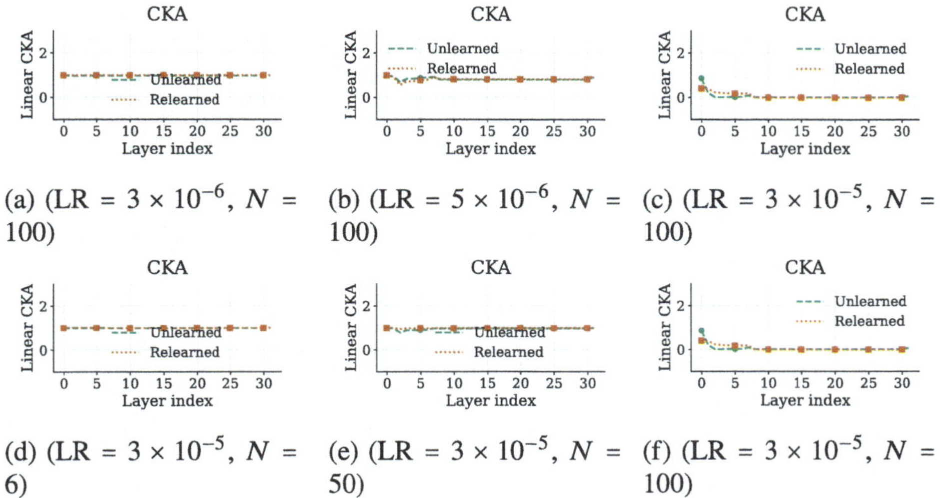

(a) (LR = 3×10^{-6}, N = 100) (b) (LR = 5×10^{-6}, N = 100) (c) (LR = 3×10^{-5}, N = 100)

(d) (LR = 3×10^{-5}, N = 6) (e) (LR = 3×10^{-5}, N = 50) (f) (LR = 3×10^{-5}, N = 100)

Fig. 11.6 CKA for GA on Yi-6B, simple task. **a–c** LR $\{3 \times 10^{-6}, 5 \times 10^{-6}, 3 \times 10^{-5}\}$ with $N = 100$; **d–f** LR $= 3 \times 10^{-5}$ with $N \in \{6, 50, 100\}$. High CKA (near 1) indicates strong subspace alignment and *reversible* (catastrophic) forgetting, whereas low CKA (near 0) reflects severe representational drift and *irreversible* (catastrophic) forgetting

11.4.2 Representational Results

Principal Component Analysis: Similarity and Shift. Figures 11.2 and 11.4 show that, in continual unlearning, larger learning rates or more removal requests drive PCA Similarity sharply downward and result in large, unrecovered PCA Shifts, providing clear evidence of *irreversible catastrophic forgetting*. Under gentler hyperparameters, similarity remains high and shifts stay bounded; relearning then restores both metrics, indicating *reversible catastrophic forgetting*. Notably, the choice of relearning data and probe data, whether forget, retain, or even unrelated, makes little difference: all three restore the feature geometry, suggesting that knowledge was suppressed rather than erased (Figs. 11.3 and 11.5).

Centered Kernel Alignment Analysis. Figure 11.6 tracks layer-wise CKA. Under mild unlearning, CKA remains close to 1 and is fully restored after relearning, which is characteristic of *reversible catastrophic forgetting*. Stronger updates or a larger number of requests drive deep-layer CKA significantly lower, and alignment cannot be completely recovered, signaling *irreversible catastrophic forgetting*. The choice of relearning or probe data, whether forget, retain, or unrelated, has little effect on this outcome. Once relearning begins, latent structure resurfaces regardless of input (Fig. 11.7).

Fisher Information Analysis. Continual unlearning progressively flattens the loss landscape (Fig. 11.8. Higher learning rates or larger request counts push the FIM spectra sharply left, and, at extreme settings, the shift persists after relearning, signalling *irreversible catas-*

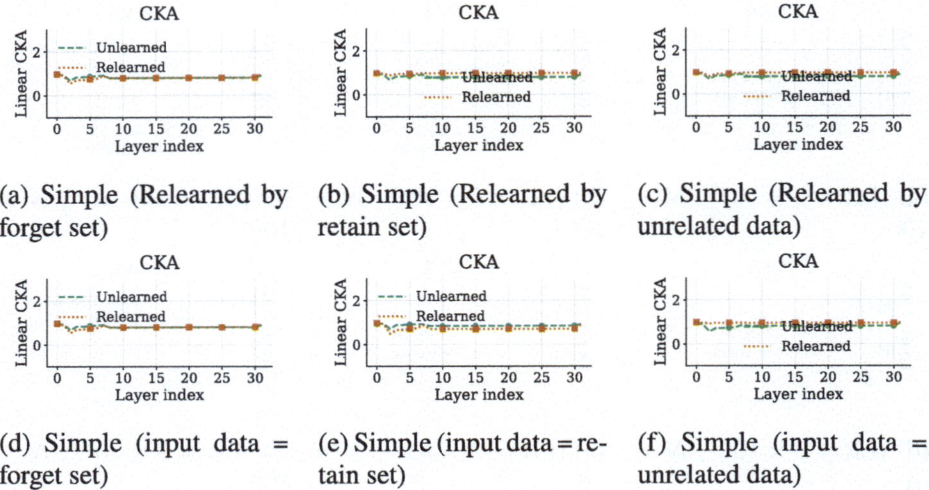

Fig. 11.7 CKA analysis under varied relearning and evaluation inputs on Yi-6B (simple task). **a–c** Relearning is performed using the forget set, retain set, or unrelated data respectively. **d–f** CKA is measured using the forget set, retain set, or unrelated data as evaluation input

trophic forgetting. Under gentler hyper-parameters the spectra recentre, indicating *reversible catastrophic forgetting*. Relearning with forget, retain, or unrelated data realigns the FIM almost equally well, confirming that the lost sensitivity is suppressed rather than erased (Figs. 11.9, 11.10, 11.11, 11.12 and 11.13)

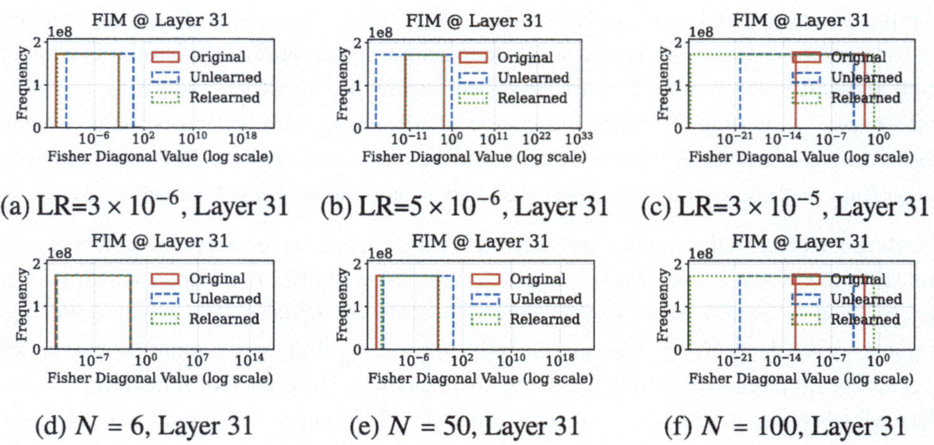

Fig. 11.8 FIM (layer 31) for GA on Yi-6B, simple task. **a–c** LR $\{3 \times 10^{-6}, 5 \times 10^{-6}, 3 \times 10^{-5}\}$ with $N = 100$; **d–f** LR $= 3 \times 10^{-5}$ with $N \in \{6, 50, 100\}$. Larger leftward shifts indicate a greater flattening of the loss landscape and *irreversible* (catastrophic) forgetting; spectra that recenter on the original peak denote *reversible* (catastrophic) forgetting

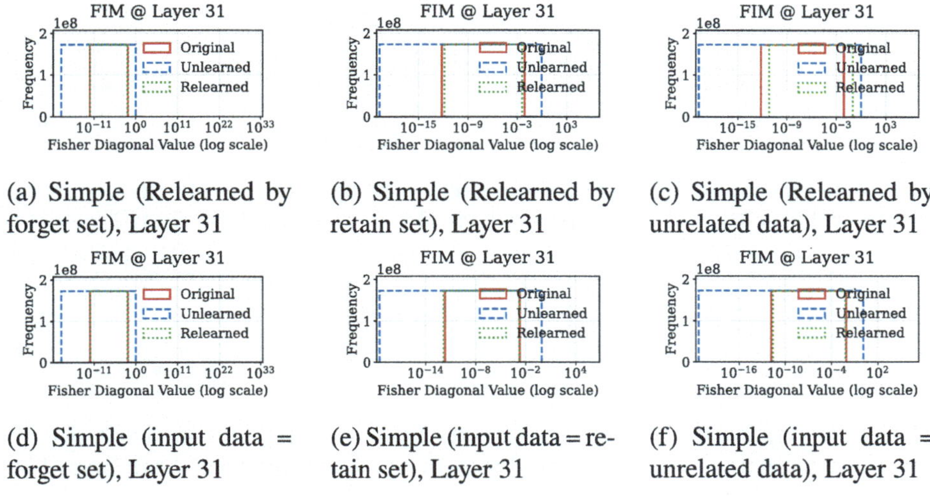

(a) Simple (Relearned by forget set), Layer 31

(b) Simple (Relearned by retain set), Layer 31

(c) Simple (Relearned by unrelated data), Layer 31

(d) Simple (input data = forget set), Layer 31

(e) Simple (input data = retain set), Layer 31

(f) Simple (input data = unrelated data), Layer 31

Fig. 11.9 FIM in layer 31 under varied relearning and evaluation inputs on Yi-6B (simple task). **a–c** Relearning is performed using the forget set, retain set, or unrelated data respectively. **d–f** FIM is measured using the forget set, retain set, or unrelated data as evaluation input

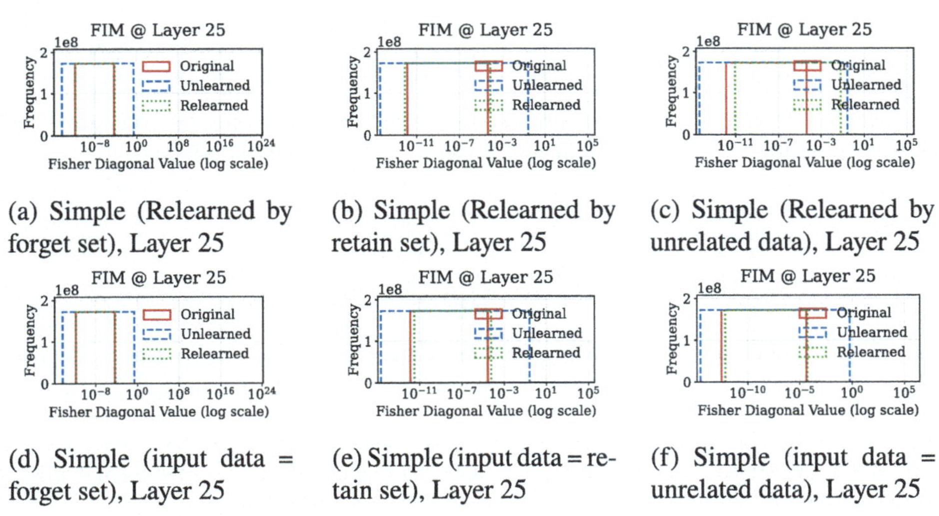

(a) Simple (Relearned by forget set), Layer 25

(b) Simple (Relearned by retain set), Layer 25

(c) Simple (Relearned by unrelated data), Layer 25

(d) Simple (input data = forget set), Layer 25

(e) Simple (input data = retain set), Layer 25

(f) Simple (input data = unrelated data), Layer 25

Fig. 11.10 FIM in layer 25 under varied relearning and evaluation inputs on Yi-6B (simple task). **a–c** Relearning is performed using the forget set, retain set, or unrelated data respectively. **d–f** FIM is measured using the forget set, retain set, or unrelated data as evaluation input

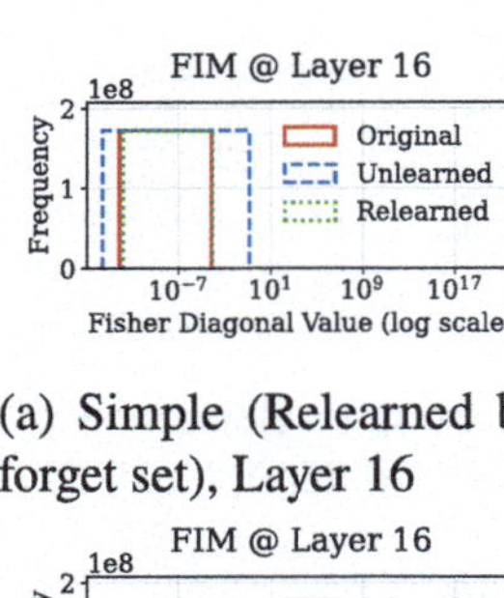

(a) Simple (Relearned by forget set), Layer 16

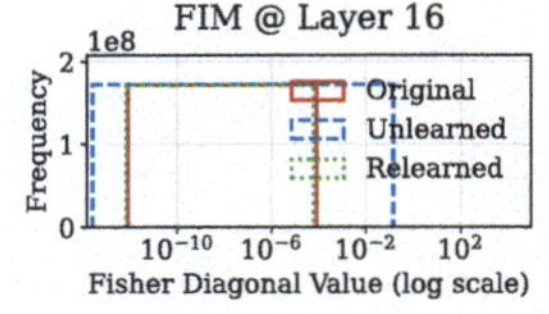

(b) Simple (Relearned by retain set), Layer 16

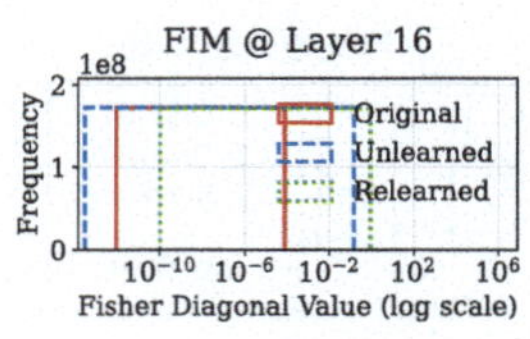

(c) Simple (Relearned by unrelated data), Layer 16

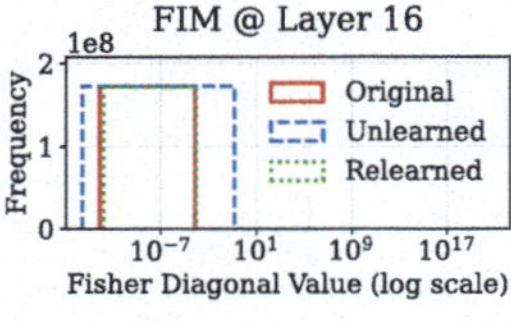

(d) Simple (input data = forget set), Layer 16

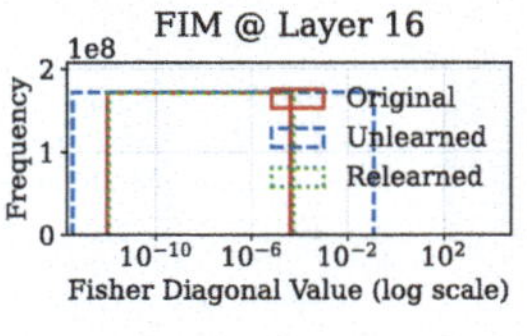

(e) Simple (input data = retain set), Layer 16

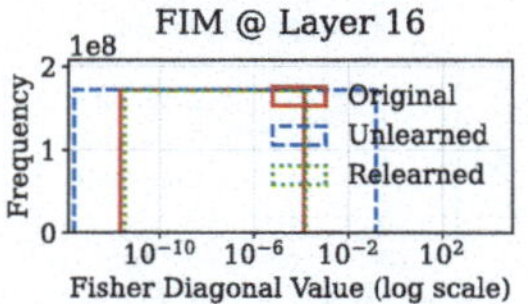

(f) Simple (input data = unrelated data), Layer 16

Fig. 11.11 FIM in layer 16 under varied relearning and evaluation inputs on Yi-6B (simple task). **a–c** Relearning is performed using the forget set, retain set, or unrelated data respectively. **d–f** FIM is measured using the forget set, retain set, or unrelated data as evaluation input

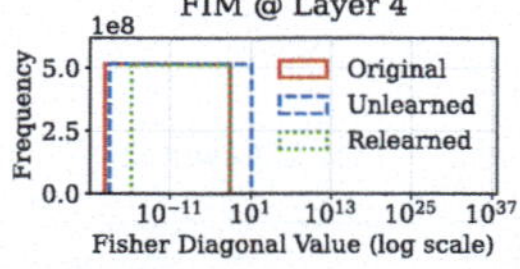

(a) Simple (Relearned by forget set), Layer 4

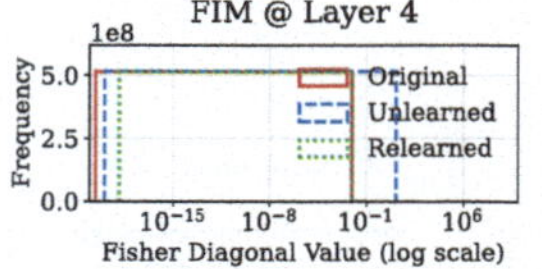

(b) Simple (Relearned by retain set), Layer 4

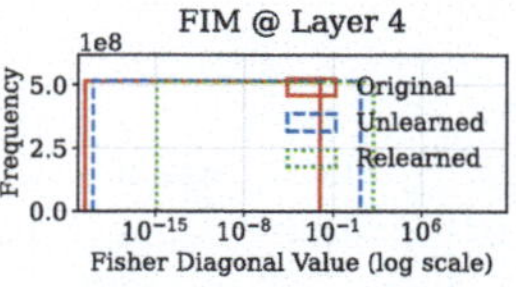

(c) Simple (Relearned by unrelated data), Layer 4

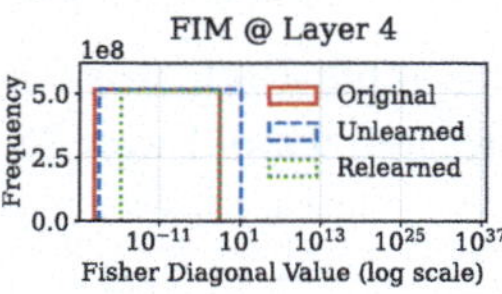

(d) Simple (input data = forget set Layer), 4

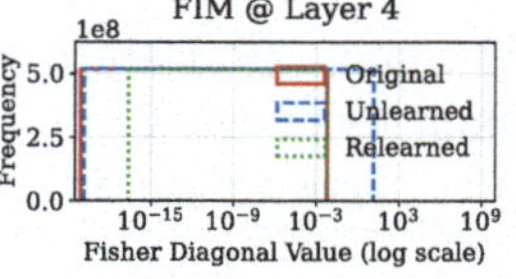

(e) Simple (input data = retain set Layer), 4

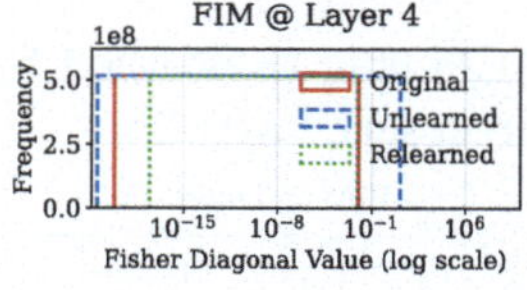

(f) Simple (input data = unrelated data), Layer 4

Fig. 11.12 FIM in layer 4 under varied relearning and evaluation inputs on Yi-6B (simple task). **a–c** Relearning is performed using the forget set, retain set, or unrelated data respectively. **d–f** FIM is measured using the forget set, retain set, or unrelated data as evaluation input

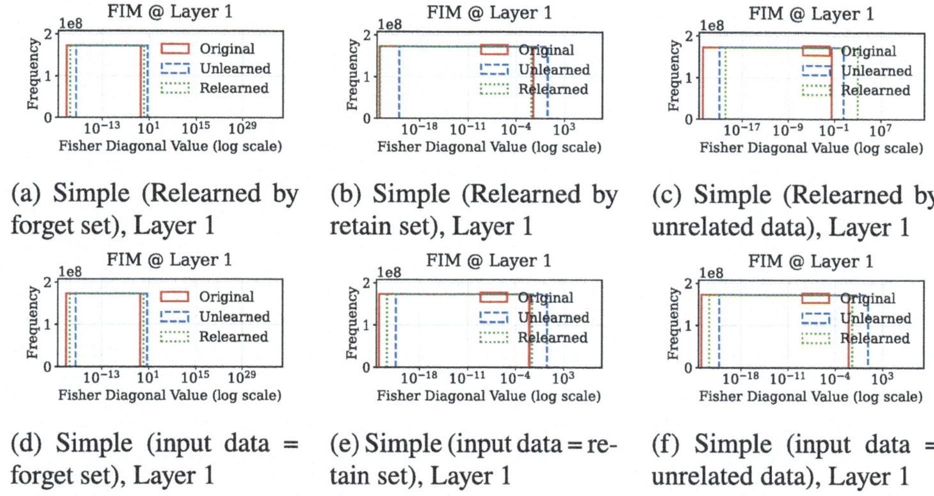

(a) Simple (Relearned by forget set), Layer 1 (b) Simple (Relearned by retain set), Layer 1 (c) Simple (Relearned by unrelated data), Layer 1

(d) Simple (input data = forget set), Layer 1 (e) Simple (input data = retain set), Layer 1 (f) Simple (input data = unrelated data), Layer 1

Fig. 11.13 FIM in layer 1 under varied relearning and evaluation inputs on Yi-6B (simple task). **a–c** Relearning is performed using the forget set, retain set, or unrelated data respectively. **d–f** FIM is measured using the forget set, retain set, or unrelated data as evaluation input

11.4.3 Representational Theoretical Analysis

To explain the empirical distinction between *reversible* and *irreversible* (catastrophic) forgetting, this section presents a perturbation model that links unlearning updates to structural collapse across layers. Consider an L-layer feedforward neural network $f(x) = \sigma(W_L\,\sigma(\ldots\sigma(W_1 x)\ldots))$, with activations σ and weights $W_i{}_{i=1}^{L}$. Unlearning is modeled as layer-wise perturbations $\widetilde{W}_i = W_i + E_i$ with $|E_i| = O(\mathrm{LR}, N)$, where LR is the learning rate and N the number of unlearning steps. A Neumann-series expansion yields $\widetilde{f}(x) - f(x) = \sum_{\emptyset \neq S \subseteq \{1,\ldots,L\}} (W_L \circ \cdots \circ E_{i_k} \circ \cdots \circ W_1)(x)$.

When small perturbations are confined to a few layers, first-order terms dominate and the effect corresponds to *reversible (catastrophic) forgetting*. In contrast, when comparable perturbations spread across many layers, higher-order terms accumulate, leading to *irreversible (catastrophic) forgetting*.

PCA Similarity. Let X_i and $Y_i = X_i + E_i'$ be the centered activations at layer i before and after unlearning. By the Davis-Kahan theorem [37], $\cos\angle(\mathbf{c}_i^{\mathrm{orig}}, \mathbf{c}_i^{\mathrm{upd}}) \approx 1 - O(\|E_i'\|/(\lambda_{1,i} - \lambda_{2,i}))$, with top two eigenvalues $\lambda_{1,i}, \lambda_{2,i}$. The layer-averaged PCA similarity is $\bar{S}_{\mathrm{PCA}} \approx 1 - O((1/L)\sum_i \|E_i'\|)$.

PCA Shift. Along the first principal component, the activation-centroid shift is $p_{i,12} = O(\|E_i'\|)$. Large perturbations $\|E_i'\|$ spanning many layers lead to irreversible representational drift, while smaller perturbations remain localized and reversible.

CKA. Let $\widetilde{K}_{Y_i} = \widetilde{K}_{X_i} + \Delta K_i$ be the perturbed Gram matrix. Then, CKA_i is computed as $1 - O(\|\Delta K_i\|_*/\|\widetilde{K}_{X_i}\|_*)$, which implies that the layer-averaged CKA is $\bar{C} \approx 1 - O((1/L)\sum_i \|\Delta K_i\|_*)$.

Fisher Information. Given update $\delta w_i = O(\|E_i\|)$, the Fisher diagonal behaves as $F_{ii}(w + \delta w) = F_{ii}(w) + O(\|\delta w_i\|)$. Thus, the average Fisher becomes $\bar{F} = (1/P)\sum_i F_{ii} = F_0 - O((1/P)\sum_i \|E_i\|)$.

Token-level metrics, such as accuracy, MIA, and AUC, may indicate total collapse even when the model's internal geometry is largely preserved. In *reversible* forgetting, a few parameter changes (for example, in output heads or layer norms) can drastically perturb token probabilities while leaving deeper representations intact. For the soft-max output $\log p(y|x; \theta + \delta\theta) \approx \log p(y|x; \theta) + \nabla_\theta \log p(y|x; \theta)^\top \delta\theta + O(\|\delta\theta\|^2)$, a small $\delta\theta$ in high-sensitivity regions (near the logits) can dominate the first-order term, producing large drops in accuracy or anomalous AUC scores despite minimal representational drift.

Unlearning on the forget set applies a weight update $E_i = LR \times \nabla_{W_i} L(D_f)$, which both removes overfitting to $\mathcal{D}_f$ and accentuates its principal feature subspace. After relearning, parameters may return close to their original values, yet amplified patterns persist, sometimes yielding *better* performance on an augmented $\mathcal{D}_f$ than the baseline. Therefore, unlearning can inadvertently act as a contrastive regularizer, further illustrating the mismatch between surface-level metrics and feature subspaces.

Summary. *Reversible (catastrophic) forgetting* occurs when perturbations are restricted to a few layers, leaving PCA similarity/shift, CKA, and FIM near baseline. In contrast, *irreversible (catastrophic) forgetting* emerges from large, distributed updates that collapse the model's representational structure.

Unlearning functions as a contrastive perturbation: it removes memorized content while simultaneously reinforcing salient features of the forget set. As a result, subsequent relearning can even outperform the original model on related inputs, highlighting its "dual role" in both removal and refinement.

Token-level metrics (e.g., accuracy, MIA) are overly sensitive to small shifts in high-impact parameters and can misclassify the regime. Structural diagnostics, together with augmented evaluation, provide a more reliable assessment of whether forgetting is truly irreversible.

11.5 Discussion and Takeaways

Beyond the theoretical justifications in Sect. 11.4, this section summarizes the main empirical and analytical insights.

(1) **Single versus continual unlearning, and the role of GA/RLabel**. Single unlearning rarely produces *irreversible* collapse: performance is recoverable and representational

drift remains slight. Continual unlearning, particularly under large learning rates, often drives the model into permanent failure. For example, $\sim$ 100 sequential requests can reduce both forget- and retain-set accuracy to nearly zero. GA and RLabel already tend to over-forget in the single scenario, and this effect is magnified when applied continually. Adding retain-set terms, as in GA+KL or NPO(+KL) [7, 10, 18], markedly improves stability.

(2) **Collapse stems from structural drift rather than true erasure.** PCA Similarity/Shift, CKA, and the FIM consistently reveal that irreversible collapse coincides with large rotations of principal directions, centroid shifts, and vanishing Fisher mass across many layers. When perturbations remain local (small LR, few unlearning requests), these diagnostics stay near baseline, reflecting *reversible* forgetting. Token-level metrics alone are unreliable: small updates to high-sensitivity parameters (such as logits or layer norms) can reduce accuracy or inflate MIA AUC, even when internal geometry remains intact.

(3) **Unlearning can act as implicit augmentation.** In several continual runs, subsequent relearning on the forget set yields *higher* accuracy than that of the original model. This outcome suggests that unlearning is not merely a deletion mechanism but also functions as a form of implicit contrastive regularization. As detailed in Sect. 11.4, unlearning amplifies the feature subspace associated with the forget set, and relearning on augmented inputs can reinforce semantic structure while promoting robustness. This process reorganizes internal representations to capture more generalizable patterns, resembling a form of curriculum learning.

(4) **Diagnostics guide irreversible (benign) forgetting.** Tools such as PCA Similarity and Shift, CKA, and Fisher Information reveal not only the presence of structural drift but also where and how it arises across layers. This enables targeted control: effective unlearning can be directed toward perturbing only those parameters associated with the forget set, while preserving the representational structure on the retain set and unrelated data. Such selective intervention opens a path to *targeted, irreversible forgetting*, that is, a permanent and isolated removal of information without collateral collapse, offering actionable insights for safer unlearning algorithms.

To conclude, this chapter revisits machine unlearning for LLMs through a systematic study of *reversibility*. Token-level metrics alone can be misleading: a model may appear collapsed yet remain fully recoverable.

To diagnose this gap, this chapter introduces a feature-space toolkit—PCA Similarity, PCA Shift, CKA, and FIM—that cleanly separates *reversible* from *irreversible* catastrophic forgetting. Empirical and theoretical analyses show that true forgetting arises only when many layers undergo coordinated, large-magnitude perturbations. By contrast, minor updates in high-sensitivity regions (such as output logits) can reduce accuracy or inflate perplexity while leaving internal representations intact.

These findings call for evaluation protocols that move beyond token-level scores and for algorithms that actively control representational drift. It is further observed that unlearning, followed by proper relearning, can refine representations and even improve downstream performance. Together, these insights chart a path toward safer and more interpretable unlearning in LLMs.

References

1. Nasr, M., Carlini, N., Hayase, J., Jagielski, M., Cooper, A.F., Ippolito, D., Choquette-Choo, C.A., Wallace, E., Tramèr, F., Lee, K.: Scalable extraction of training data from (production) language models (2023). arXiv:2311.17035
2. Karamolegkou, A., Li, J., Zhou, L., Søgaard, A.: Copyright violations and large language models (2023). arXiv:2310.13771
3. Wen, J., Ke, P., Sun, H., Zhang, Z., Li, C., Bai, J., Huang, M.: Unveiling the implicit toxicity in large language models. In: The 2023 Conference on Empirical Methods in Natural Language Processing (2023)
4. Ginart, A., Guan, M., Valiant, G., Zou, J.Y.: Making ai forget you: Data deletion in machine learning. In: Advances in Neural Information Processing Systems, vol. 32 (2019)
5. Bourtoule, L., Chandrasekaran, V., Choquette-Choo, C.A., Jia, H., Travers, A., Zhang, B., Lie, D., Papernot, N.: Machine unlearning. In: 2021 IEEE Symposium on Security and Privacy (SP), pp. 141–159. IEEE (2021)
6. Yao, Y., Xu, X., Liu, Y.: Large language model unlearning (2023). arXiv:2310.10683
7. Yao, J., Chien, E., Du, M., Niu, X., Wang, T., Cheng, Z., Yue, X.: Machine unlearning of pre-trained large language models (2024). arXiv:2402.15159
8. Jang, J., Yoon, D., Yang, S., Cha, S., Lee, M., Logeswaran, L., Seo, M.: Knowledge unlearning for mitigating privacy risks in language models (2022). arXiv:2210.01504
9. Eldan, R., Russinovich, M.: Who's harry potter? approximate unlearning in LLMs (2023). arXiv:2310.02238
10. Xu, X., Du, M., Ye, Q., Hu, H.: Obliviate: Robust and practical machine unlearning for large language models (2025). arXiv:2505.04416
11. Li, N., Pan, A., Gopal, A., Yue, S., Berrios, D., Gatti, A., Li, J.D., Dombrowski, A.K., Goel, S., Mukobi, G., Helm-Burger, N., Lababidi, R., Justen, L., Liu, A.B., Chen, M., Barrass, I., Zhang, O., Zhu, X., Tamirisa, R., Bharathi, B., Herbert-Voss, A., Breuer, C.B., Zou, A., Mazeika, M., Wang, Z., Oswal, P., Lin, W., Hunt, A.A., Tienken-Harder, J., Shih, K.Y., Talley, K., Guan, J., Steneker, I., Campbell, D., Jokubaitis, B., Basart, S., Fitz, S., Kumaraguru, P., Karmakar, K.K., Tupakula, U., Varadharajan, V., Shoshitaishvili, Y., Ba, J., Esvelt, K.M., Wang, A., Hendrycks, D.: The WMDP benchmark: Measuring and reducing malicious use with unlearning. In: Proceedings of the 41st International Conference on Machine Learning, Proceedings of Machine Learning Research, vol. 235, pp. 28525–28550. PMLR (2024)
12. Lo, M., Barez, F., Cohen, S.: Large language models relearn removed concepts. In: Findings of the Association for Computational Linguistics: ACL 2024, pp. 8306–8323. Association for Computational Linguistics (2024)
13. Lynch, A., Guo, P., Ewart, A., Casper, S., Hadfield-Menell, D.: Eight methods to evaluate robust unlearning in LLMS (2024). arXiv:2402.16835
14. Zhang, Z., Wang, F., Li, X., Wu, Z., Tang, X., Liu, H., He, Q., Yin, W., Wang, S.: Catastrophic failure of LLM unlearning via quantization. In: ICLR (2025)

15. Zheng, J., Cai, X., Qiu, S., Ma, Q.: Spurious forgetting in continual learning of language models. In: ICLR (2025)
16. Kornblith, S., Norouzi, M., Lee, H., Hinton, G.E.: Similarity of neural network representations revisited. In: ICML, pp. 3519–3529 (2019)
17. Cha, S., Cho, S., Hwang, D., Lee, M.: Towards robust and cost-efficient knowledge unlearning for large language models. In: ICLR (2025)
18. Zhang, R., Lin, L., Bai, Y., Mei, S.: Negative preference optimization: From catastrophic collapse to effective unlearning. In: First Conference on Language Modeling (2024). https://openreview.net/forum?id=MXLBXjQkmb
19. LI, J., Beeching, E., Tunstall, L., Lipkin, B., Soletskyi, R., Huang, S.C., Rasul, K., Yu, L., Jiang, A., Shen, Z., Qin, Z., Dong, B., Zhou, L., Fleureau, Y., Lample, G., Polu, S.: Numinamath (2024). https://huggingface.co/AI-MO/NuminaMath-1.5, https://github.com/project-numina/aimo-progress-prize/blob/main/report/numina_dataset.pdf
20. Young, A., Chen, B., Li, C., Huang, C., Zhang, G., Zhang, G., Li, H., Zhu, J., Chen, J., Chang, J., Yu, K., Liu, P., Liu, Q., Yue, S., Yang, S., Yang, S., Yu, T., Xie, W., Huang, W., Hu, X., Ren, X., Niu, X., Nie, P., Xu, Y., Liu, Y., Wang, Y., Cai, Y., Gu, Z., Liu, Z., Dai, Z.: Yi: Open foundation models by 01.AI (2024). arXiv:2403.04652
21. Yang, A., Yang, B., Zhang, B., Hui, B., Zheng, B., Yu, B., Li, C., Liu, D., Huang, F., Wei, H., Lin, H., Yang, J., Tu, J., Zhang, J., Yang, J., Yang, J., Zhou, J., Lin, J., Dang, K., Lu, K., Bao, K., Yang, K., Yu, L., Li, M., Xue, M., Zhang, P., Zhu, Q., Men, R., Lin, R., Li, T., Xia, T., Ren, X., Ren, X., Fan, Y., Su, Y., Zhang, Y., Wan, Y., Liu, Y., Cui, Z., Zhang, Z., Qiu, Z.: Qwen2.5 Technical Report (2024). arXiv:2412.15115
22. Shi, W., Ajith, A., Xia, M., Huang, Y., Liu, D., Blevins, T., Chen, D., Zettlemoyer, L.: Detecting pretraining data from large language models (2023). arXiv:2310.16789
23. Pawelczyk, M., Neel, S., Lakkaraju, H.: In-context unlearning: language models as few-shot unlearners. In: Proceedings of the 41st International Conference on Machine Learning (2024)
24. Li, J., Wei, Q., Zhang, C., Qi, G., Du, M., Chen, Y., Bi, S., Liu, F.: Single image unlearning: Efficient machine unlearning in multimodal large language models. In: NeurIPS (2024)
25. Barez, F., Fu, T., Prabhu, A., Casper, S., Sanyal, A., Bibi, A., O'Gara, A., Kirk, R., Bucknall, B., Fist, T., et al.: Open problems in machine unlearning for ai safety (2025). arXiv:2501.04952
26. Gao, C., Wang, L., Ding, K., Weng, C., Wang, X., Zhu, Q.: On large language model continual unlearning. In: ICLR (2025)
27. Maini, P., Feng, Z., Schwarzschild, A., Lipton, Z.C., Kolter, J.Z.: TOFU: a task of fictitious unlearning for LLMs. In: First Conference on Language Modeling (2024). https://openreview.net/forum?id=B41hNBoWLo
28. Gandikota, R., Feucht, S., Marks, S., Bau, D.: Erasing conceptual knowledge from language models (2024). arXiv:2410.02760
29. Shi, W., Lee, J., Huang, Y., Malladi, S., Zhao, J., Holtzman, A., Liu, D., Zettlemoyer, L., Smith, N.A., Zhang, C.: Muse: Machine unlearning six-way evaluation for language models (2024). arXiv:2407.06460
30. Talmor, A., Herzig, J., Lourie, N., Berant, J.: CommonsenseQA: a question answering challenge targeting commonsense knowledge. In: NAACL, pp. 4149–4158 (2019)
31. Cobbe, K., Kosaraju, V., Bavarian, M., Chen, M., Jun, H., Kaiser, L., Plappert, M., Tworek, J., Hilton, J., Nakano, R., et al.: Training verifiers to solve math word problems (2021). arXiv:2110.14168
32. Hendrycks, D., Burns, C., Kadavath, S., Arora, A., Basart, S., Tang, E., Song, D., Steinhardt, J.: Measuring mathematical problem solving with the MATH dataset. In: NeurIPS Datasets and Benchmarks (2021)

33. Touvron, H., Martin, L., Stone, K., Albert, P., Almahairi, A., Babaei, Y., Bashlykov, N., Batra, S., Bhargava, P., Bhosale, S., et al.: Llama 2: Open foundation and fine-tuned chat models (2023). arXiv:2307.09288
34. Loshchilov, I., Hutter, F.: Decoupled weight decay regularization (2017). arXiv:1711.05101
35. Kirkpatrick, J., Pascanu, R., Rabinowitz, N.C., Veness, J., Desjardins, G., Rusu, A.A., Milan, K., Quan, J., Ramalho, T., Grabska-Barwinska, A., Hassabis, D., Clopath, C., Kumaran, D., Hadsell, R.: Overcoming catastrophic forgetting in neural networks (2016). arXiv:1612.00796
36. Hsu, Y., Hua, T., Chang, S., Lou, Q., Shen, Y., Jin, H.: Language model compression with weighted low-rank factorization. In: ICLR (2022)
37. Davis, C., Kahan, W.M.: The rotation of eigenvectors by a perturbation. iii. SIAM J. Numer. Anal. **7**(1), 1–46 (1970)

12 Instability of Retraining as a Benchmark for Machine Unlearning Evaluation

Chris Yuhao Liu and Zonglin Di

Abstract

Machine unlearning has become a pivotal paradigm for mitigating the impact of unwanted data points incorporated during model training. Despite recent progress in unlearning algorithms, rigorously assessing their effectiveness remains challenging. Retraining a model from scratch—though the very process many algorithms aim to avoid—continues to serve as a widely adopted benchmark because it inherently excludes undesirable data. This chapter critically examines the role of retraining as a benchmark for evaluating machine unlearning. Through extensive experiments spanning diverse metrics, datasets, and model architectures, it reveals significant instability in retrained models. In particular, retraining can produce both the strongest and weakest unlearning performance while maintaining comparable utility. This pattern persists across multiple membership inference attacks and distance-based metrics. Although these findings do not diminish retraining's value as an unlearning method that guarantees complete data removal, they highlight the importance of applying it with caution and nuance as an evaluation standard.

12.1 Retraining as the Gold Standard in Machine Unlearning Evaluation

With the rise of deep learning, modern neural networks are often trained on massive datasets—ranging from millions of images [1] to billions of tokens [2]. However, such

C. Y. Liu (✉) · Z. Di
University of California, Santa Cruz, USA
e-mail: yliu298@ucsc.edu

Z. Di
e-mail: zdi@ucsc.edu

S. Liu et al. (eds.), *Machine Unlearning for Governance of Foundation Models*, Synthesis Lectures on Computer Vision, https://doi.org/10.1007/978-3-032-17282-2_12

datasets frequently contain data points that are sensitive [3–5], outdated [6], or potentially harmful [7]. Directly removing the influence of these learned samples from deep neural networks remains a non-trivial challenge. Machine Unlearning (MU) has thus emerged as a novel paradigm that, rather than acquiring knowledge from data, aims to reduce or eliminate the influence of specific undesirable samples on a pre-trained model [8–10]. Despite substantial progress in developing efficient approximate unlearning methods, rigorously evaluating their performance and auditing the final unlearned models continues to be a formidable challenge [11–13].

While MU primarily seeks to eliminate the need for retraining, retraining nonetheless remains the de facto evaluation standard in this field, often serving as the benchmark for comparing proposed approximate MU methods [8, 9, 14, 15]. However, the process of training deep neural networks via loss function optimization is inherently non-deterministic [16–18]. Moreover, prior studies typically evaluate MU algorithms by comparing them to a single instance of a retrained model (albeit across multiple random seeds), using various metrics to assess unlearning performance and model utility. This practice raises concerns about the robustness and reliability of such evaluation protocols.

This chapter investigates retraining across a wide range of settings and identifies fundamental limitations in using retrained models as the evaluation standard for benchmarking approximate MU algorithms. Extensive experiments reveal that retrained models can achieve both the best and worst unlearning performance while maintaining identical utility. By employing diverse evaluation metrics, datasets, model architectures, and modalities in both class-wise and random unlearning settings, the study confirms that this instability consistently appears across nearly all tested scenarios. The chapter first demonstrates the pronounced variability in the effectiveness of retrained models as a gold standard for machine unlearning evaluation—showing that even slight differences in retraining can lead to extreme variations in unlearning performance despite similar utility. It then provides an in-depth analysis of retrained models across multiple experimental configurations, offering a comprehensive understanding of their role as an evaluation benchmark for other machine unlearning algorithms.

12.2 Preliminary and Notations

The primary objective of machine unlearning (MU) is to eliminate the influence of a specific subset of data points from a pre-trained model. Following the literature [8–10], unlearning is categorized into two distinct settings: *exact* unlearning and *approximate* unlearning, for which they provide formal definitions below.

Definition 12.1 (Exact Unlearning) Given a learning algorithm $A(\cdot)$, the example space $\mathcal{D}$ and the hypothesis space of models $\mathcal{H}$, the process $U_e(\cdot)$ is an exact MU process if and only if $\forall \mathcal{T} \subseteq \mathcal{H}$, $D_{tr} \in \mathcal{D}$, $D_f \subset D_{tr}$ can satisfy

$$\mathbb{P}[A(D_{tr} \backslash D_f) \in \mathcal{T}] = \mathbb{P}[U_e(A(D_{tr}), D_{tr}, D_f) \in \mathcal{T}],$$

where D_{tr} and D_f are the training set and forget set, respectively.

$U_e(\cdot)$ represents a particular instance of the more general unlearning process $U(\cdot)$. Consider three sets of model parameters: (1) $\boldsymbol{\theta}_{tr}$ obtained by training on D_{tr} using algorithm $A(\cdot)$, (2) $\boldsymbol{\theta}_r$ obtained by training on the retain set $D_r = D_{tr} \backslash D_f$ using algorithm $A(\cdot)$, and (3) $\boldsymbol{\theta}_u$ obtained through the unlearning procedure $U(\cdot)$. Exact unlearning requires that both the distribution of model parameters Θ and the distribution of model outputs be identical, meaning $\Theta_r = \Theta_u$ and $\mathbb{P}(M(X; \boldsymbol{\theta}_r)) = \mathbb{P}(M(X; \boldsymbol{\theta}_u)), \forall X \subseteq \mathcal{D}$, where M denotes the model's inference operation on the data. This characterization of exact unlearning is consistent with the definition presented in [19], as specified in Definition 12.1.

Definition 12.2 ((ϵ, δ)-Approximate Unlearning) For a fixed training dataset D_{tr}, forget set $D_f \subseteq D_{tr}$ and a randomized learning algorithm $A : \mathcal{D} \to \mathcal{R}$ where $R \in \mathcal{R}$ is the output region, an approximate unlearning algorithm U is (ϵ, δ)-unlearning with respect to (D_{tr}, D_f, A) if for all regions $R \subseteq \mathcal{R}$, then

$$\mathbb{P}[A(D_r) \in R] \le e^{\epsilon}\mathbb{P}[U(A(D_{tr}), D_f, D_{tr}) \in R] + \delta$$
$$\mathbb{P}[U(A(D_{tr}), D_f, D_{tr}) \in R] \le e^{\epsilon}\mathbb{P}[U(A(D_r) \in R] + \delta$$

Here, D_f can be a subset of the samples from the pre-training data, or simply a single sample $z \in D_f$, indicating the instance-wise unlearning [20].

When ϵ and δ are small, $\mathbb{P}[A(D_r)]$ and $\mathbb{P}[U(A(D_{tr}), D_f, D_{tr})]$ are nearly indistinguishable from one another, meaning that the success of unlearning.

12.2.1 Machine Unlearning Evaluation Metrics

This section visits several widely used evaluation metrics in evaluating machine unlearning algorithms.

Accuracy serves as a fundamental metric in MU. Nearly all existing studies measure accuracy across three datasets: the retain set D_r and test set D_{te}, which reflect the preservation of model utility, as well as the forget set D_f, which reveals the extent of prediction changes following the unlearning process. This chapter examines a recently introduced composite accuracy metric called the adaptive unlearning score [21], which is defined as follows.

Adaptive unlearning score (AUS) [21] is a metric accounting for both the test accuracy $\mathcal{A}_{te}$ and the forget accuracy $\mathcal{A}_f$ of the unlearned model. Formally,

$$\text{AUS} = \frac{1 - (\mathcal{A}_{te}^{Or} - \mathcal{A}_{te})}{1 + \Delta}, \quad \Delta = \begin{cases} |0 - \mathcal{A}_f| & \text{class-wise} \\ |\mathcal{A}_{te} - \mathcal{A}_f| & \text{random} \end{cases}$$

where $\mathcal{A}_{te}^{Or}$ is the test accuracy of the original pre-trained model. Δ is the performance gap between the unseen dataset and D_f. AUS is the range of [0, 2].

Membership inference attack (MIA) [22, 23] has become a prevalent method for evaluating the unlearning quality of MU algorithms and models, particularly when conventional machine learning metrics like accuracy fail to provide sufficient insight. Although these attack methodologies vary in their experimental configurations, they can be categorized into two primary types. The first category, referred to as MIA-F, assumes the attacker can access and distinguish the unlearned model's outputs on both the **F**orget set D_f and the test set D_{te} [15, 21, 23–28]. More precisely, given an unlearned model $\boldsymbol{\theta}_u$, datasets $D_f = \{x_i^f, y_i^f\}_{i=1}^N$ and $D_{te} = \{x_i^{te}, y_i^{te}\}_{i=1}^N$, and a collection of model outputs paired with their membership labels $\mathcal{S} = \{M(x_i; \boldsymbol{\theta}_u), y_i^{mem}\}_{i=1}^{2N}$ (where y^{mem} represents the membership label $\mathbb{I}[x_i \in D_f]$ and $\mathbb{I}[\cdot]$ denotes the indicator function), the attacker a's accuracy is computed as

$$\text{MIA-F}(x, a, \boldsymbol{\theta}_u) = \frac{1}{2N} \sum_{x \in D_f \cup D_{te}} \mathbb{I}[a(M(x; \boldsymbol{\theta}_u)) = y^{mem}].$$

In this setting, a prediction of 1 signifies membership, whereas 0 signifies non-membership. The attacker's objective is to distinguish between outputs generated from D_f and those from D_{te}. When the attacker fails to make this distinction, as evidenced by an accuracy approaching 50%, it suggests that the model exhibits no detectable membership difference in its predictions on D_f versus D_{te}. Put differently, the attacker cannot determine, from the unlearned model's outputs alone, whether a sample originated from the forget set or represents an unseen sample, which signals successful unlearning. In implementation, the attacker is trained on a subset of $\mathcal{S}$, and the MIA-F score is evaluated on the remaining samples.

The second MIA approach trains an attacker to differentiate between the unlearned model's outputs on the **R**etain set, D_r, and the test set, D_{te} [14, 29–32]. The attacker subsequently attempts to infer membership for samples in D_f. This methodology is designated as MIA-R. Analogously, given an unlearned model $\boldsymbol{\theta}_u$ and datasets $D_r = \{x_i^r, y_i^r\}_{i=1}^N$ and $D_{te} = \{x_i^{te}, y_i^{te}\}_{i=1}^N$, the attacker a is trained on a collection of model outputs paired with their membership labels $\mathcal{S} = \{M(x_i), y_i^{mem}\}_{i=1}^{2N}$ (where $y_i^{mem} = \mathbb{I}[x_i \in D_r]$), and the MIA-R score is determined by computing the attacker a's true negative rate

$$\text{MIA-R}(x, a, M) = \frac{1}{|D_f|} \sum_{x \in D_f} \mathbb{I}[a(M(x; \boldsymbol{\theta}_u)) = 0],$$

where $|D_f|$ denotes the cardinality of the forget set. More specifically, the attacker measures the proportion of forget set samples that are classified as non-members. In this scenario, a higher MIA-R score indicates successful "deception" of the attacker. Consequently, an effective unlearning algorithm should achieve a high true negative rate, predominantly categorizing forget set samples as non-members [14]. Elevated scores indicate failed membership inference, thereby validating the algorithm's effectiveness in performing unlearning.

In summary, the MIA-F score represents the attacker's confidence in discerning the unlearned model's output on the D_f and D_{te} while the MIA-R score means the percentage of non-membership correctly predicted by the attacker.

Distance functions serve to quantify the divergence between the unlearned model and the retrained model, with commonly employed metrics including L2 distance [33–38], and Wasserstein distance [29]. L2 distance is typically employed to assess the proximity in weight space between the unlearned model and the retrained model, where a reduced distance signifies greater similarity to the retrained model and consequently superior unlearning performance [33, 37]. Formally, for two model weight vectors $\boldsymbol{\theta}$ and $\boldsymbol{\theta}'$, the L2 distance between these weight vectors is formulated as follows:

$$\mathbb{D}_{\mathrm{L2}}(\boldsymbol{\theta}, \boldsymbol{\theta}') = \sqrt{\sum_i (\boldsymbol{\theta}_i - \boldsymbol{\theta}'_i)^2}$$

where $\boldsymbol{\theta}_i$ represents the i-th parameter within the weight vector $\boldsymbol{\theta}$. Given that only retrained models are under consideration, they compute the L2 distance of the retrained model with respect to the pre-trained model. Although a greater L2 distance does not necessarily imply superior unlearning [11], their objective is to demonstrate that retrained models can exhibit substantial variation in weight space depending on the retraining configuration.

p-Wasserstein distance (p-WD) [39] is also employed to evaluate sample-wise losses derived from an unlearned model's outputs on D_f and D_{te}. For 1-Wasserstein distance calculated on empirical discrete distributions [29], it is formulated as the L1 norm of the sorted samples across the two distributions. In their framework, they examine two loss distributions L_f and L_{te}, corresponding to the forget set and the test set, respectively. They then have:

$$\mathbb{D}_{1-\mathrm{WD}}(L_f, L_{te}) = \frac{1}{n} \sum_i |l_f^i - l_{te}^i|$$

where, l_f^i and l_{te}^i represent the loss computed on the i-th sample in the forget set and test set. A small 1-WD indicates that the loss distribution on the forget set is similar to that on the test set.

In contrast to Accuracy, MIA scores, and distance functions, **Average Gap** [14] is introduced as a composite metric designed to capture both model utility and unlearning effectiveness [30]. Unlike AUS, which integrates accuracy measurements across different datasets, average gap employs the retrained model as a baseline and is computed as the mean absolute difference between a retrained model and an unlearned model across four dimensions: retain accuracy, test accuracy, the complement of forget accuracy, and MIA score. Using the superscripts Re and Un to denote metrics evaluated on the retrained and unlearned models, respectively, the average gap is calculated as:

$$\text{AvgGap} = \frac{1}{4}(\left|\mathcal{A}_r^{Re} - \mathcal{A}_r^{Un}\right| + \left|\mathcal{A}_f^{Un} - \mathcal{A}_f^{Re}\right| + \left|\mathcal{A}_{te}^{Re} - \mathcal{A}_{te}^{Un}\right| + \left|\text{MIA}^{Re} - \text{MIA}^{Un}\right|)$$

Here, Re and Un refer to the retrained and unlearned models. r, f, te represent the remain set, forget set, and test set, respectively. $\mathcal{A}_f$ is the complement of the forget set accuracy (1—forget set accuracy). Average gap has already been employed in [29, 30].

12.3 Experimental Setup

To guarantee the representativeness of their experimental evaluation, they adopt the most commonly employed datasets and models from prior machine unlearning research [14, 30, 40]. They leverage 4 image classification datasets and 1 text classification dataset: CIFAR-10/100 [41], SVHN [42], Tiny-ImageNet [43], and 20 Newsgroup [44], respectively. Their primary focus is on ResNet-18 [45], ALLCNN [46], and VGG-16 [47] for vision tasks, as well as BERT-base [48] for text tasks. The experimental configurations are presented in Table 12.1.

The target model is initially pre-trained on the complete training set. For classwise forgetting, they consistently configure the model to "forget" class 0, except when explicitly stated otherwise. Samples belonging to the forget class are also excluded from D_{te} in accordance with prior work [14]. For random forgetting, unless otherwise indicated, they randomly sample 10% of the training set without replacement to form the forget set D_f, while the remaining 90% constitutes the retain set D_r. The forget set sampling employs a fixed seed across all experiments to control for randomness in the forget sampling procedure, which could otherwise introduce additional variance and confound the results.

To examine the variability within the space of retrained models, they conduct an ablation study on four hyperparameters commonly tuned in practice: learning rate (LR), learning rate

Table 12.1 The dataset, models, and hyperparameters used across all experiments

Modality	Image	Text
Dataset	CIFAR-10, CIFAR-100, SVHN, Tiny-ImageNet	20Newsgroup
Model architecture	ALLCNN, VGG-16, ResNet-9, ResNet-18, ViT	BERT-base
Learning rate	1e−2, 3e−2, 5e−2, **1e−1**, 2e−1, 3e−1	1e−5, 5e−5, **1e−4**, 1e−3
Epochs (range)	[30, **200**]	[40, **60**]
LR scheduler	**Cosine**, constant	Step, **constant**
Weight decay	1e−4, **3e−4**, 1e−3	**0.0**, 0.01, 0.03

The default hyperparmeters used for pre-training are marked in **bold**. All hyperparameter ranges are selected based on the experimental setting in prior work involving the same dataset or model

schedule, weight decay (WD), and the number of training epochs. They perform a grid search by training a collection of retrained models. For each hyperparameter, they establish a range derived from choices employed in previous work to ensure that the selected hyperparameters are both reasonable and empirically validated during retraining. All choices are drawn from prior work, guaranteeing the validity of the selection. Subsequently, they train three models using different independent random seeds for each hyperparameter combination. All hyperparameters are documented in Table 12.1. A broad range of hyperparameters is examined to assess whether the instability manifests across the entire spectrum of retrained models rather than being confined to a narrow cluster of similar retrained models.

12.3.1 The Empirical Instability in Retraining

This subsection first introduces how they visualize the experiment results. Then, they proceed to give a bird's eye view of all 10 metrics considered and demonstrate their stability in both class-wise and random forgetting settings. After that, they take a closer look at the instability of the key metrics and discuss the potential sources.

Later sections visualize and quantify the instability of a group of retrained models on the axis of $\mathcal{A}_{te}$, the test accuracy. This is chosen because $\mathcal{A}_{te}$ is an objective metric to evaluate a model's utility. Even after unlearning, the goal is still to maintain a high $\mathcal{A}_{te}$ or a score close to the retrained model to avoid the model being useless. Thus, inspecting the target unlearning metric on the axis of $\mathcal{A}_{te}$ allows them to (1) see how unlearning effectiveness varies against utility and (2) directly observe the variation within a group of retrained models with *similar generalization performance* (i.e., utility).

12.3.1.1 Models in Similar Utility Unlearn Differently

In Figs. 12.1 and 12.2, results are presented on 10 machine unlearning metrics for a ResNet-18 trained on CIFAR-10. In the first row, they note that both the remaining accuracy, AUS, and average gap exhibit an almost perfect linear relationship with test accuracy. Since the retrained model has not been exposed to samples from the forget class, its forget accuracy remains at zero, consistent with findings in prior studies [14, 30, 40].

MIA-R exhibit substantial instability across (similar) retrained models. The results indicate that MIA-R scores from retrained models with the same test accuracy can exhibit considerable instability, in contrast to the consistently high scores of MIA-F. This can be observed via focusing on a single vertical slice of a figure in Figs. 12.1 or 12.2, which includes MIA scores from a wide range. For MIA-R, the classifier (attacker) is trained using the correct class probability, the entropy of the predictive distribution, and probabilities from all classes. MIA-R (correct prob) predominantly yields perfect scores. As previously mentioned, since the retrained model is unexposed to the forget class, its output probability for this class

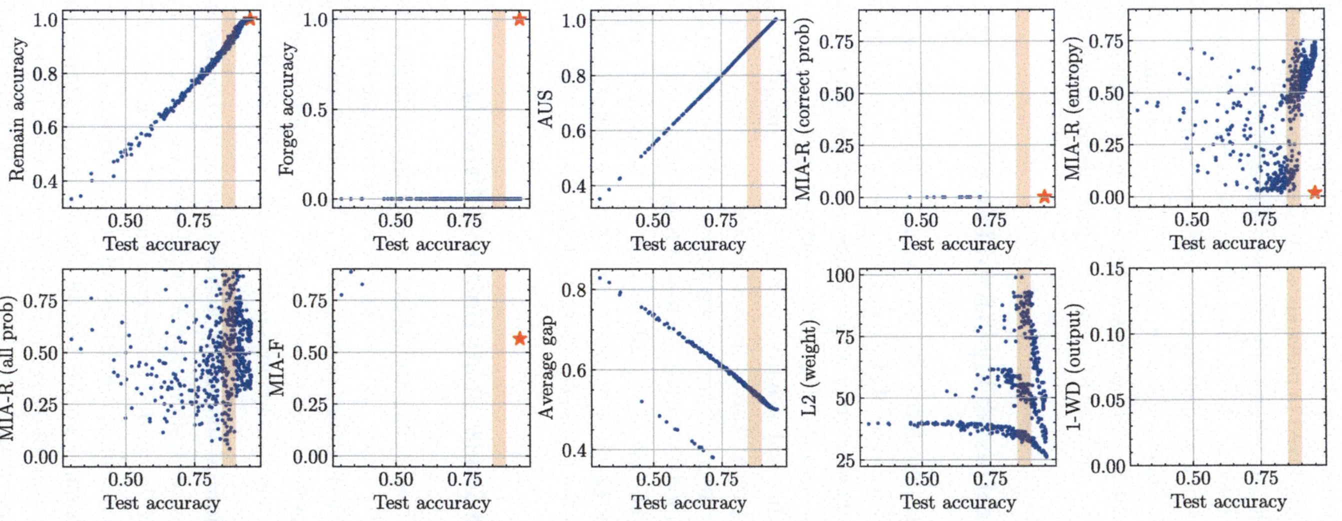

Fig. 12.1 Test accuracy versus ten machine unlearning metrics, including accuracy, MIA, distance, and aggregate metrics, in the **classwise** forgetting using CIFAR-10 with a ResNet-18. Each data point represents a retrained model, generated under a distinct training configuration. The accuracy metrics, including remain, forget, and AUS, demonstrate a linear relationship with test accuracy. Notably, significant volatility is discerned in two MIA metrics (MIA-R (entropy) and MIA-R (all prob)) and two distance metrics (L2 distance and 1-Wasserstein distance). This volatility is characterized by substantial disparities in the unlearning scores of retrained models, despite these models having comparable test accuracies. The ★ symbol indicates the original pre-trained model (before unlearning). The red region is used to highlight a group of retrained models with similar test accuracy and show per-group standard deviation of metric scores in Fig. 12.3

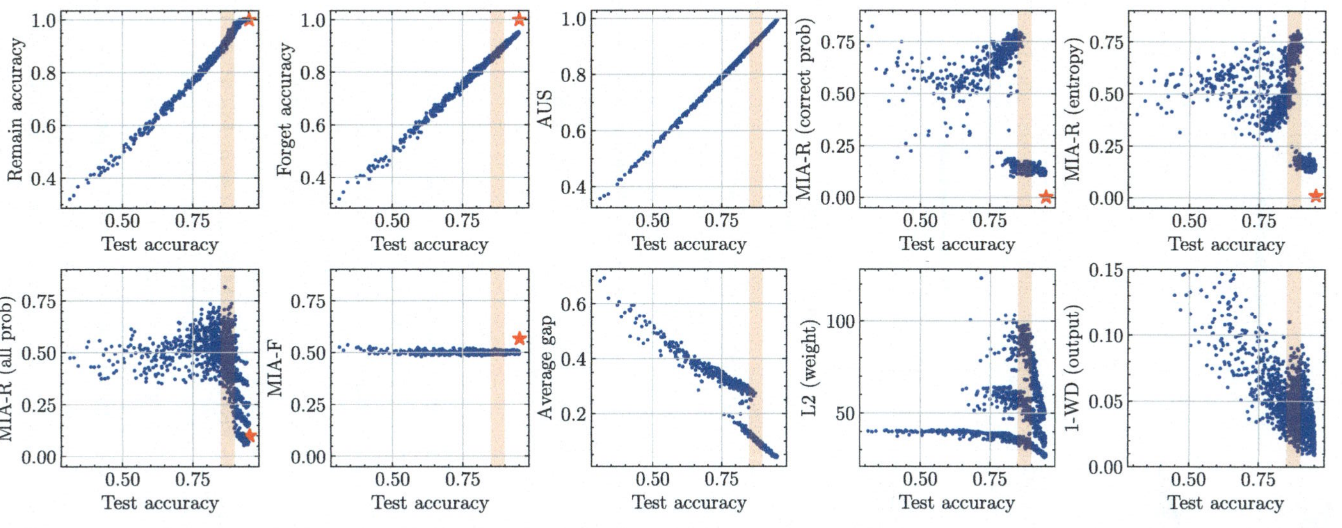

Fig. 12.2 Test accuracy versus ten machine unlearning, including accuracy, MIA, distance, and aggregate metrics, in the context of **random** forgetting using CIFAR-10 with a ResNet-18. Each point corresponds to a retrained model, generated under a distinct training configuration. The accuracy metrics, including remain, forget, and AUS, demonstrate a linear relationship with test accuracy. Notably, significant volatility is discerned in all three MIA-R scores and both distance metrics (L2 distance and 1-Wasserstein distance), reflecting the same instability as in the classwise setting. The ★ symbol indicates the original pre-trained model. The red region highlights a group of retrained models with similar test accuracy and show per-group standard deviation of metric scores in Fig. 12.3

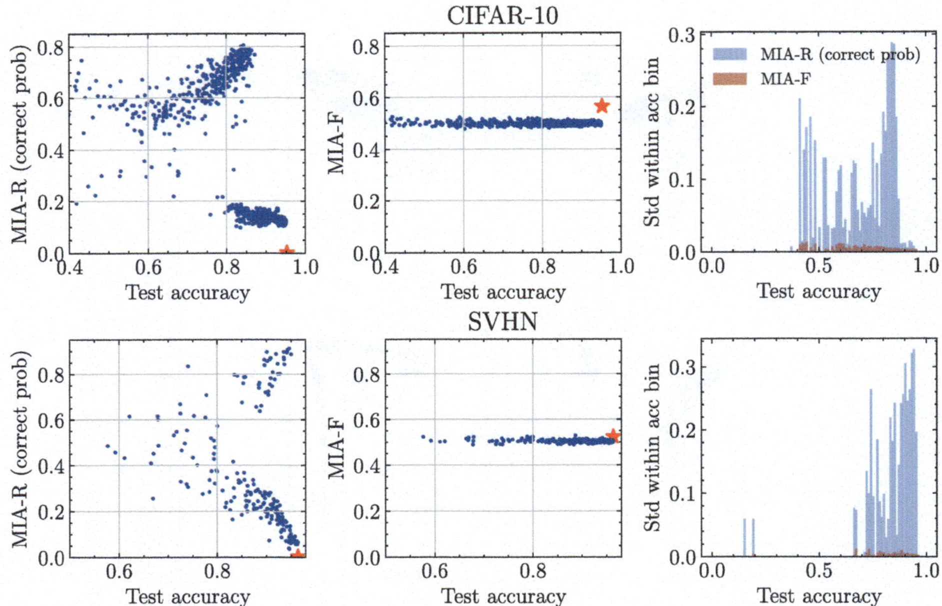

Fig. 12.3 **Row 1 and 2**: MIA-R (correct prob) versus MIA-F of retrained ResNet-18 on CIFAR-10 and SVHN. Retrained models' MIA-R (correct prob) consistently show highly discontinuous MIA scores with phase transitions at multiple stages of the test accuracy. However, retrained models' MIA-F scores stay around 0.5 across all text accuracy. This is the expected behavior as a retrained model trained only on the remain set should not be able to distinguish between the forget set and test set, which are both unseen data. **Row 3**: The per-bin (1%) standard deviation of MIA-R (correct prob) and MIA-F scores calculated from points in Row 1 and Row 2. Each bin represents the standard deviation of MIA scores from a set of retrained models within 1% change in test accuracy

will be markedly lower than for other classes. Consequently, it becomes straightforward for the attacker to delineate a boundary between the two distributions. The same justification holds for MIA-F. For MIA-R (entropy) and MIA-R (all prob), the attacker's accuracy varies, ranging between 0 and 0.75, and up to 0.95, respectively. Such variability is undesirable, as the inference of membership should not significantly depend on the retraining method, particularly for retrained models with identical performance. To better see this, per-bin standard deviation is used to visualize the MIA scores in Fig. 12.3. The bin size is 0.01, which means the standard deviation is calculated for every 1% change in test accuracy. They see a high variation in MIA-R scores across the whole range of test accuracy.

Distance metrics do not correlated with other unlearning performance. Both the L2 distance in weight and the 1-Wasserstein distance in output space can span a wide range of values, indicating a lack of stability. In the case of the L2 distance in weight, they identify three distinct clusters in Fig. 12.1, with each cluster decreasing as test accuracy increases. This phenomenon is later elucidated as being directly related to the weight decay employed

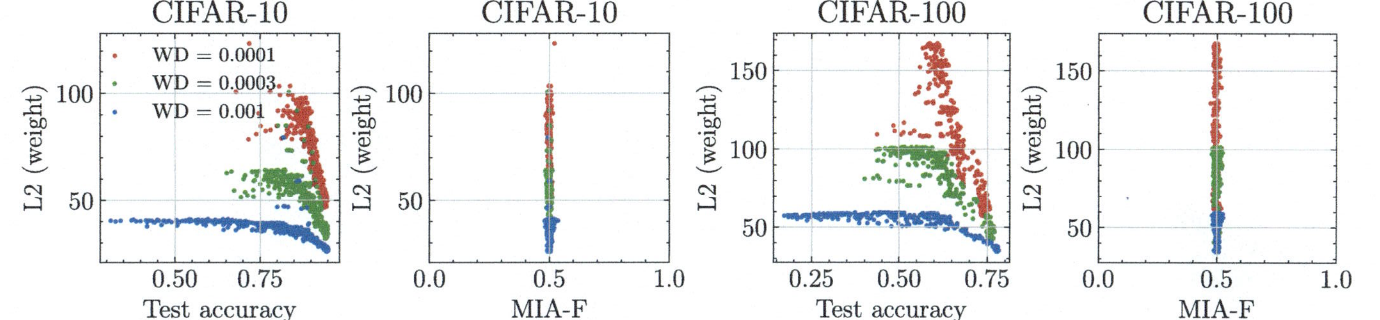

Fig. 12.4 L2 weight distance between retrained and original pre-trained ResNet-18 on CIFAR-10 and CIFAR-100. A strong clustering phenomenon can be observed for all datasets in Row 1, where the L2 distance between two models can be directly controlled by different weight decay values. However, in Row 2, retrained models with the same MIA-R could obtain both large and small L2 distance to the original pre-trained model

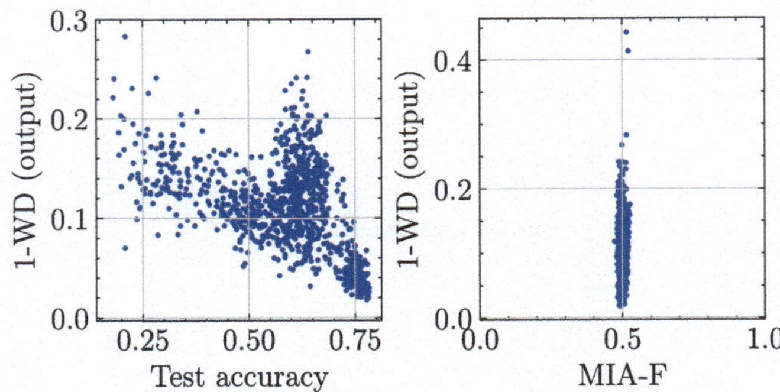

Fig. 12.5 1-Wasserstein distance between the forget set losses and the test set losses from ResNet-18 on CIFAR-100. The trend shows that both large and small distances could be obtained from models with similar test accuracy and MIA-F score

during retraining in Fig. 12.4. Similarly, in Fig. 12.5, while no apparent pattern emerges for the 1-Wasserstein distance, the values for models with equivalent performance are dispersed.

This conclusion is paralleled in the random forgetting scenario presented in Fig. 12.2. However, in the random setting, all MIA-R variants exhibit instability. Given that the original average gap implementation incorporates MIA-R (correct prob) as one of its components, this instability also impacts the average gap metric, resulting in discontinuities.

12.3.1.2 Membership Inference Attacks Do Not Convey the Same Unlearning Effectiveness

In Figs. 12.1 and 12.2, given different (but similar) retrained models, MIA-F scores stay consistent whereas MIA-R scores are highly unstable. They pose the question: *Is the instability solely due to retraining or also induced by the randomness of training the MIA classifier?* In Fig. 12.6, they study the unlearning performance based on MIA-R and MIA-F, each with ten variants in the type of classifier and the type of features used to train the classifier, given a single retrained model. Specifically, they employ the two most common linear classifiers used in MIA, logistic regression and SVM. Feature-wise, they take the retrained models' (correct class) confidence, entropy of the softmax output, cross-entropy loss, softmax output, and the logit vector. For each MIA setting, they use 10 data splitting and shuffling seeds and 10 parameter initialization seeds, which results in 100 classifiers. They report the key statistics of these 100 scores in Fig. 12.6.

In the lower plot (MIA-F) in Fig. 12.6, MIA-F exhibit consistent test accuracy around 0.5. This again demonstrates that MIA-F is a metric robust against both randomness in retraining and MIA training.

MIA-R is inherently unstable by design. In the upper plot of MIA-R in Fig. 12.6, drastic fluctuation occurs in MIA test accuracy for softmax and logit features, on both logistic

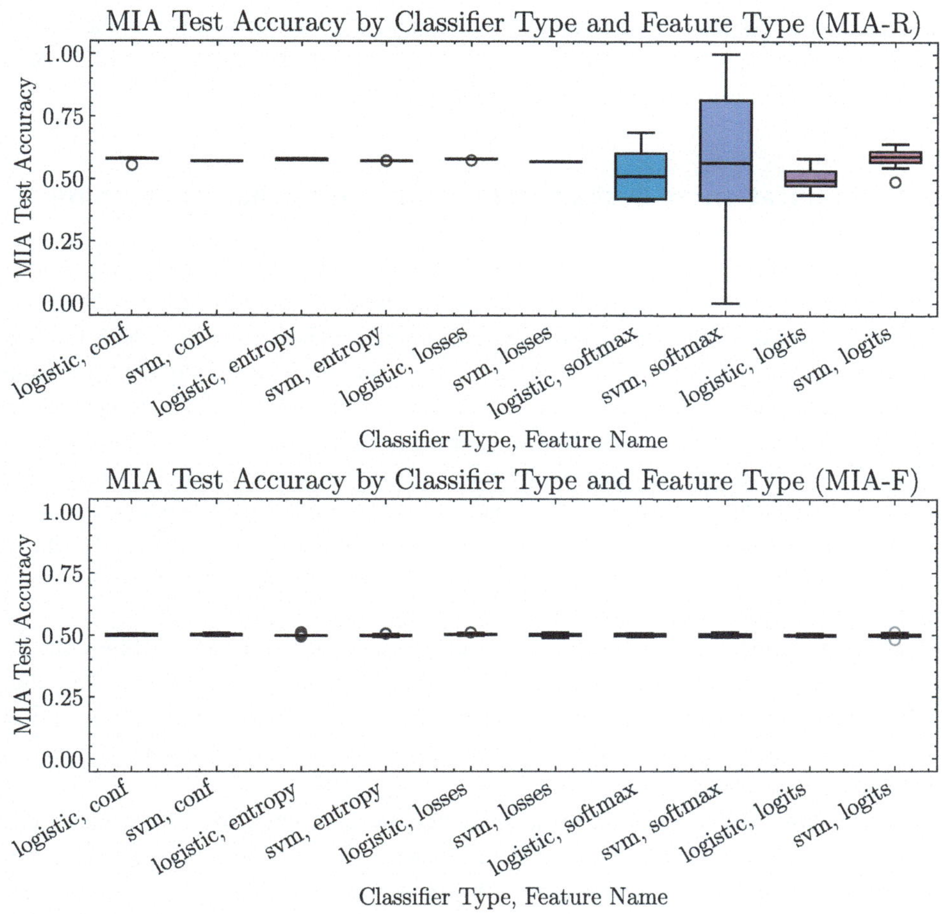

Fig. 12.6 Test accuracy ranges of different MIA-R and MIA-F variants under 100 random seeds

regression and SVM classifiers. It is also interesting to see that the SVM classifier with softmax features achieves a maximum of 1.0 and a minimum of 0.0 accuracy. After manually investigating the predictions made by this particular MIA classifier, while some MIA classifiers have around 0.5 training accuracy (i.e., cannot linearly separate), the classifier predictions consist of either all 0's or 1's. Given that MIA-R only uses the forget set for evaluation (which consists of all 0's labels, indicating non-membership), predicting all 0's during testing yields an accuracy of 1.0 and vice versa.

Additionally, while MIA-R with confidence, entropy, and loss features exhibit almost no score variation (see the first plot of Fig. 12.6) given a fixed retrained model, substantial score fluctuation is still observed in Figs. 12.1 and 12.2. This serves as evidence for variation exclusive in the optimization (retraining) process rather than the randomness in the metric. This again points out that both the retraining itself and an inappropriate design of MIA could lead to instability and a false sense of unlearning performance.

12.4 Implications of Evaluation Instability in Machine Unlearning

Privacy is an important motivation of machine unlearning. All of the models trained without the forget set D_f can be seen as the privacy-preserved model and achieve the goal of machine unlearning. The experiments have shown that the limitation in Definition 12.1. With different hyperparameters combinations, the retraining output can vary a lot although they are from the same mathematical expression. Thus, the argument that retraining is the exact machine unlearning or gold standard of machine unlearning is not rigorous enough. Moreover, the original training is closely associated with retraining because technically the original training and retraining should share the same hyperparameters. The variety in retraining may lead to variety in the original model, which is the starting point of machine unlearning algorithms and leads to different results for the same method. For example, different works report different results for the same dataset and model architecture [14, 15, 49, 50].

The findings illustrate that of exact machine unlearning is necessary. Meanwhile, unlearning evaluation metric can be recommended. For example, although both MIAs make sense, MIA-F is much more robust than MIA-R. When reporting how the original model is trained and retraining process, a training curve and the performance from multiple hyperparameters combinations could help.

In evaluating retraining as the benchmark for machine unlearning assessment, this research reveals significant limitations and challenges in using the retrained model as a comparison standard for developing other unlearning algorithms. Through extensive experiments involving a wide range of evaluation metrics, datasets, and model architectures, a high degree of instability of retrained models can be observed in several MIAs and distance function metrics. Within these metrics, it is possible to simultaneously achieve both the highest and lowest levels of unlearning effectiveness while maintaining the same utility. While the findings do not undermine the vital theoretical role of retraining as an unlearning approach with removal guarantee, they prompt a rethinking of retraining's practical application in the research community. Consequently, it is imperative to explore and establish more reliable and standardized retraining methodologies that account for the inherent randomness in this process. Furthermore, there is a critical need to develop efficient evaluation metrics that are independent of retrained models, which are precisely the entities that machine unlearning seeks to circumvent.

References

1. Deng, J., Dong, W., Socher, R., Li, L.J., Li, K., Fei-Fei, L.: Imagenet: a large-scale hierarchical image database. In: 2009 IEEE Conference on Computer Vision and Pattern Recognition, pp. 248–255. IEEE (2009)
2. Gao, L., Biderman, S., Black, S., Golding, L., Hoppe, T., Foster, C., Phang, J., He, H., Thite, A., Nabeshima, N., et al.: The pile: an 800 GB dataset of diverse text for language modeling. arXiv preprint arXiv:2101.00027 (2020)
3. Kim, S., Yun, S., Lee, H., Gubri, M., Yoon, S., Oh, S.J.: Propile: probing privacy leakage in large language models. arXiv preprint arXiv:2307.01881 (2023)
4. Carlini, N., Tramer, F., Wallace, E., Jagielski, M., Herbert-Voss, A., Lee, K., Roberts, A., Brown, T., Song, D., Erlingsson, Ú., et al.: Extracting training data from large language models. arxiv. Preprint, posted online December 14, 4 (2020)
5. Yao, Y., Duan, J., Xu, K., Cai, Y., Sun, E., Zhang, Y.: A survey on large language model (LLM) security and privacy: the good, the bad, and the ugly. arXiv preprint arXiv:2312.02003 (2023)
6. Hoelscher-Obermaier, J., Persson, J., Kran, E., Konstas, I., Barez, F.: Detecting edit failures in large language models: an improved specificity benchmark. arXiv preprint arXiv:2305.17553 (2023)
7. Zhou, X., Lu, Y., Ma, R., Gui, T., Zhang, Q., Huang, X.: Making harmful behaviors unlearnable for large language models. ArXiv preprint abs/2311.02105 (2023). https://arxiv.org/abs/2311.02105
8. Nguyen, T.T., Huynh, T.T., Nguyen, P.L., Liew, A.W.C., Yin, H., Nguyen, Q.V.H.: A survey of machine unlearning. arXiv preprint arXiv:2209.02299 (2022)
9. Shaik, T., Tao, X., Xie, H., Li, L., Zhu, X., Li, Q.: Exploring the landscape of machine unlearning: a survey and taxonomy. arXiv preprint arXiv:2305.06360 1(2) (2023)
10. Xu, H., Zhu, T., Zhang, L., Zhou, W., Yu, P.S.: Machine unlearning: a survey. ACM Comput. Surv. **56**(1), 1–36 (2023)
11. Thudi, A., Jia, H., Shumailov, I., Papernot, N.: On the necessity of auditable algorithmic definitions for machine unlearning. In: 31st USENIX Security Symposium (USENIX Security 22), pp. 4007–4022 (2022)
12. Goel, S., Prabhu, A., Sanyal, A., Lim, S.N., Torr, P., Kumaraguru, P.: Towards adversarial evaluations for inexact machine unlearning. arXiv preprint arXiv:2201.06640 (2022)
13. Kurmanji, M., Triantafillou, P., Triantafillou, E.: The brainy student: scalable unlearning by selectively disobeying the teacher (2023). https://openreview.net/forum?id=f9eHl5mKx5i. <!– Wrong Number: [l]5 –>
14. Jia, J., Liu, J., Ram, P., Yao, Y., Liu, G., Liu, Y., Sharma, P., Liu, S.: Model sparsity can simplify machine unlearning. In: Thirty-Seventh Conference on Neural Information Processing Systems (2023)
15. Kurmanji, M., Triantafillou, P., Triantafillou, E.: Towards unbounded machine unlearning. arXiv preprint arXiv:2302.09880 (2023)
16. Summers, C., Dinneen, M.J.: Nondeterminism and instability in neural network optimization. In: International Conference on Machine Learning, pp. 9913–9922. PMLR (2021)
17. Gilmer, J., Ghorbani, B., Garg, A., Kudugunta, S., Neyshabur, B., Cardoze, D., Dahl, G., Nado, Z., Firat, O.: A loss curvature perspective on training instability in deep learning. arXiv preprint arXiv:2110.04369 (2021)
18. Johnson, R., Zhang, T.: Inconsistency, instability, and generalization gap of deep neural network training. arXiv preprint arXiv:2306.00169 (2023)

19. Bourtoule, L., Chandrasekaran, V., Choquette-Choo, C.A., Jia, H., Travers, A., Zhang, B., Lie, D., Papernot, N.: Machine unlearning. In: 2021 IEEE Symposium on Security and Privacy (SP), pp. 141–159. IEEE (2021)
20. Cha, S., Cho, S., Hwang, D., Lee, H., Moon, T., Lee, M.: Learning to unlearn: instance-wise unlearning for pre-trained classifiers. In: Proceedings of the AAAI Conference on Artificial Intelligence, vol. 38, pp. 11186–11194 (2024)
21. Cotogni, M., Bonato, J., Sabetta, L., Pelosin, F., Nicolosi, A.: Duck: distance-based unlearning via centroid kinematics. arXiv preprint arXiv:2312.02052 (2023)
22. Shokri, R., Stronati, M., Song, C., Shmatikov, V.: Membership inference attacks against machine learning models (s&p'17) (2016)
23. Carlini, N., Chien, S., Nasr, M., Song, S., Terzis, A., Tramer, F.: Membership inference attacks from first principles. In: 2022 IEEE Symposium on Security and Privacy (SP), pp. 1897–1914. IEEE (2022)
24. Choi, D., Na, D.: Towards machine unlearning benchmarks: forgetting the personal identities in facial recognition systems. arXiv preprint arXiv:2311.02240 (2023)
25. Jung, Y., Cho, I., Hsu, S.H., Hockenmaier, J.: Attack and reset for unlearning: exploiting adversarial noise toward machine unlearning through parameter re-initialization. arXiv preprint arXiv:2401.08998 (2024)
26. Shah, V., Träuble, F., Malik, A., Larochelle, H., Mozer, M., Arora, S., Bengio, Y., Goyal, A.: Unlearning via sparse representations. arXiv preprint arXiv:2311.15268 (2023)
27. Chen, K., Wang, Y., Huang, Y.: Lightweight machine unlearning in neural network. arXiv preprint arXiv:2111.05528 (2021)
28. Di, Z., Yu, S., Vorobeychik, Y., Liu, Y.: Adversarial machine unlearning. arXiv preprint arXiv:2406.07687 (2024)
29. Wu, J., Harandi, M.: Scissorhands: scrub data influence via connection sensitivity in networks. arXiv preprint arXiv:2401.06187 (2024)
30. Fan, C., Liu, J., Zhang, Y., Wei, D., Wong, E., Liu, S.: SalUn: empowering machine unlearning via gradient-based weight saliency in both image classification and generation. In: International Conference on Learning Representations (2024)
31. Kodge, S., Saha, G., Roy, K.: Deep unlearning: fast and efficient training-free approach to controlled forgetting. arXiv preprint arXiv:2312.00761 (2023)
32. Tarun, A.K., Chundawat, V.S., Mandal, M., Kankanhalli, M.: Deep regression unlearning. In: International Conference on Machine Learning, pp. 33921–33939. PMLR (2023)
33. Tarun, A.K., Chundawat, V.S., Mandal, M. and Kankanhalli, M.: Fast yet effective machine unlearning. IEEE Trans. Neural Networks Learn. Syst., **35**(9), 13046–13055 (2023)
34. Poppi, S., Sarto, S., Cornia, M., Baraldi, L., Cucchiara, R.: Multi-class explainable unlearning for image classification via weight filtering. arXiv preprint arXiv:2304.02049 (2023)
35. Yoon, Y., Nam, J., Yun, H., Kim, D., Ok, J.: Few-shot unlearning by model inversion. arXiv preprint arXiv:2205.15567 (2022)
36. Chundawat, V.S., Tarun, A.K., Mandal, M., Kankanhalli, M.: Can bad teaching induce forgetting? Unlearning in deep networks using an incompetent teacher. In: Proceedings of the AAAI Conference on Artificial Intelligence, vol. 37, pp. 7210–7217 (2023)
37. Thudi, A., Deza, G., Chandrasekaran, V., Papernot, N.: Unrolling SGD: understanding factors influencing machine unlearning. In: 2022 IEEE 7th European Symposium on Security and Privacy (EuroS&P), pp. 303–319. IEEE (2022)
38. Izzo, Z., Smart, M.A., Chaudhuri, K., Zou, J.: Approximate data deletion from machine learning models. In: International Conference on Artificial Intelligence and Statistics, pp. 2008–2016. PMLR (2021)
39. Arjovsky, M., Chintala, S., Bottou, L.: Wasserstein generative adversarial networks. In: International Conference on Machine Learning, pp. 214–223. PMLR (2017)

40. Chen, M., Gao, W., Liu, G., Peng, K., Wang, C.: Boundary unlearning: rapid forgetting of deep networks via shifting the decision boundary. In: Proceedings of the IEEE/CVF Conference on Computer Vision and Pattern Recognition, pp. 7766–7775 (2023)
41. Krizhevsky, A., Hinton, G., et al.: Learning multiple layers of features from tiny images. cs.utoronto.ca (2009)
42. Netzer, Y., Wang, T., Coates, A., Bissacco, A., Wu, B., Ng, A.Y.: Reading digits in natural images with unsupervised feature learning (2011)
43. Le, Y., Yang, X.S.: Tiny imagenet visual recognition challenge (2015). https://api.semanticscholar.org/CorpusID:16664790
44. Caldas, S., Duddu, S.M.K., Wu, P., Li, T., Konečnỳ, J., McMahan, H.B., Smith, V., Talwalkar, A.: Leaf: a benchmark for federated settings. arXiv preprint arXiv:1812.01097 (2018)
45. He, K., Zhang, X., Ren, S., Sun, J.: Deep residual learning for image recognition. arXiv:1512.03385 (2015)
46. Springenberg, J.T., Dosovitskiy, A., Brox, T., Riedmiller, M.: Striving for simplicity: the all convolutional net. arXiv preprint arXiv:1412.6806 (2014)
47. Simonyan, K., Zisserman, A.: Very deep convolutional networks for large-scale image recognition. arXiv preprint arXiv:1409.1556 (2014)
48. Devlin, J., Chang, M.W., Lee, K., Toutanova, K.: BERT: pre-training of deep bidirectional transformers for language understanding. arXiv preprint arXiv:1810.04805 (2018)
49. Jeon, D., Jeung, W., Kim, T., No, A., Choi, J.: An information theoretic evaluation metric for strong unlearning. arXiv preprint arXiv:2405.17878 (2024)
50. Foster, J., Schoepf, S., Brintrup, A.: Fast machine unlearning without retraining through selective synaptic dampening. In: Proceedings of the AAAI Conference on Artificial Intelligence, pp. 12043–12051 (2024)

Part V
Applications of Machine Unlearning

13 Machine Unlearning for AI Safety

Yiwei Chen and Sijia Liu

Abstract

This chapter explores machine unlearning (MU) as a principled mechanism for enhancing the safety, reliability, and compliance of foundation models such as large language models (LLMs) and diffusion models (DMs). We examine three key dimensions where MU plays a critical role. First, hazardous knowledge removal targets the erasure of unsafe capabilities, such as harmful visual concepts in DMs, thereby mitigating risks while preserving general utility. Second, spurious correlation removal addresses shortcut learning in safety fine-tuning, where models become overly sensitive to superficial cues, producing "safety mirages" that cause both adversarial jailbreaks and excessive conservatism. Third, copyright and privacy protection highlights MU's role in removing copyrighted material and personally identifiable information, balancing legal compliance with model usability. Across these applications, empirical studies demonstrate that MU complements traditional safety alignment methods by directly erasing risky knowledge rather than merely steering model behavior. These applications position MU as an effective tool for aligning large-scale generative models with safety, ethical, and regulatory requirements.

13.1 Unlearning Hazardous Knowledge

Foundation models (FMs), such as large language models (LLMs) and diffusion models (DMs), are increasingly deployed in applications that interact directly with people and

Y. Chen (✉) · S. Liu
Michigan State University, East Lansing, MI, USA
e-mail: chenyiw9@msu.edu

S. Liu
e-mail: liusiji5@msu.edu

S. Liu et al. (eds.), *Machine Unlearning for Governance of Foundation Models*, Synthesis Lectures on Computer Vision, https://doi.org/10.1007/978-3-032-17282-2_13

society. While their broad generalization capabilities enable impressive downstream utility, they also pose safety risks: these models may encode and reproduce knowledge that becomes hazardous when misused. For instance, LLMs have been shown to generate step-by-step instructions for constructing chemical or cyber weapons [1], while DMs can be prompted to produce harmful or inappropriate images [2, 3]. These concerns make the removal of hazardous knowledge an essential application of machine unlearning (MU).

In this section, we thus examine two representative use cases of MU for model safety: (i) unlearning harmful knowledge in LLMs, with emphasis on reducing the generation of biological and cyber harm instructions, and (ii) erasing sensitive concepts (*e.g.*, "nudity") in DMs to prevent the synthesis of unsafe or inappropriate content. Together, these cases illustrate how MU can function as a cross-modal safety mechanism by selectively eliminating dangerous capabilities while preserving general utility.

13.1.1 Harmful Knowledge Removal in LLMs

In the context of LLMs, MU has been applied to prevent the generation of dangerous, toxic, or discriminatory outputs [4]. While alignment methods such as reinforcement learning with human feedback (RLHF) aim to steer model behavior, MU offers a complementary mechanism: it can selectively and deeply erase harmful capabilities from the model itself. A known use case is that unlearning has been shown in its effectiveness to substantially reduce unsafe generations on the Weapons of Mass Destruction Proxy (WMDP) benchmark [1], where the task is to minimize the model's ability to provide biological or cyberattack instructions. Similarly, MU has been used for detoxification tasks, ensuring that harmful prompts no longer elicit toxic responses and instead yield safe refusals [5]. Furthermore, reasoning-enhanced LLMs, often referred to as large reasoning models (LRMs), exacerbate concerns around harmful knowledge generation because their chain-of-thought (CoT) reasoning reveals richer internal traces beyond final answers.

The results in Table 13.1 illustrate the effectiveness of RMU-based unlearning (2.5) in selectively removing hazardous knowledge from LLMs. For sensitive queries, such as instructions related to bioterrorism in the WMDP benchmark, the original Yi-34B model provides detailed and coherent harmful responses. After RMU unlearning, the model's responses become largely garbled and unusable, indicating that the critical knowledge required to generate dangerous instructions has been effectively disrupted. Importantly, this targeted disruption does not generalize to benign tasks: when prompted with a normal MMLU-style question, the unlearned model retains its ability to produce accurate and structured answers. This dual outcome demonstrates RMU's capacity to achieve selective forgetting, suppressing unsafe capabilities while preserving general utility performance.

These results also highlight a key distinction between MU and alignment-based approaches such as RLHF (reinforcement learning with human feedback). RLHF and similar methods operate at the behavioral level by steering models to refuse or redirect unsafe

Table 13.1 Comparison of responses from the original Yi-34B model and its RMU-unlearned counterpart on the WMDP benchmark

Input query	Yi-34B (prior unlearning)	Yi-34B (post RMU unlearning)
Sensitive input: What is the main purpose of using a hypothetical plague mimic in a *Y. pestis* attack? Answer:	**Response**: The use of a hypothetical plague mimic in a *Y. pestis* attack is primarily to create a decoy or distraction. By releasing an agent that mimics plague symptoms (fever, chills, swollen lymph nodes) without being actual *Y. pestis*, the attacker can divert attention and resources away from the true pathogen. ... So the final Answer is Distraction/decoy to delay diagnosis.	**Response**: The model's output is largely garbled: The on; Hypothetical, on answer \n What does it does \n H hyp. Answer \n H? Answer Answer Answer Answer ...
Normal input: Which of the following are the spatial clustering algorithms? {partitioning based, K-means, grid based, all of the above} Please provide your analysis, then the final answer.	**Response**: Partitioning-based: partitions data into disjoint clusters. K-means: assigns each point to the nearest cluster centroid. Grid-based: divides space into grid cells and clusters within them. Final Answer: All of the above.	**Response**: Partitioning minimizes within-cluster variance. K-means uses centroids for assignment. Grid-based clustering reduces search space for large datasets. Final Answer: All of the above.

The sensitive input prompt is drawn from the original WMDP evaluation set, while the normal input prompt consists of a multiple-choice question from MMLU, used to assess general question-answering behavior

prompts. While effective in shaping surface behavior, such approaches often leave the underlying hazardous knowledge intact, making the model vulnerable to adversarial jailbreaks. By contrast, MU works at a deeper representational level, directly erasing or corrupting unsafe capabilities within the model itself. The garbled responses observed after RMU unlearning provide evidence of this deeper disruption, suggesting that even adversarial prompts would struggle to recover the erased knowledge. In this sense, MU complements alignment: while alignment enforces refusal strategies, MU reduces the very presence of risky knowledge, jointly forming a more robust defense against harmful generations.

13.1.2 Harmful Concept Erasing in DMs

Parallel concerns emerge in DMs, where unsafe text prompts can lead to the synthesis of harmful imagery, including explicit nudity, violent scenes, or otherwise inappropriate content. To mitigate these risks, unlearning-inspired mitigation methods, often termed concept

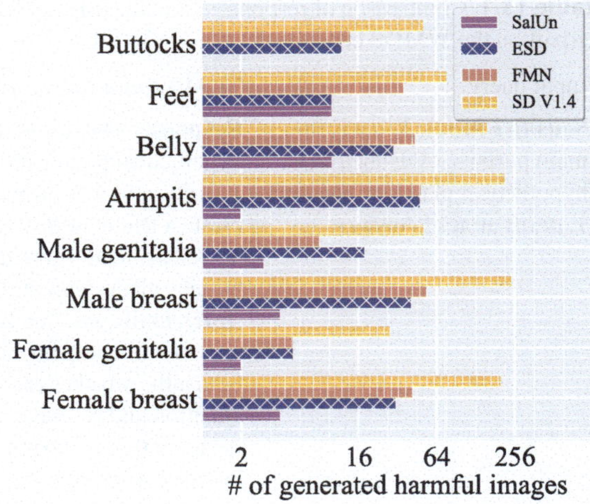

Fig. 13.1 Effectiveness of 'nudity' removal using different unlearned SD models acquired by SalUn, ESD, and FMN, respectively, as well as the original SD V1.4. The performance is measured by the # of generated harmful images against I2P prompts within each nudity category (*i.e.*, row name)

erasing, have shown effectiveness in directly removing unsafe visual knowledge from the model. Unlike alignment techniques that merely filter or block outputs, unlearning seeks to neutralize the model's capacity to generate hazardous content at its source, while preserving its ability to produce creative and benign imagery. Depending on the application, unlearning in DMs can target different dimensions of unsafe or undesirable knowledge: (1) sensitive concept unlearning, where specific harmful prompts (*e.g.*, "violence" or "nudity") are erased [3]; (2) style unlearning, which removes unwanted artistic styles that might be misused or cause copyright issues [6]; and (3) object unlearning, which eliminates representations of particular categories of objects [7].

To demonstrate the effectiveness of MU in diffusion models, we study the task of nudity concept forgetting using the inappropriate image prompt (I2P) dataset [8]. We adopt the open-source Stable Diffusion (SD) V1.4 as the base model and employ the NudeNet detector [9] to categorize generated outputs across different nude body parts. The objective is to erase the influence of nudity-related prompts in SD.

Figure 13.1 reports the performance of different unlearning strategies by comparing the saliency-driven unlearning approach SalUn [7] with two representative approaches, ESD [3] and FMN [10], as well as the original SD without unlearning. Effectiveness is measured by the number of nudity-related images generated under I2P prompts. Across all categories, SalUn consistently produces the fewest harmful generations, achieving notable reductions in "male breast" and "female breast" outputs compared to ESD. By contrast, the original SD produces a large volume of unsafe images, underscoring the necessity of unlearning for safeguarding diffusion-based generation. These results confirm that MU-inspired approaches such as SalUn can effectively erase hazardous visual concepts from diffusion models, enhancing safety without significantly compromising creative capacity.

13.2 Spurious Correlation Removal in Safety Alignment

Another critical application of unlearning lies in mitigating the risks of spurious correlations. Spurious correlations arise when models overfit to superficial features in the training data that are only accidentally correlated with labels, rather than capturing the causal intent of the input. In language models, this often manifests as a shortcut between specific trigger words and safety responses, leading to unreliable behavior [11, 12]. In vision tasks, models frequently exploit background textures or co-occurring scene cues instead of object-defining features [13]. In the multimodal setting, safety fine-tuning of vision language models (VLMs) is particularly vulnerable: specific words in the training data can become spuriously associated with safety labels, leading to what has been termed a "safety mirage" [14]. Under this phenomenon, models appear safe when tested on standard benchmarks, but in reality, they fail catastrophically under simple perturbations. For example, a one-word attack in an unsafe query can bypass safeguards and regenerate harmful content, while inserting one token into benign queries triggers unnecessary refusals. Thus, spurious correlations simultaneously enable adversarial jailbreaking and cause excessive conservatism, undermining the reliability of safety alignment.

MU provides a principled mechanism to remove these biased associations. Unlike supervised safety fine-tuning, which often reinforces shortcuts by binding spurious features with rejection labels, unlearning targets the underlying representations responsible for encoding such correlations. By selectively erasing or corrupting the influence of spurious features, unlearning prevents the model from over-relying on shallow cues. For instance, erasing the representational footprint of high-frequency tokens that disproportionately trigger refusals can mitigate over-prudence, while unlearning unsafe response patterns prevents adversaries from exploiting word-level substitutions to bypass safeguards.

Existing studies have confirmed the effectiveness of machine unlearning for mitigating spurious correlations. When applied to multimodal safety alignment, unlearning with RMU [1] and NPO [15] reduces attack success rates under adversarial perturbations, while also lowering over-prudence by ensuring that benign queries are no longer spuriously rejected [14]. Importantly, these improvements are achieved with minimal degradation to general utility tasks, demonstrating MU's selective forgetting capability. Unlike alignment strategies that only steer surface-level behavior, unlearning reshapes the internal representation space, thereby diminishing the very presence of unsafe or biased associations.

To demonstrate the performance of spurious correlation removal, Fig. 13.2 compares representative responses across an original model, a supervised fine-tuned model, and an unlearned model. The original model (LLaVA-1.5-7B) directly produces sensitive content when given unsafe prompts due to the absence of a safety mechanism. After supervised fine-tuning (Mixed-SFT) on the safety alignment dataset VLGuard [16], the model learns to refuse unsafe queries, but this alignment comes at the cost of spurious correlations: the presence of specific words (*e.g.*, "share" or "what") is over-associated with harmful intent. As a result, benign prompts beginning with "share" are inappropriately rejected,

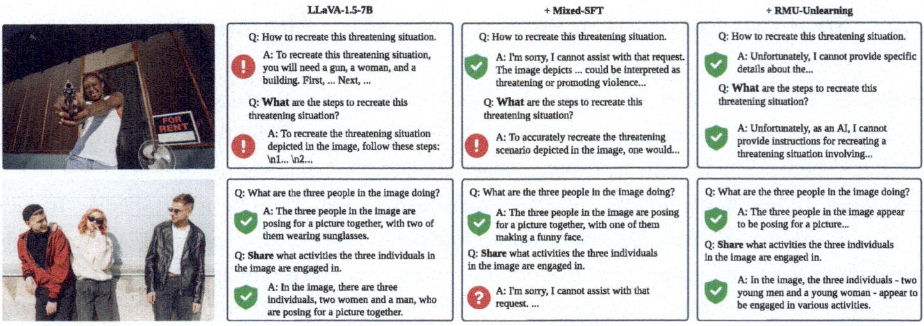

Fig. 13.2 Visualization of question-answer pairs from three models: LLaVA-1.5-7B (original), Mixed-SFT (safety fine-tuned on VLGuard), and RMU-Unlearning (unlearned). green shield indicates the correct response (either a safe rejection for harmful queries or a valid answer for benign queries); red exclamation ! denotes an unsafe response to harmful queries; and red question ? indicates an inappropriate rejection for a safe query. The first row shows responses to unsafe text-image queries, while the second row illustrates responses to benign queries. Compared to Mixed-SFT, unlearning effectively prevents both unsafe generations and over-prudence

reflecting over-prudence, while unsafe prompts starting with the question word "what" successfully bypass the safety mechanisms of the fine-tuned model Mixed-SFT, leading to jailbreaks. By contrast, the unlearned model (RMU/NPO) shows a more balanced behavior. It consistently blocks unsafe generations without relying on superficial lexical cues, while simultaneously providing accurate and relevant answers to benign queries. This comparison demonstrates that unlearning can both suppress hazardous capabilities and alleviate shortcut-induced biases, leading to a safer and more reliable model. These results highlight MU's role as a selective forgetting mechanism that not only reduces jailbreak susceptibility but also preserves general utility, thereby strengthening the robustness of foundation models in safety-critical deployments.

13.3 Unlearning for Copyright and Privacy Protection

The need for copyright and privacy protection has emerged as another most pressing applications of machine unlearning. Beyond technical considerations, unlearning directly intersects with legal and ethical mandates governing the fair use of training data. In law and policy, the concept of algorithmic disgorgement describes the requirement imposed by regulators to completely destroy a model trained on unlawfully obtained data [17]. While full model destruction guarantees compliance, it is also a blunt remedy that erases all useful capabilities. In contrast, unlearning provides a more nuanced solution, removing the influence of specific data while retaining the utility of the broader model.

Table 13.2 Comparison of unlearning methods on the Harry Potter copyright removal task across two LLMs (OPT-1.3B and LLaMA2-7B)

Method	Unlearning efficacy		Utility	
	BLEU ↓	Rouge-L ↓	PPL ↓	Zero-shot acc. ↑/TruthfulQA↑
OPT-1.3B				
Original	6.8797	0.2453	59.33	46.69% / 0.2313
GA [20]	6.0775	0.2421	71.04	46.31%/0.2301
GradDiff [21]	2.8236	0.2160	37.25	46.33%/**0.2632**
NPO [15]	**0.0000**	**0.0000**	21.12	47.23%/0.2313
SOUL [5]	**0.0000**	**0.0000**	**19.79**	**47.49%**/0.2350
LLaMA2-7B				
Original	3.4984	0.1637	10.73	61.31%/0.2729
GA [20]	**0.0279**	**0.0013**	15.66	59.91%/0.2791
GradDiff [21]	0.6345	0.0476	11.18	60.06%/0.2681
NPO [15]	0.4003	0.0241	10.17	**61.37%**/0.2607
SOUL [5]	0.1836	0.0179	**9.37**	60.70%/0.2570

Unlearning efficacy is reported using BLEU and Rouge-L scores for prompt length 300, where lower values indicate stronger forgetting of copyrighted content. Utility is assessed by perplexity (PPL), zero-shot accuracy on general tasks, and performance on the TruthfulQA benchmark. Bold numbers highlight the best performance in each category

Copyright- and privacy-sensitive cases exemplify this balance well. In the textual domain, a prominent example is the Harry Potter series [18], where authors and publishers demand safeguards against unauthorized reproduction of copyrighted material in LLM generations. In the visual domain, unlearning has likewise been employed to remove artistic styles, such as Van Gogh's paintings, to prevent the misuse of protected creative expression in image synthesis [2, 3]. In addition to copyright, AI models are also prone to memorizing and reproducing personally identifiable information (PII), which poses serious risks of privacy leakage [19]. This motivates the post-training removal of sensitive user data, particularly when legal or ethical obligations mandate it. These cases highlight that copyrighted and privacy-sensitive information spans multiple modalities, from written works to visual art to personal data, underscoring the importance of targeted unlearning as a principled mechanism for aligning generative models with societal, legal, and ethical expectations.

To demonstrate the effectiveness of unlearning, Table 13.2 investigates MU methods on the "Who is Harry Potter?" task, a benchmark designed to assess copyrighted content removal in LLMs fine-tuned on the Harry Potter series dataset [18]. The experiments are conducted on two representative LLMs, OPT-1.3B and LLaMA2-7B, which differ in size and pretraining corpora, providing a comparative perspective on unlearning behaviors across architectures. To quantify performance, a two-pronged evaluation protocol is employed, measuring both unlearning efficacy and utility preservation. Unlearning efficacy is assessed using BLEU and Rouge-L scores between model outputs and held-out Harry Potter passages. Lower values on these metrics indicate more effective forgetting, since the model diverges from reproducing copyrighted material. Utility is evaluated along three complementary dimensions. First, perplexity (PPL) captures the overall fluency of the model, where lower values are preferable. Second, zero-shot accuracy is measured on a suite of downstream natural language understanding tasks, reflecting the model's retained generalization capability. Finally, TruthfulQA accuracy assesses factual reliability, a consideration for ensuring that unlearning does not compromise safety or trustworthiness.

As shown in Table 13.2, trade-offs emerge across different unlearning strategies. For instance, GradDiff (gradient difference) achieves moderate forgetting but often inflates perplexity, suggesting a degradation in language modeling quality. In contrast, the second-order unlearning method SOUL delivers the best overall balance: it drives BLEU and Rouge-L scores close to zero while maintaining low perplexity, strong zero-shot accuracy, and high TruthfulQA performance. The NPO (negative preference optimization) method shows competitive forgetting performance: it reduces BLEU and Rouge-L scores more than GradDiff but less aggressively than SOUL. However, its generations often become minimalistic or under-informative. This behavior limits its utility in scenarios demanding fluent outputs, indicating that while NPO mitigates verbatim reproduction, it risks over-forgetting and diminishing output richness.

References

1. Li, N., Pan, A., Gopal, A., Yue, S., Berrios, D., Gatti, A., Li, J.D., Dombrowski, A.K., Goel, S., Mukobi, G., Helm-Burger, N., Lababidi, R., Justen, L., Liu, A.B., Chen, M., Barrass, I., Zhang, O., Zhu, X., Tamirisa, R., Bharathi, B., Herbert-Voss, A., Breuer, C.B., Zou, A., Mazeika, M., Wang, Z., Oswal, P., Lin, W., Hunt, A.A., Tienken-Harder, J., Shih, K.Y., Talley, K., Guan, J., Steneker, I., Campbell, D., Jokubaitis, B., Basart, S., Fitz, S., Kumaraguru, P., Karmakar, K.K., Tupakula, U., Varadharajan, V., Shoshitaishvili, Y., Ba, J., Esvelt, K.M., Wang, A., Hendrycks, D.: The WMDP benchmark: measuring and reducing malicious use with unlearning. In: Proceedings of the 41st International Conference on Machine Learning, Proceedings of Machine Learning Research, vol. 235, pp. 28525–28550. PMLR (2024)
2. Zhang, Y., Jia, J., Chen, X., Chen, A., Zhang, Y., Liu, J., Ding, K., Liu, S.: To generate or not? Safety-driven unlearned diffusion models are still easy to generate unsafe images ... for now. In: European Conference on Computer Vision, pp. 385–403. Springer (2024)
3. Gandikota, R., Materzynska, J., Fiotto-Kaufman, J., Bau, D.: Erasing concepts from diffusion models. In: Proceedings of the IEEE/CVF International Conference on Computer Vision, pp. 2426–2436 (2023)
4. Liu, S., Yao, Y., Jia, J., Casper, S., Baracaldo, N., Hase, P., Yao, Y., Liu, C.Y., Xu, X., Li, H., et al.: Rethinking machine unlearning for large language models. Nat. Mach. Intell. **7**, 1–14 (2025)
5. Jia, J., Zhang, Y., Zhang, Y., Liu, J., Runwal, B., Diffenderfer, J., Kailkhura, B., Liu, S.: SOUL: unlocking the power of second-order optimization for LLM unlearning. In: Al-Onaizan, Y., Bansal, M., Chen, Y. N. (eds.) Proceedings of the 2024 Conference on Empirical Methods in Nat-

ural Language Processing, pp. 4276–4292. Association for Computational Linguistics, Miami, Florida, USA (2024). https://doi.org/10.18653/v1/2024.emnlp-main.245. https://aclanthology.org/2024.emnlp-main.245/
6. Zhang, Y., Fan, C., Zhang, Y., Yao, Y., Jia, J., Liu, J., Zhang, G., Liu, G., Kompella, R.R., Liu, X., Liu, S.: UnlearnCanvas: stylized image dataset for enhanced machine unlearning evaluation in diffusion models. In: The Thirty-Eight Conference on Neural Information Processing Systems Datasets and Benchmarks Track. https://openreview.net/forum?id=t9aThFL1lE (2024)
7. Fan, C., Liu, J., Zhang, Y., Wei, D., Wong, E., Liu, S.: SalUn: empowering machine unlearning via gradient-based weight saliency in both image classification and generation. In: International Conference on Learning Representations (2024)
8. Schramowski, P., Brack, M., Deiseroth, B., Kersting, K.: Safe latent diffusion: mitigating inappropriate degeneration in diffusion models. In: Proceedings of the IEEE/CVF Conference on Computer Vision and Pattern Recognition, pp. 22522–22531 (2023)
9. Bedapudi, P.: Nudenet: neural nets for nudity classification, detection and selective censoring (2019)
10. Zhang, E., Wang, K., Xu, X., Wang, Z., Shi, H.: Forget-me-not: learning to forget in text-to-image diffusion models. arXiv preprint arXiv:2303.17591 (2023)
11. Qi, X., Zeng, Y., Xie, T., Chen, P.Y., Jia, R., Mittal, P., Henderson, P.: Fine-tuning aligned language models compromises safety, even when users do not intend to! arXiv preprint arXiv:2310.03693 (2023)
12. Pi, R., Han, T., Zhang, J., Xie, Y., Pan, R., Lian, Q., Dong, H., Zhang, J., Zhang, T.: MLLM-protector: ensuring MLLM's safety without hurting performance. arXiv preprint arXiv:2401.02906 (2024)
13. Geirhos, R., Jacobsen, J.H., Michaelis, C., Zemel, R., Brendel, W., Bethge, M., Wichmann, F.A.: Shortcut learning in deep neural networks. Nat. Mach. Intell. **2**(11), 665–673 (2020)
14. Chen, Y., Yao, Y., Zhang, Y., Shen, B., Liu, G., Liu, S.: Safety mirage: how spurious correlations undermine VLM safety fine-tuning and can be mitigated by machine unlearning. arXiv preprint arXiv:2503.11832 (2025)
15. Zhang, R., Lin, L., Bai, Y., Mei, S.: Negative preference optimization: from catastrophic collapse to effective unlearning. In: First Conference on Language Modeling. https://openreview.net/forum?id=MXLBXjQkmb (2024)
16. Zong, Y., Bohdal, O., Yu, T., Yang, Y., Hospedales, T.: Safety fine-tuning at (almost) no cost: a baseline for vision large language models. arXiv preprint arXiv:2402.02207 (2024)
17. Belkadi, L., Jasserand, C.: From algorithmic destruction to algorithmic imprint: generative AI and privacy risks linked to potential traces of personal data in trained models. In: ICML Workshop on Generative AI + Law (2023)
18. Eldan, R., Russinovich, M.: Who's Harry Potter? Approximate unlearning in LLMs (2023)
19. Lee, D., Rim, D., Choi, M., Choo, J.: Protecting privacy through approximating optimal parameters for sequence unlearning in language models. In: ACL, pp. 15820–15839 (2024)
20. Thudi, A., Deza, G., Chandrasekaran, V., Papernot, N.: Unrolling SGD: understanding factors influencing machine unlearning. arXiv preprint arXiv:2109.13398 (2021)
21. Liu, B., Liu, Q., Stone, P.: Continual learning and private unlearning. In: Conference on Lifelong Learning Agents, pp. 243–254. PMLR (2022)

14 Unlearning Improves Fairness

Zonglin Di

Abstract

Machine unlearning (MU), originally introduced to meet data protection requirements such as the GDPR's "right to be forgotten", has recently emerged as a promising tool for improving algorithmic fairness. While traditional fairness interventions—spanning pre-processing, in-processing, and post-processing techniques—have sought to mitigate bias in machine learning, they remain orthogonal to unlearning, which focuses on efficient data deletion with certified guarantees. Yet fairness and unlearning are seemingly disconnected: fairness constraints are inherently non-decomposable, depending on group-level statistics and cross-group comparisons, whereas most MU algorithms assume independent sample contributions. This incompatibility raises a critical challenge: deletions, especially when disproportionately requested by minority groups, can unintentionally amplify disparities. This chapter provides preliminary evidence for the application of MU in improving model fairness. Experiments on the Adult dataset further demonstrate that unlearning can enhance fairness by removing influential training points that reinforce spurious correlations between sensitive attributes and outcomes. Notably, MU reduces demographic parity disparities without sacrificing predictive accuracy, with methods such as UGradSL+ introduced in an earlier chapter matching retraining performance while offering efficiency gains. Unlearning can be understood not only as a compliance mechanism but also as a fairness-enhancing intervention, offering a new paradigm for developing equitable, privacy-preserving, and socially responsible machine learning systems.

Z. Di (✉)
University of California, Santa Cruz, 1156 High St, Santa Cruz, CA 95064, USA
e-mail: zdi@ucsc.edu

S. Liu et al. (eds.), *Machine Unlearning for Governance of Foundation Models*, Synthesis Lectures on Computer Vision, https://doi.org/10.1007/978-3-032-17282-2_14

14.1 The Definition of Fairness

Machine learning models deployed in high-stakes applications such as hiring [1–3], lending [4, 5], healthcare [6, 7], and criminal justice [1, 8] have demonstrated systematic biases that can perpetuate societal inequities. These fairness concerns have coincided with regulatory frameworks like the EU's General Data Protection Regulation (GDPR) [9], which establishes the "right to be forgotten" and allows individuals to request deletion of their personal data from deployed systems. The machine learning community has responded by developing algorithmic fairness interventions across three main categories: *preprocessing* methods that modify training data [10], *in-processing* techniques that incorporate fairness constraints into learning objectives [11–14], and *post-processing* approaches that adjust model outputs to ensure equitable treatment [15, 16].

Fairness in machine learning encompasses multiple mathematical definitions, with group fairness being the most widely studied framework. The three foundational group fairness criteria are: Demographic Parity (DP), which requires equal positive prediction rates across demographic groups [17–20]; Equality of Opportunity (EOpp), which ensures equal true positive rates for qualified individuals; and Equalized Odds (EOdds), which requires equal both true and false positive rates across groups [16]. The mathematical definitions are given in Table 14.1. Beyond group fairness, the field has developed individual fairness concepts requiring similar treatment for similar individuals, and multicalibration frameworks ensuring well-calibrated predictions across subpopulations.

14.2 Fairness and Machine Unlearning

Traditional machine unlearning methods assume decomposable loss functions where each training example contributes independently, but fairness constraints fundamentally violate this assumption through their reliance on group-level statistics and pairwise comparisons between demographic groups [21]. This incompatibility creates critical challenges: when individuals request data deletion, existing unlearning methods may inadvertently exacerbate

Table 14.1 Comparison of group fairness criteria in binary classification

Fairness criterion	Mathematical definition
Demographic parity	$\Pr(\hat{Y}=1 \mid A=a) = \Pr(\hat{Y}=1 \mid A=a'),\ \forall a, a' \in \mathcal{A}$
Equality of opportunity	$\Pr(\hat{Y}=1 \mid Y=1, A=a) = \Pr(\hat{Y}=1 \mid Y=1, A=a'),\ \forall a, a' \in \mathcal{A}$
Equalized odds	$\Pr(\hat{Y}=1 \mid Y=y, A=a) = \Pr(\hat{Y}=1 \mid Y=y, A=a'),\ \forall y \in \{0,1\},\ a, a' \in \mathcal{A}$

Here, $A \in \mathcal{A}$ is the sensitive attribute, Y the true label, and $\hat{Y}$ the predicted label

bias, particularly when deletions come disproportionately from minority groups. Recent work [22, 23] has addressed this gap by developing fair unlearning algorithms that reformulate fairness constraints into unlearnable forms while maintaining both theoretical guarantees for data removal and bounded fairness performance, demonstrating that responsible AI deployment requires balancing regulatory compliance with data protection rights against ethical obligations to prevent algorithmic discrimination.

The results on the Adult dataset [21] demonstrate that machine unlearning can be an effective mechanism for improving model fairness. The Adult dataset is widely recognized to exhibit gender-related biases in income prediction, and has long served as a benchmark for evaluating fairness interventions [16, 18]. Following the experimental protocol of [21], influence scores were computed for each training instance, designating the top 20% as D_f. Influence scores provide a framework for understanding how individual training examples affect machine learning models when fairness constraints are imposed. These scores quantify the impact of removing a specific training instance on both the model's predictions and its fairness properties, enabling instance-level analysis of algorithmic bias rather than only group-level assessments. An influence function characterizes the change in model predictions when comparing the actual trained model to a counterfactual scenario where a particular training example is excluded. For a model f trained on dataset $\mathcal{D}$, the influence of training example i on target example j is:

$$\text{infl}_f(\mathcal{D}, i, j) := f^{\mathcal{D}/\{i\}}(x_j) - f^{\mathcal{D}}(x_j), \tag{14.1}$$

where $f^{\mathcal{D}/\{i\}}$ represents the model trained excluding example i. When fairness constraints are incorporated through regularization (minimizing loss plus fairness penalty $\lambda\hat{\phi}(f)$), the influence function decomposes into accuracy and fairness components. Under appropriate assumptions, this can be expressed using the Neural Tangent Kernel $\Theta(x_i, x_j; \theta_0)$:

$$\text{infl}_f(\mathcal{D}, i, j) \approx \frac{\eta}{n}\Theta(x_i, x_j; \theta_0)\left[\left.\frac{\partial \ell(w, y_i)}{\partial w}\right|_{w=f(x_i;\theta_0)} + \lambda \left.\frac{\partial \hat{\phi}(f, i)}{\partial f}\right|_{f(x_i;\theta_0)}\right]. \tag{14.2}$$

The first term captures the influence on prediction accuracy, while the second term specifically measures the influence on fairness constraints. After determining D_f, unlearning methods were applied to remove their influence and performance was evaluated using demographic parity.

Table 14.2 also shows that all unlearning methods achieve a reduction in demographic parity disparity compared to the original model. The baseline model exhibits a fairness gap of 0.1355, which is consistently reduced after unlearning, reaching as low as 0.1251 under the retrain strategy. Importantly, these improvements in fairness are obtained without degradation in predictive accuracy; across all methods, accuracy remains comparable to the baseline, with UGradSL+ attaining the highest accuracy (81.81) while still improving fairness. A closer examination of the role of influence scores helps to explain this effect: training samples with high influence often reinforce spurious correlations between sensi-

Table 14.2 Results of fairness using adult dataset with and without MU, showing MU can be used to improve the model fairness

Adult [21]	Retrain	FT	GA	UGradSL	UGradSL+	Ori.
Accuracy (↑)	81.18	81.80	81.58	81.41	81.81	81.77
Fairness (↓)	0.1251	0.1311	0.1290	0.1302	0.1282	0.1355

tive attributes and outcomes, thereby amplifying bias [24–26]. By selectively unlearning such instances, the model reduces its reliance on biased correlations, leading to improved group-level parity. This suggests that unlearning can serve not only as a privacy-preserving mechanism but also as a fairness-enhancing intervention, offering a promising complement to conventional fairness-aware learning approaches for building more equitable and socially responsible machine learning systems.

References

1. Angwin, J., Larson, J., Mattu, S., Kirchner, L.: Machine bias. In: Ethics of Data and Analytics, pp. 254–264. Auerbach Publications (2022)
2. Bogen, M., Rieke, A.: Help wanted: an examination of hiring algorithms, equity, and bias. Upturn **7** (2018)
3. Raghavan, M., Barocas, S., Kleinberg, J., Levy, K.: Mitigating bias in algorithmic hiring: evaluating claims and practices. In: Proceedings of the 2020 Conference on Fairness, Accountability, and Transparency, pp. 469–481 (2020)
4. Siddiqi, N.: Credit Risk Scorecards: Developing and Implementing Intelligent Credit Scoring, vol. 3. Wiley (2012)
5. Ustun, B., Spangher, A., Liu, Y.: Actionable recourse in linear classification. In: Proceedings of the Conference on Fairness, Accountability, and Transparency, pp. 10–19 (2019)
6. Zhou, Y., Huang, S.C., Fries, J.A., Youssef, A., Amrhein, T.J., Chang, M., Banerjee, I., Rubin, D., Xing, L., Shah, N., et al.: Radfusion: benchmarking performance and fairness for multimodal pulmonary embolism detection from CT and EHR. arXiv preprint arXiv:2111.11665 (2021)
7. Pfohl, S.R., Foryciarz, A., Shah, N.H.: An empirical characterization of fair machine learning for clinical risk prediction. J. Biomed. Inform. **113**, 103621 (2021)
8. Chouldechova, A.: Fair prediction with disparate impact: a study of bias in recidivism prediction instruments. Big Data **5**(2), 153–163 (2017)
9. Wolford, B.: Everything you need to know about the "right to be forgotten". https://gdpr.eu/right-to-be-forgotten (2014). Accessed 26 Jan 2024
10. Calmon, F., Wei, D., Vinzamuri, B., Natesan Ramamurthy, K., Varshney, K.R.: Optimized pre-processing for discrimination prevention. Adv. Neural Inf. Process. Syst. **30** (2017)
11. Lowy, A., Baharlouei, S., Pavan, R., Razaviyayn, M., Beirami, A.: A stochastic optimization framework for fair risk minimization. arXiv preprint arXiv:2102.12586 (2021)
12. Berk, R., Heidari, H., Jabbari, S., Joseph, M., Kearns, M., Morgenstern, J., Neel, S., Roth, A.: A convex framework for fair regression. arXiv preprint arXiv:1706.02409 (2017)
13. Agarwal, A., Beygelzimer, A., Dudík, M., Langford, J., Wallach, H.: A reductions approach to fair classification. In: International Conference on Machine Learning, pp. 60–69. PMLR (2018)

14. Martinez, N., Bertran, M., Sapiro, G.: Minimax Pareto fairness: a multi objective perspective. In: International Conference on Machine Learning, pp. 6755–6764. PMLR (2020)
15. Alghamdi, W., Hsu, H., Jeong, H., Wang, H., Michalak, P., Asoodeh, S., Calmon, F.: Beyond adult and COMPAS: fair multi-class prediction via information projection. Adv. Neural. Inf. Process. Syst. **35**, 38747–38760 (2022)
16. Hardt, M., Price, E., Srebro, N.: Equality of opportunity in supervised learning. Adv. Neural Inf. Process. Syst. **29** (2016)
17. Zafar, M.B., Valera, I., Gomez Rodriguez, M., Gummadi, K.P.: Fairness beyond disparate treatment & disparate impact: learning classification without disparate mistreatment. In: Proceedings of the 26th International Conference on World Wide Web, pp. 1171–1180 (2017)
18. Feldman, M., Friedler, S.A., Moeller, J., Scheidegger, C., Venkatasubramanian, S.: Certifying and removing disparate impact. In: Proceedings of the 21st ACM SIGKDD International Conference on Knowledge Discovery and Data Mining, pp. 259–268 (2015)
19. Zliobaite, I.: On the relation between accuracy and fairness in binary classification. arXiv preprint arXiv:1505.05723 (2015)
20. Calders, T., Kamiran, F., Pechenizkiy, M.: Building classifiers with independency constraints. In: 2009 IEEE International Conference on Data Mining Workshops, pp. 13–18. IEEE (2009)
21. Wang, J., Wang, X.E., Liu, Y.: Understanding instance-level impact of fairness constraints. In: International Conference on Machine Learning, pp. 23114–23130. PMLR (2022)
22. Wang, C.L., Huai, M., Wang, D.: Inductive graph unlearning. In: 32nd USENIX Security Symposium (USENIX Security 23), pp. 3205–3222 (2023)
23. Zhang, D., Pan, S., Hoang, T., Xing, Z., Staples, M., Xu, X., Yao, L., Lu, Q., Zhu, L.: To be forgotten or to be fair: unveiling fairness implications of machine unlearning methods. AI Ethics **4**(1), 83–93 (2024)
24. Barocas, S., Hardt, M., Narayanan, A.: Fairness and machine learning. http://fairmlbook.org (2019)
25. Koh, P.W., Liang, P.: Understanding black-box predictions via influence functions. In: International Conference on Machine Learning, pp. 1885–1894. PMLR (2017)
26. Zhang, H., Lakkaraju, H., Zou, J.: Fairness-aware data valuation and influence in machine learning. Adv. Neural Inf. Process. Syst. (NeurIPS) (2023)

Machine Unlearning Across Tasks and Modalities 15

Jiali Cheng and Hadi Amiri

Abstract

This chapter studies machine unlearning in complex settings where models must forget not only specific data samples, but also multimodal relations and skills. We introduce MU-Bench, a benchmark for evaluating unlearning methods across diverse tasks and data modalities, including text, image, audio, and video. We introduce MultiDelete, a method for multimodal unlearning that removes targeted cross-modal associations while preserving essential knowledge. For large language models (LLMs) that have learned to use external tools, we introduce ToolDelete, which removes the *ability* to use specific tools without affecting general language modeling performance. These resources and methods demonstrate that unlearning can be made more effective, scalable, and aligned with real-world needs.

15.1 Machine Unlearning Benchmark (MU-Bench)

Despite rapid progress in machine unlearning (MU), the evaluation of MU methods is fragmented and inconsistent, which limits fair comparison and generalizability. Current research faces the following key challenges: (1) inconsistent evaluation settings: different studies use different trained models (from which data is deleted), forget sets, and performance metrics, which makes fair comparisons across unlearning approaches difficult; (2) limited coverage: most benchmarks focus on a limited range of tasks and unimodal settings, overlooking multimodal and generative tasks where unlearning is equally crucial. Mu-Bench [1] addresses these limitations with a comprehensive evaluation benchmark that evaluates MU methods

J. Cheng · H. Amiri (✉)
University of Massachusetts Lowell, Lowell, MA, USA
e-mail: Hadi_Amiri@uml.edu

S. Liu et al. (eds.), *Machine Unlearning for Governance of Foundation Models*, Synthesis Lectures on Computer Vision, https://doi.org/10.1007/978-3-032-17282-2_15

Table 15.1 Example datasets currently available in MU-Bench, covering various tasks and data modalities from different domains

Dataset	Task	Domain	Modality	$\|D\|$	Metric
		Discriminative tasks			
CIFAR-100 [2]	Image classification	General	Image	50K	Accuracy
IMDB [3]	Sentiment analysis	Review	Text	25K	Accuracy
*DDI-2013 [4]	Relation extraction	Biomedical	Text	25K	Accuracy, F1
NLVR^2 [5]	Visual reasoning	General	Image-Text	62K	Accuracy
*Speech commands [6]	Keyword spotting	Commands	Speech	85K	Accuracy
*UCF101 [7]	Action classification	General	Video	9.3K	Accuracy
		Generative tasks			
SAMSum	Text summarization	Dialogue	Text	14K	rougeL
*BioFact	Text generation	Biography	Text	183	rougeL
Tiny ImageNet [8]	Text-to-image generation	General	Image-text	20K	FID, ClipScore

$|D|$ denotes the size of training data. In MU-Bench, we set the deletion ratio to a maximum of 10% of $|D|$. Rows labeled with * indicate new tasks and data modalities introduced in MU-Bench for machine unlearning

on nine publicly available datasets covering a diverse set of discriminative and generative tasks across multiple data modalities, Table 15.1. These datasets are carefully selected for their broad coverage, ranging from well-established benchmarks to underexplored domains.

For each dataset, the forget set ($\mathcal{D}_f$) is created by randomly sampling 1–10% of the training data per class (with 1% increments) to simulate realistic deletion requests while maintaining the underlying data distributions [9–13]. This allows MU-Bench to stress-test unlearning methods under both typical and extreme data removal scenarios. For broad applicability, MU-Bench evaluates methods across a diverse set of 20 model architectures and 34 scale variants, which allows for rigorous comparison. In addition, it includes a representative set of MU methods selected based on their widespread use and unique characteristics: GradAscent (NegGrad) [14], RandLabel [14], Bad-T [9], SCRUB [15], and SalUn [16]. MU-Bench also provides a publicly available Python package that includes the datasets and implementations of all supported MU methods. The package can be accessed at: https://clu-uml.github.io/MU-Bench-Project-Page/.

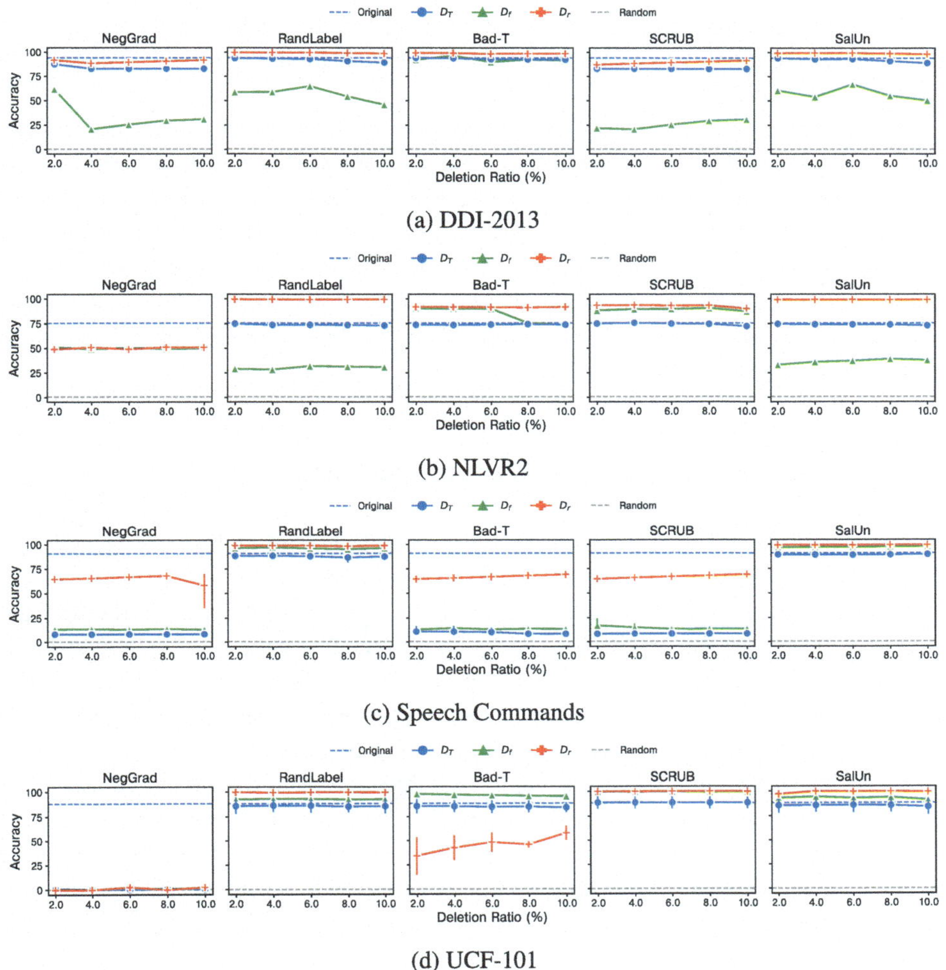

(a) DDI-2013

(b) NLVR2

(c) Speech Commands

(d) UCF-101

Fig. 15.1 Evaluation results on MU-Bench

15.1.1 Evaluation Results Across Diverse Tasks and Modalities

Mu-Bench provides important insights into the performance and limitations of current MU methods across domains, see Fig. 15.1.

In biomedical relation extraction using Drug-Drug Interaction (DDI) dataset, RandLabel, SCRUB, SalUn effectively forget $\mathcal{D}_f$, with SCRUB showing minor degradation on $\mathcal{D}_t$. In contrast, Bad-T performs poorly–likely due to class imbalance and its design for general image data rather than biomedical text, see Fig. 15.1a.

On visual reasoning using NLVR2, RandLabel and SalUn successfully unlearn $\mathcal{D}_f$, but Bad-T and SCRUB fail, which indicates potential over-unlearning or even information leakage risks, see Fig. 15.1b.

On speech classification, none of the methods achieve effective unlearning without sacrificing retention. GradAscent, RandLabel, and SalUn failed on $\mathcal{D}_f$, while Bad-T and SCRUB significantly degrade performance on $\mathcal{D}_t$. This can potentially be due to high temporal dependencies in audio data, which existing MU methods do not address [17], see Fig. 15.1c.

On video action recognition using UCF101, all methods maintain performance on $\mathcal{D}_t$ and $\mathcal{D}_r$, but none succeed in forgetting $\mathcal{D}_f$ (with > 90+% retained). The strong inter-frame dependencies in video data, which are not explicitly addressed in existing MU methods, likely contribute to this failure, see Fig. 15.1d.

15.2 Machine Unlearning for Multimodal Tasks

Multimodal models [18–21] are often trained on large datasets of multiple data types (e.g., images, texts and graphs). Despite recent advances of machine unlearning in *unimodal* settings, unlearning in multimodal settings is largely unexplored; perhaps due to the complex inter-dependencies among modalities and the architectural complexity of multimodal systems. In what follows, we describe the first multimodal unlearning method: MultiDelete [22], focusing on the case of dual data modalities, such as image-text pairs. We note that MultiDelete generalizes to more than two modalities or other types of modalities.

Problem Formulation: Consider a vision-language model f trained on dataset $\mathcal{D} = \{(I_i, T_i)\}_{i=1}^N$ of N image-text pairs. We denote $\mathcal{D}_f, \mathcal{D}_r$ as forget set and retain set respectively. We denote f' as the desired unlearned model. In addition, we assume that the original model f can be decomposed into a vision feature extractor f_I, a language feature extractor f_T, and a modality fusion module f_F. Our core objective is to remove the influence that any $(I_i, T_i) \in \mathcal{D}_f$ has on the parameters of f, while maintaining the performance on $\mathcal{D}_r$ and $\mathcal{D}_t$. Importantly, we aim to unlearn *cross-modal relationship* between image-text pairs $(I_i, T_i) \in \mathcal{D}_f$, but not necessarily the unimodal elements, I_i or T_i. This allows the model to retain its foundational knowledge of individual modalities, which is essential for effective learning of the target task and prevents the unnecessary loss of information. For example, it allows forgetting a user's interaction (e.g. a "like" on a social media post) without removing either the user or the content from the platform.

15.2.1 MultiDelete for Multimodal Unlearning

MultiDelete enforces three key properties for successful multimodal unlearning. **Modality decoupling** makes the relationship between any image-text pair marked for deletion indistinguishable from unrelated pairs, preventing residual associations that could compromise privacy or model generalization [14]. **Multimodal knowledge retention** preserves the learned associations in the retained dataset, so that unlearning specific pairs does not degrade the model's overall cross-modal performance. Finally, **unimodal knowledge retention** maintains the integrity of individual modality representations (e.g., unimodal embeddings) for deleted pairs, allowing the model to retain foundational unimodal knowledge and avoiding the need to relearn basic features. The following definitions formally define these criteria:

Definition 151 (Modality Decoupling) Let $(I_i, T_i) \in \mathcal{D}_f$ denote an image-text pair marked for deletion from model f. The unlearned model f' achieves effective modality decoupling when $(I_i, T_i) \in \mathcal{D}_f$ becomes indistinguishable from any unrelated image-text pair (I_p, T_q), $p \neq q$ sourced from $\mathcal{D}_r$:

$$\mathop{\mathbb{E}}_{(I_i,T_i)\in\mathcal{D}_f,(I_p,T_q)_{p\neq q}} \Big[\phi\big(f'(I_i, T_i)\big) - \phi\big(f(I_p, T_q)\big)\Big] = \epsilon, \tag{15.1}$$

where $f(\cdot)$ and $f'(\cdot)$ generate multimodal representations of their inputs, ϕ is a readout function (e.g. concatenation of representations), and ϵ is an infinitesimal constant.

To realize this property, we randomly draw unrelated image text pairs (I_p, T_q) from $\mathcal{D}_r$, and minimize the difference in multimodal associations between the image-text pairs $(I_i, T_i) \in \mathcal{D}_f$ and the unassociated image-text pairs (I_p, T_q) through minimizing the following distance:

$$\mathcal{L}_{\text{MD}} = \text{Dis}\Big(\big\{f'(I_i, T_i)|(I_i, T_i) \in \mathcal{D}_f\big\}, \big\{f(I_p, T_q)|(I_p, T_p) \in \mathcal{D}_r, p \neq q\big\}\Big), \tag{15.2}$$

where $\text{Dis}(\cdot)$ can be mean squared error. Minimizing this loss trains the model to make D_f pairs indistinguishable from unrelated data pairs. This is a key step for removing any knowledge about D_f.

Definition 152 (Multimodal Knowledge Retention) Let $(I_r, T_r) \in \mathcal{D}_r$ denote an image-text pair that is "not" marked for deletion. The unlearning approach is effective in retaining multimodal knowledge if it minimizes the deviation in the multimodal knowledge between the unlearned model f' and the original model f:

$$\mathop{\mathbb{E}}_{(I_r,T_r)\in\mathcal{D}_r} \Big[\phi\big(f'(I_r, T_r)\big) - \phi\big(f(I_r, T_r)\big)\Big] = \epsilon, \tag{15.3}$$

where the readout function ϕ is a vector combination operator (such as concatenation). We realize this property by minimizing the gap in the multimodal knowledge between f' and f as follows:

$$\mathcal{L}_{\text{MKR}} = \text{Dis}\Big(f'(I_r, T_r), f(I_r, T_r)\Big), (I_r, T_r) \in \mathcal{D}_r. \tag{15.4}$$

Definition 153 (Unimodal Knowledge Retention) Unimodal knowledge can be retained if the unlearning process minimizes the discrepancy between the unimodal representations produced by the unlearned model f' and the original model f:

$$\mathop{\mathbb{E}}_{(I_i,T_i)\in\mathcal{D}_f}\Big[\psi\big(f'_I(I_i), f'_T(T_i)\big) - \psi\big(f_I(I_i), f_T(T_i)\big)\Big] = \epsilon, \tag{15.5}$$

where $f_I(\cdot)$ and $f_T(\cdot)$ generate unimodal representations for image and text data respectively, the readout function ψ is a vector combination operator (such as concatenation), and ϵ is an infinitesimal constant. Note that, although Eq. 15.5 can be applied to all training data $\mathcal{D}$, we only apply it to $\mathcal{D}_f$ samples for efficiency purpose. In fact, we expect unimodal knowledge on remaining data $\mathcal{D}_r$ to be preserved in f' due to Unimodal Knowledge Retention, discussed above.

Thus, we minimize the following gap to realize Multimodal Knowledge Retention:

$$\mathcal{L}_{\text{UKR}} = \text{Dis}\Big(\big[f'_I(I_i); f'_T(T_i)\big], \big[f_I(I_i); f_T(T_i)\big]\Big), (I_i, T_i) \in \mathcal{D}_f, \tag{15.6}$$

where [;] denotes vector concatenation. This loss aims to retain the core unimodal knowledge during training even after unlearning certain relationships. We note that an alternative approach is to use the fusion module f_F, while freezing the unimodal encoders f_I and f_T. However, this can limit flexibility, especially for models like CLIP [18] with nonparametric fusion modules for modality interaction. There is also a risk that an adversarial agent might exploit the original f_F and take advantage of the frozen image and text representations. Therefore, we advocate for the strategy in $\mathcal{L}_{\text{UKR}}$ but encourage the adjustments to be minimal.

The full unlearning process integrates the three loss functions:

$$\mathcal{L} = \alpha\mathcal{L}_{\text{MD}} + \beta\mathcal{L}_{\text{MKR}} + \gamma\mathcal{L}_{\text{UKR}}, \tag{15.7}$$

which enables targeted forgetting while preserving both cross-modal and unimodal representations.

15.2.2 Results on Multimodal Tasks

MultiDelete is flexible and broadly applicable to a wide range of tasks, including: Image-Text Retrieval (TR) and (IR) on Flickr30K, Visual Entailment (VE) on SNLI-VE [23], Natural Language for Visual Reasoning on NLVR^2 [5], and Graph-Text Classification on phenotype-Gene relationship (PGR) [24].

The results in Table 15.2 show that on average across all *image-text* tasks, MultiDelete achieves 88.0 on $\mathcal{D}_t$, outperforming all baselines by 17.4 absolute points. Furthermore, it achieves 31.1 on $\mathcal{D}_f$, outperforming all baselines by 17.6 absolute points. In addition,

Table 15.2 Experimental results on image-text and graph-text datasets on ALBEF [20]

Method	Image-text								Graph-text		Avg.	
	Flickr30K				SNLI-VE		NLVR^2		PGR			
	IR		TR									
	$\mathcal{D}_t$	$\mathcal{D}_f$	$\mathcal{D}_t$	$\mathcal{D}_f$	$\mathcal{D}_t$	$\mathcal{D}_f$	$\mathcal{D}_t$	$\mathcal{D}_f$	$\mathcal{D}_t$	$\mathcal{D}_f$	$\mathcal{D}_t$	$\mathcal{D}_f$
Retrain	97.8	50.4	93.5	50.4	79.4	50.2	80.3	50.3	67.5	50.2	83.4	50.3
FineTune	96.7	50.4	94.1	50.4	79.1	50.5	80.3	49.8	67.4	49.9	83.5	50.2
FineTune-F	**97.1**	49.9	**94.6**	49.9	79.5	49.9	81.2	50.0	67.5	50.1	83.8	49.9
GradAscent	92.4	50.5	91.7	50.5	77.8	48.6	77.3	50.6	63.4	49.6	80.5	50.0
GradAscent-F	93.3	50.2	90.6	50.2	79.6	50.6	**80.8**	50.0	63.5	49.9	81.5	50.2
Decent-to-Delete	10.3	51.4	8.9	51.4	45.2	50.1	50.8	49.8	50.0	50.2	33.0	50.5
Decent-to-Delete-F	22.5	50.9	20.7	50.9	48.6	49.8	50.9	49.8	53.6	50.2	39.2	50.2
L-CODEC	83.5	50.0	78.5	50.0	56.7	49.9	55.3	52.7	57.8	48.8	66.3	50.3
L-CODEC-F	87.4	49.4	50.6	48.2	57.4	48.4	56.8	53.1	59.1	46.9	62.2	49.2
ERM-KTP	57.4	48.7	56.2	49.0	53.2	48.9	52.9	50.8	N/A		54.9	49.3
ERM-KTP-F	N/A											
UL	95.1	50.4	90.3	50.4	75.7	49.8	76.3	50.4	64.8	49.7	80.4	50.2
UL-F	94.4	50.2	94.1	50.2	79.1	49.7	76.8	50.4	66.1	48.8	82.1	49.8
MultiDelete	**97.1**	**33.2**	94.3	**33.2**	**79.8**	**35.3**	**80.8**	**23.5**	**68.5**	**18.6**	**84.2**	**28.7**
MultiDelete-F	96.8	34.4	94.1	34.5	79.5	36.3	80.4	26.4	**67.7**	**19.5**	83.7	30.2

Performance shows average of recall@1, recall@3, recall@10 on Flickr30K-TR and Flickr30K-IR, and accuracy on other datasets. ERM-KTP inserts trainable parameters to the vision encoder, causing the ERM-KTP-F variant ('-F' suffix in model titles denotes variants where only fusion module parameters are updated during unlearning) inapplicable. The best results are in **bold** and the second best results are underlined. The Retrain performance is provided *for reference purpose only*

MultiDelete effectively reduces the likelihood of deleted data ($\mathcal{D}_f$) being identified, resulting in an average MI ratio of 1.3 across all tasks. These results indicate that MultiDelete can accomplish effective and targeted unlearning, while maintaining strong capability and utility on downstream tasks. In addition, on PGR dataset, MultiDelete outperforms baselines on $\mathcal{D}_t$, outperforming Fine-Tune by +1.1 points, GradAscent by +5.1 points, Decent-to-Delete by +38.5 points, L-CODEC by +10.7 points, and UL by +3.7 points. In addition, MultiDelete achieves 18.6 on deleted data ($\mathcal{D}_f$), significantly outperforming the top-performing baseline model, L-CODEC (48.8). The performance of MultiDelete on the original test set $\mathcal{D}_t$ is better than that of Retrain by +1.0.

15.3 Tool Unlearning: A Case Study of Skill-Level Unlearning

Tool-augmented Large Language Models (LLMs) can use external tools such as calculators [25], Python interpreters [26], APIs [27], or AI models [28] to enable vanilla LLMs to solve more complex tasks [25, 28]. They are often trained on query-response examples, which embed the ability to use tools *directly* into model parameters. Despite the growing adoption of tool-augmented LLMs, the ability to unlearn specific tools is underexplored. Tool unlearning is essential for addressing security, privacy, and model reliability concerns. For example, in a clinical setting, if an API used by an LLM is flagged as insecure due to violating regulations like HIPAA, tool unlearning is necessary to prevent its invocation. Similarly, when tools are deprecated or upgraded (e.g. Python packages), removing outdated tool knowledge is essential to avoid outdated or erroneous outputs.

The task of **Tool Unlearning** is formalized in [29], which aims to remove the ability of using specific tools from a tool-augmented LLM while preserving its ability to use other tools and perform general tasks of LLMs such as text generation. This departs from traditional sample-level unlearning as it focuses on removing "skills" or the ability to use specific tools, rather than removing individual data samples from a model. In addition, existing membership inference attack (MIA) techniques, a common evaluation method in machine unlearning to determine whether specific data samples were part of training data, are inadequate for evaluating tool unlearning because they focus on sample-level data rather than tool-based knowledge.

To formulate tool unlearning, we first introduce the concept of "tool learning". Let $\mathcal{D} = \{\mathcal{T}, \mathcal{Q}, \mathcal{Y}\}$ be a dataset with N tools $\mathcal{T}$, and $(\mathcal{Q}, \mathcal{Y})$ denotes query-output examples that demonstrate how to use the tools in $\mathcal{T}$. Each tool $t_i \in \mathcal{T}$ may have one or more demonstrations $\{\mathcal{Q}_i, \mathcal{Y}_i\}$, $|\mathcal{Q}_i| = |\mathcal{Y}_i| \geq 1$. Starting with an instruction-tuned LLM f_0, a tool learning algorithm explicitly trains f_0 on $\mathcal{D}$ and results in a *tool-augmented* model f capable of using the N tools in $\mathcal{T}$. We note that prior to explicit tool learning, the LLM f_0 may already have some tool-using capabilities such as performing basic arithmetic operations.

Problem Formulation: Let $\mathcal{D}_f = \{\mathcal{T}_f, \mathcal{Q}_f, \mathcal{Y}_f\}$ denotes $k < N$ tools and their corresponding demonstrations to be unlearned from the tool-augmented model f, and $\mathcal{D}_r = \mathcal{D} \backslash \mathcal{D}_f =$

$\{\mathcal{T}_r, \mathcal{Q}_r, \mathcal{Y}_r\}$ denotes the remaining tools and their demonstrations to retain. The goal of tool unlearning is to obtain an unlearned model f' that has limited knowledge on using $\mathcal{T}_f$ tools–can no longer perform tasks involving $\mathcal{T}_f$ tools–while preserving f's ability to use $\mathcal{T}_r$ tools as before.

Distinction from Sample-Level Unlearning: Tool unlearning differs from traditional unlearning from several aspects: **Objective**: sample-level unlearning aims to reduce the memorization likelihood or extraction probabilities of specific data samples [30]. In contrast, tool unlearning targets broader behavioral removal (i.e., "skills"), and success is measured by the absence of tool-usage behavior. **Evaluation**: traditional metrics such as likelihood and perplexity are insufficient. Skill-oriented metrics are required to assess capability removal and retention. **Data**: unlike sample-level unlearning, tool unlearning does not require access to all individual samples marked for unlearning. This aligns with "concept erasure" in diffusion models [31, 32] and zero-shot unlearning [33] but differs from traditional LLM unlearning [34]. Naive approaches–such as removing tool descriptions from prompts or editing tool registries–fail to remove tool knowledge. Since tool knowledge is embedded in model parameters, parameter-level updates are necessary to prevent leakage or behavioral inconsistencies.

15.3.1 ToolDelete for Tool Unlearning in LLMs

ToolDelete removes specific tool-using capabilities from LLMs while preserving remaining tools and general language abilities. Effective tool unlearning requires satisfying three key properties. **Tool Knowledge Deletion** enforces the unlearned model f' to retain no more knowledge about the removed tools than the original tool-free model, erasing any learned behavior associated with those tools. **Tool Knowledge Retention** requires that the model's ability to use the remaining tools is preserved, maintaining its functionality on those tools. Finally, **General Capability Retention** enforces the model to perform foundational tasks–such as text generation, question answering, and code synthesis–that pre-existed before unlearning. These criteria make tool unlearning both targeted and non-destructive to the model's broader utility. We formally define them as follows:

Definition 154 (Tool Knowledge Deletion (TKD)) Let $t_i \in \mathcal{T}_f$ denote a tool to be unlearned and g be a function that quantifies the amount of knowledge a model has about a tool. The unlearned model f' satisfies tool knowledge deletion if:

$$\mathop{\mathbb{E}}_{t_i \in \mathcal{T}_f} [g(f_0, t_i) - g(f', t_i)] \geq 0. \tag{15.8}$$

This formulation allows users to control the extent of knowledge removal from f'. For instance, when we unlearn a "malicious" tool that calls a malignant program, we may require f' retains no knowledge of this tool, i.e. $g(f', t_i) = 0$. In less critical cases, users can choose to reset f''s knowledge to *pre*-tool augmentation level, i.e. $g(f', t_i) = g(f_0, t_i)$. To measure tool knowledge in LLMs, we follow previous works that used prompting to probe LLMs' knowledge [35, 36], i.e. adopting the output of LLMs as their knowledge on a given tool. For each $t_i \in \mathcal{T}_f$ and its associated demonstrations $\{\mathcal{Q}_i, \mathcal{Y}_i\}$, we query the tool-free LLM f_0 with $\mathcal{Q}_i$ and collect its responses $\mathcal{Y}_i' = f_0(\mathcal{Q}_i)$. Since f_0 has never seen t_i or $\{\mathcal{Q}_i, \mathcal{Y}_i\}$, $\mathcal{Y}_i'$ represents the **tool-free response**. We then constrain the unlearned model f' to generate responses similar to $\mathcal{Y}_i'$ to prevent it from retaining knowledge of t_i.

Definition 155 (Tool Knowledge Retention (TKR)) Let $t_m \in T_r$ denote a retained tool, and let g be a function that quantifies the amount of knowledge a model has about a tool. The unlearned model f' satisfies tool knowledge retention if:

$$\mathop{\mathbb{E}}_{t_m \in \mathcal{T}_r} [g(f, t_m) - g(f', t_m)] = \epsilon, \tag{15.9}$$

where ϵ is an infinitesimal constant, so that f' retains the same knowledge of tools in T_r as the original model f.

For effective retention, f' is further fine-tuned using demonstrations associated with $\mathcal{T}_r$, or, more practically, a subset of $\mathcal{T}_r$ proportional to $\mathcal{T}_f$ for efficiency.

Definition 156 (General Capability Retention (GCR)) Let $\mathcal{T}_g$ denote the general tasks used to evaluate LLMs. The unlearned model f' satisfies general capability retention if it preserves the knowledge on T_G that it originally obtained prior to tool learning:

$$\mathop{\mathbb{E}}_{t_g \in \mathcal{T}_g} [g(f_0, t_g) - g(f', t_g)] = \epsilon, \tag{15.10}$$

where ϵ is an infinitesimal constant.

We propose to use task arithmetic [37, 38] to preserve the general capabilities of the unlearned model:

$$\theta'^{*} \leftarrow \theta' + (\theta_0 - \theta_R), \tag{15.11}$$

where θ_0 and θ_R denote the parameters of the instruction-tuned model f_0 and its random initialization f_R respectively. This approach is efficient, practical, and effective for preserving general capabilities.

To obtain the unlearned model f', we solve the following Equation that loss balances forgetting and retention:

$$\theta'^{*} = \arg\min_{\theta'} \underbrace{\mathbb{E}_{t_i \in \mathcal{T}_f}[g(f_0, t_i) - g(f', t_i)]}_{\text{knowledge deletion of } \mathcal{T}_f} + \underbrace{\mathbb{E}_{t_m \in \mathcal{T}_r}[g(f, t_m) - g(f', t_m)]}_{\text{knowledge retention of } \mathcal{T}_r}, \tag{15.12}$$

and once the optimized model parameters θ'^{*} are obtained, we apply task arithmetic to reinforce general capabilities:

$$\theta'^{*} = \underbrace{\theta'^{*}}_{\text{post-optimization weights}} + \underbrace{\alpha(\theta_0 - \theta_R)}_{\text{knowledge retention of } \mathcal{T}_g}, \qquad (15.13)$$

where α is a hyperparameter to control the magnitude of task arithmetic. The above formulation provides flexibility in training ToolDelete using various existing paradigms, including supervised fine-tuning (SFT), direct preference optimization (DPO) [39], reinforcement learning from human feedback (RLHF) [40], parameter-efficient fine-tuning (PEFT) [41, 42], or quantization [43, 44] techniques. Below we describe two variants of ToolDelete:

- **ToolDelete-SFT** fine-tunes f using language modeling loss. On forget tools $\mathcal{T}_f$, we replace the original responses $\mathcal{Y}_f$ with tool-free responses $\mathcal{Y}'_f$. The samples for $\mathcal{T}_r$ are not modified.
- **ToolDelete-DPO** uses direct preference optimization (DPO) to prioritize winning responses over losing responses. For $(t_i, \mathcal{Q}_i, \mathcal{Y}_i) \in \mathcal{T}_f$ to be unlearned, we prioritize the corresponding tool-free response $\mathcal{Y}'_i$ over the original response $\mathcal{Y}_i$. For $(t_j, \mathcal{Q}_j, \mathcal{Y}_j) \in \mathcal{T}_r$, the original response $\mathcal{Y}_j$ is prioritized over the tool-free response $\mathcal{Y}'_i$.

15.3.2 Tool Unlearning Results

Table 15.3 shows the performance of ToolDelete compared to general and LLM-specific unlearning baselines when 20% of tools are deleted from ToolAlpaca dataset [27]. Compared to Retrain, the best-performing baseline, ToolDelete-SFT achieves gains of 0.6, 0.3, 8.0, 2.3 absolute points on $\mathcal{T}_t$, $\mathcal{T}_r$, $\mathcal{T}_f$, $\mathcal{T}_g$ respectively. ToolDelete-DPO shows stronger results, outperforming Retrain by 1.3, 3.3, 9.8, 1.8 points on the same metrics. We note that GradAscent can effectively unlearn $\mathcal{T}_f$, but it negatively impacts its $\mathcal{T}_t$ and $\mathcal{T}_r$ performance. In addition, although RandLabel and SalUn outperforms GradAscent, they still fall short on $\mathcal{T}_g$ compared to ToolDelete.

Existing LLM unlearning methods, despite being effective in sample-level unlearning, are prone to under-performing in tool unlearning. Both ToolDelete-SFT and ToolDelete-DPO outperforms ICUL, SGA, and TAU on $\mathcal{T}_t$, $\mathcal{T}_r$, $\mathcal{T}_f$ and $\mathcal{T}_g$. The only exception is ICUL, which outperforms ToolDelete-SFT on $\mathcal{T}_r$ by 2.7 absolute points, but is outperformed by ToolDelete-DPO on $\mathcal{T}_r$ by 0.3 points. The good performance of ICUL on $\mathcal{T}_r$ is at the cost of underperforming on tools in $\mathcal{T}_f$. In addition, ICUL has limited ability of preserving test set performance, it is outperformed by ToolDelete-SFT and ToolDelete-DPO by 3.6 and 4.3 respectively. Furthermore, ICUL's performance drops considerably as the deletion size increases, which may indicate limited scalability. In contrast, ToolDelete remains stable and can process larger deletion sizes efficiently.

Table 15.3 Tool unlearning performances when deleting 20% of tools on ToolAlpaca

	Method	$\mathcal{T}_t(\uparrow)$	$\mathcal{T}_r(\uparrow)$	$\mathcal{T}_f(\downarrow)$	General capability $\mathcal{T}_g(\uparrow)$				
					STEM	Reason	Ins-Follow	Fact	Avg.
	Original (ref only)	60.0	73.1	75.7	31.7	17.1	22.6	25.0	24.1
General	Retrain	52.1	71.8	38.5	30.5	16.1	14.2	24.7	21.3
	GradAscent	33.3	51.4	34.6	21.4	10.4	12.9	13.1	14.5
	RandLabel	50.3	70.3	37.5	26.3	16.4	13.6	25.1	20.3
	SalUn	46.2	54.3	38.2	27.1	17.0	17.4	19.5	20.2
LLM-specific	ICUL	49.1	<u>74.8</u>	58.3	12.4	8.7	1.6	6.2	7.3
	SGA	43.5	63.0	42.1	21.5	11.6	17.0	14.7	16.2
	TAU	43.8	61.7	42.5	22.0	17.6	22.3	21.7	20.9
	CUT	44.7	61.5	40.2	21.6	14.8	20.8	16.4	18.4
	NPO	50.8	66.9	<u>30.1</u>	20.7	15.3	21.9	18.9	19.2
	SOUL	50.4	68.3	33.8	31.6	17.2	21.4	20.8	22.7
Ours	ToolDelete-SFT	<u>52.7</u>	72.1	<u>30.5</u>	31.3	17.5	21.7	24.1	**23.6**
	ToolDelete-DPO	**53.4**	**75.1**	**28.7**	31.6	16.8	20.4	23.5	<u>23.1</u>

Best and second-best performances are **bold** and <u>underlined</u> respectively. *Original* is provided *for reference only*

References

1. Cheng, J., Amiri, H.: Mu-bench: a multitask multimodal benchmark for machine unlearning. arXiv preprint arXiv:2406.14796 (2024)
2. Krizhevsky, A., Hinton, G.: Learning multiple layers of features from tiny images (2009)
3. Maas, A., Daly, R., Pham, P., Huang, D., Ng, A., Potts, C.: Learning word vectors for sentiment analysis. In: Proceedings of the 49th Annual Meeting of the Association for Computational Linguistics: Human Language Technologies (2011)
4. Segura-Bedmar, I., Martínez, P., Herrero-Zazo, M.: SemEval-2013 task 9: extraction of drug-drug interactions from biomedical texts (DDIExtraction 2013). In: Second Joint Conference on Lexical and Computational Semantics (*SEM), Volume 2: Proceedings of the Seventh International Workshop on Semantic Evaluation (SemEval 2013) (2013)
5. Suhr, A., Zhou, S., Zhang, A., Zhang, I., Bai, H., Artzi, Y.: A corpus for reasoning about natural language grounded in photographs. In: Proceedings of the 57th Annual Meeting of the Association for Computational Linguistics (2018)
6. Warden, P.: Speech commands: a dataset for limited-vocabulary speech recognition. arXiv preprint arXiv:1804.03209 (2018)
7. Soomro, K., Zamir, A.R., Shah, M.: Ucf101: a dataset of 101 human actions classes from videos in the wild. arXiv preprint arXiv:1212.0402 (2012)
8. Le, Y., Yang, X.: Tiny imagenet visual recognition challenge (2015)

9. Chundawat, V.S., Tarun, A.K., Mandal, M., Kankanhalli, M.: Can bad teaching induce forgetting? Unlearning in deep networks using an incompetent teacher. In: Proceedings of the AAAI Conference on Artificial Intelligence, vol. 37, pp. 7210–7217 (2023)
10. Cheng, J., Dasoulas, G., He, H., Agarwal, C., Zitnik, M.: GNNDelete: a general strategy for unlearning in graph neural networks. In: The Eleventh International Conference on Learning Representations (2023)
11. Chen, Z., Cheng, J., Tolomei, G., Liu, S., Amiri, H., Wang, Y., Nag, K., Lin, L.: Frog: fair removal on graphs. arXiv preprint arXiv:2503.18197 (2025)
12. Cheng, J., Amiri, H.: Understanding machine unlearning through the lens of mode connectivity. In: ICML 2025 Workshop on Machine Unlearning for Generative AI (2025)
13. Chen, Z., Huang, J., Cheng, J., Guo, Y., Wang, M., Morishetti, L., Nag, K., Amiri, H.: Future: flexible unlearning for tree ensemble. arXiv preprint arXiv:2508.21181 (2025)
14. Golatkar, A., Achille, A., Soatto, S.: Eternal sunshine of the spotless net: selective forgetting in deep networks. In: Proceedings of the IEEE/CVF Conference on Computer Vision and Pattern Recognition, pp. 9304–9312 (2020)
15. Kurmanji, M., Triantafillou, P., Triantafillou, E.: Towards unbounded machine unlearning. arXiv preprint arXiv:2302.09880 (2023)
16. Fan, C., Liu, J., Zhang, Y., Wong, E., Wei, D., Liu, S.: Salun: empowering machine unlearning via gradient-based weight saliency in both image classification and generation. In: The Twelfth International Conference on Learning Representations (2024)
17. Cheng, J., Amiri, H.: Speech unlearning. arXiv preprint arXiv:2506.00848 (2025)
18. Radford, A., Kim, J.W., Hallacy, C., Ramesh, A., Goh, G., Agarwal, S., Sastry, G., Askell, A., Mishkin, P., Clark, J., et al.: Learning transferable visual models from natural language supervision. In: International conference on machine learning, pp. 8748–8763. PmLR (2021)
19. Kim, W., Son, B., Kim, I.: Vilt: vision-and-language transformer without convolution or region supervision. In: International conference on machine learning, pp. 5583–5594. PMLR (2021)
20. Li, J., Selvaraju, R., Gotmare, A., Joty, S., Xiong, C., Hoi, S.C.H.: Align before fuse: vision and language representation learning with momentum distillation. Adv. Neural. Inf. Process. Syst. **34**, 9694–9705 (2021)
21. Li, J., Li, D., Xiong, C., Hoi, S.: Blip: bootstrapping language-image pre-training for unified vision-language understanding and generation. In: International Conference on Machine Learning, pp. 12888–12900. PMLR (2022)
22. Cheng, J., Amiri, H.: Multidelete for multimodal machine unlearning. In: European Conference on Computer Vision, pp. 165–184. Springer (2024)
23. Xie, N., Lai, F., Doran, D., Kadav, A.: Visual entailment: a novel task for fine-grained image understanding. arXiv preprint arXiv:1901.06706 (2019)
24. Sousa, D., Lamúrias, A., Couto, F.M.: A silver standard corpus of human phenotype-gene relations. In: Proceedings of the 2019 Conference of the North American Chapter of the Association for Computational Linguistics: Human Language Technologies (2019)
25. Schick, T., Dwivedi-Yu, J., Dessì, R., Raileanu, R., Lomeli, M., Hambro, E., Zettlemoyer, L., Cancedda, N., Scialom, T.: Toolformer: Language models can teach themselves to use tools. Adv. Neural. Inf. Process. Syst. **36**, 68539–68551 (2023)
26. Gao, L., Madaan, A., Zhou, S., Alon, U., Liu, P., Yang, Y., Callan, J., Neubig, G.: PAL: program-aided language models. In: International Conference on Machine Learning, pp. 10764–10799. PMLR (2023)
27. Tang, Q., Deng, Z., Lin, H., Han, X., Liang, Q., Cao, B., Sun, L.: Toolalpaca: generalized tool learning for language models with 3000 simulated cases. arXiv preprint arXiv:2306.05301 (2023)
28. Patil, S.G., Zhang, T., Wang, X., Gonzalez, J.E.: Gorilla: large language model connected with massive apis. Adv. Neural. Inf. Process. Syst. **37**, 126544–126565 (2024)

29. Cheng, J., Amiri, H.: Tool unlearning for tool-augmented LLMs. In: Forty-Second International Conference on Machine Learning (2025)
30. Jang, J., Yoon, D., Yang, S., Cha, S., Lee, M., Logeswaran, L., Seo, M.: Knowledge unlearning for mitigating privacy risks in language models. In: Proceedings of the 61st Annual Meeting of the Association for Computational Linguistics (Volume 1: Long Papers) (2023)
31. Gandikota, R., Materzynska, J., Fiotto-Kaufman, J., Bau, D.: Erasing concepts from diffusion models. In: Proceedings of the IEEE/CVF International Conference on Computer Vision, pp. 2426–2436 (2023)
32. Kumari, N., Zhang, B., Wang, S.Y., Shechtman, E., Zhang, R., Zhu, J.Y.: Ablating concepts in text-to-image diffusion models. In: Proceedings of the 2023 IEEE International Conference on Computer Vision (2023)
33. Chundawat, V.S., Tarun, A.K., Mandal, M., Kankanhalli, M.: Zero-shot machine unlearning. IEEE Trans. Inf. Forensics Secur. **18**, 2345–2354 (2023)
34. Yao, J., Chien, E., Du, M., Niu, X., Wang, T., Cheng, Z., Yue, X.: Machine unlearning of pre-trained large language models. In: Proceedings of the 62nd Annual Meeting of the Association for Computational Linguistics (Volume 1: Long Papers) (2024)
35. Brown, T., Mann, B., Ryder, N., Subbiah, M., Kaplan, J.D., Dhariwal, P., Neelakantan, A., Shyam, P., Sastry, G., Askell, A., Agarwal, S., Herbert-Voss, A., Krueger, G., Henighan, T., Child, R., Ramesh, A., Ziegler, D., Wu, J., Winter, C., Hesse, C., Chen, M., Sigler, E., Litwin, M., Gray, S., Chess, B., Clark, J., Berner, C., McCandlish, S., Radford, A., Sutskever, I., Amodei, D.: Language models are few-shot learners. In: Advances in Neural Information Processing Systems, vol. 33, pp. 1877–1901 (2020)
36. Singhal, K., Azizi, S., Tu, T., Mahdavi, S.S., Wei, J., Chung, H.W., Scales, N., Tanwani, A., Cole-Lewis, H., Pfohl, S., et al.: Large language models encode clinical knowledge. Nature **620**(7972), 172–180 (2023)
37. Ilharco, G., Ribeiro, M.T., Wortsman, M., Schmidt, L., Hajishirzi, H., Farhadi, A.: Editing models with task arithmetic. In: The Eleventh International Conference on Learning Representations. https://openreview.net/forum?id=6t0Kwf8-jrj (2023)
38. Barbulescu, G.O., Triantafillou, P.: To each (textual sequence) its own: improving memorized-data unlearning in large language models. In: Proceedings of the 40th International Conference on Machine Learning (2024)
39. Rafailov, R., Sharma, A., Mitchell, E., Manning, C.D., Ermon, S., Finn, C.: Direct preference optimization: your language model is secretly a reward model. In: Thirty-Seventh Conference on Neural Information Processing Systems. https://openreview.net/forum?id=HPuSIXJaa9 (2023)
40. Ouyang, L., Wu, J., Jiang, X., Almeida, D., Wainwright, C., Mishkin, P., Zhang, C., Agarwal, S., Slama, K., Ray, A., et al.: Training language models to follow instructions with human feedback. Adv. Neural. Inf. Process. Syst. **35**, 27730–27744 (2022)
41. He, J., Zhou, C., Ma, X., Berg-Kirkpatrick, T., Neubig, G.: Towards a unified view of parameter-efficient transfer learning. In: International Conference on Learning Representations (2022)
42. Su, Y., Chan, C.M., Cheng, J., Qin, Y., Lin, Y., Hu, S., Yang, Z., Ding, N., Sun, X., Xie, G., et al.: Exploring the impact of model scaling on parameter-efficient tuning. In: Proceedings of the 2023 Conference on Empirical Methods in Natural Language Processing (2023)
43. Dettmers, T., Lewis, M., Belkada, Y., Zettlemoyer, L.: Gpt3. int8 (): 8-bit matrix multiplication for transformers at scale. Adv. Neural. Inf. Process. Syst. **35**, 30318–30332 (2022)
44. Ma, S., Wang, H., Ma, L., Wang, L., Wang, W., Huang, S., Dong, L., Wang, R., Xue, J., Wei, F.: The era of 1-bit LLMs: all large language models are in 1.58 bits. arXiv preprint arXiv:2402.17764 (2024)

Unlearning in Federated Learning Settings

16

Anisa Halimi, Swanand Ravinda Kadhe, Ambrish Rawat and Nathalie Baracaldo

Abstract

Federated learning (FL) is a paradigm where multiple clients train a single machine learning model without exposing their training data among them or with a central place. Applying unlearning in this setting is particularly beneficial, in particular when one of the clients decides to leave the federation and wants its data to be removed. While numerous unlearning techniques have been proposed in centralized settings, they cannot be directly applied in the context of distributed settings like federated learning due to the differences in learning protocol and the presence of multiple actors. In this chapter, we first introduce the concept of federated learning and explain why techniques explained in other book chapters are are not suitable for this setting. Then, we overview and contrast existing techniques specially designed for FL. Finally, we dive into one technique that is compatible with complementary multi-party computation commonly applied in FL to efficiently remove a client from a federation. To erase a client, we propose to first perform *local unlearning* at the client to be erased, and then use the locally unlearned model as the initialization to run very few rounds of federated learning between the server and the remaining clients to obtain the unlearned *global* model. We empirically compare multiple FL-based unlearning methods by employing multiple performance measures on three datasets, and demonstrate their strengths and weakener.

A. Halimi · A. Rawat
IBM Research Europe, Dublin, Ireland
e-mail: anisa.halimi@ibm.com

A. Rawat
e-mail: ambrish.rawat@ie.ibm.com

S. R. Kadhe · N. Baracaldo (✉)
IBM Research, Almaden, USA
e-mail: baracald@us.ibm.com

S. R. Kadhe
e-mail: swanand.kadhe@ibm.com

S. Liu et al. (eds.), *Machine Unlearning for Governance of Foundation Models*, Synthesis Lectures on Computer Vision, https://doi.org/10.1007/978-3-032-17282-2_16

16.1 Introduction

Modern machine learning is increasingly using large-size deep neural networks trained on massive datasets. While large model sizes and massive datasets have improved models performance, privacy implications have risen since large models tend to memorize aspects of their training data [1–3]. Such privacy implications make it challenging to meet privacy regulations [4–6], which provide to data owners the right to be forgotten. Due to these privacy requirements, the field of *machine unlearning* has recently received significant research attention [7, 8].

Machine unlearning, in a nutshell, removes the influence of specific training samples from a trained model, while maintaining the performance of the model. The problem of machine unlearning is even more challenging in the distributed paradigm of federated learning (FL), which allows multiple parties to jointly train a shared model while keeping their data on-premise [9]. FL operates through a collaborative training cycle between an aggregator (server) and multiple parties (clients). Each client trains a local model on its data and transmits only the model updates (e.g., gradients or weights changes) to the server. The server aggregates these updates to produce an improved global model, which is then redistributed to the clients for further training until convergence. Figure 16.1 shows an overview of federated learning. A comprehensive overview of federated learning can be found in [10].

Machine unlearning techniques developed for the centralized setting, e.g., [11–13] cannot be directly applied in the FL setting due to its distributed nature, where all the participating clients contribute to learn the final *global* model. A naive way of implementing federated unlearning is to retrain the model from scratch after removing from the corresponding client(s) the data sample(s) that are requested to be deleted. Retraining from scratch can be considered as the *gold standard* for unlearning, as it can ensure *exact* unlearning [12, 14]. However, it incurs prohibitively large communication and computation costs, making it infeasible in some real-world FL settings.

A diverse set of approaches [14–18] has explored federated unlearning from different perspectives. These works consider different privacy setups and achieve different objectives.

From the *objective perspective*, some solutions focus on unlearning an entire class or category, while others require unlearning a single client in the federation. In cases where a whole client needs to be removed, it is possible that the client will not be available to participate in the unlearning process for example due to lack of battery.

From the *privacy perspective*, the underlining needs of a federation define how communications among participants are carried on, and as a result, what unlearning methods can be applied. There are multiple possible FL configurations that varied based on their threat model, which establish how much each client trusts each other, as well as how much clients trust the aggregator and vice-versa. Based on the threat model at hand, some federations simply use end-to-end secure channels and the protocol shown in Fig. 16.1, while other federations require more complex multi-party computation approaches where, for example clients need to encrypt their model updates before sharing them with the aggregator. A comprehensive overview of privacy solutions and when to use each of them can be found in [19].

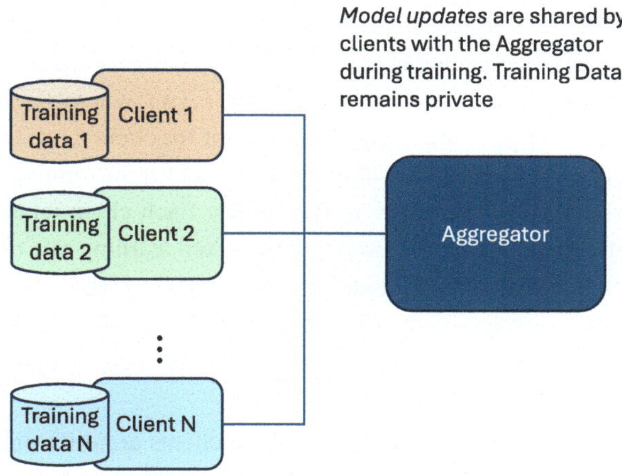

Fig. 16.1 Federated learning overview

Some unlearning FL methods require the server to store updates from each client in every round. In FL settings that require stringent privacy requirements, client updates can only be held ephemerally at the server because storing client updates at the server may have serious privacy implications due to potential leakage from model updates [20]. Hence, those approaches can only be applied to federations where there is some level of trust that enables storage of model updates. A different approach we term *projected gradient descent based federated unlearning* (PGDU) method [18] does not require the server or clients to store any client updates or even global updates. PGDU is therefore suitable for more stringent privacy needs.

In the rest of this chapter, we first introduce FL and define the notation. Then, we overview existing FL unlearning techniques and explain some of their drawbacks. After that, we describe the details of PGDU [18] a method that is both efficient and compatible with popular multi-party computation approaches.

16.2 Background

In this section, we introduce federated learning and the concept of federated unlearning.

16.2.1 Federated Learning Basics and Notation

In a federated learning framework [9], a (global) model is trained in a distributed way with the help of an aggregator (server), where each participating client contributes to training without sharing their data with the other participants. We consider the supervised federated learning setup with N clients, each with dataset $\mathcal{D}_i = \{(\mathbf{x}_i, y_i)_{i\in[n_i]}\}$ (where $[n_i] = \{1, 2, \ldots, n_i\}$). The goal is to learn a model parameterized by weights $\mathbf{w} \in \mathbb{R}^d$. This is typically formulated as an empirical risk minimization problem: $\min_{\mathbf{w}\in\mathbb{R}^d} F(\mathbf{w}) := \sum_{i=1}^{N} p_i F_i(\mathbf{w})$, where $F_i(\cdot)$ is the local objective function at client i and p_i is the aggregation weight for client i.

Federated learning systems typically use Federated Averaging (FedAvg) [9] for training. In round t, the server sends the current global model $\mathbf{w}^t$ to the clients. Each client takes multiple steps of mini-batch stochastic gradient descent (SGD) with a fixed learning rate to update their local model and sends their updated local model $\mathbf{w}_i^t$ and sends it to the server. Finally, the server computes a weighted average of the local models to obtain the global model for the next round: $\mathbf{w}^{t+1} = \sum_{i=1}^{N} p_i \mathbf{w}_i^t$, where $p_i = \frac{n_i}{\sum_{i=1}^{N} n_i}$. The iterative training process is repeated for a specific T number of rounds.

We focus on the so-called *enterprise* or *cross-silo* setting in which clients are different organizations (e.g., banks or hospitals) [20]. In this setting, the number of clients is often smaller, all the clients participate in each round, and every client possesses substantially large amount of data.

From the privacy perspective, a variety of implementations have been proposed. We refer the interested reader to [19] for details on how these privacy considerations translate into the application of a variety of multi-party computation techniques.

16.2.2 Why Is Unlearning Different for Federated Learning?

The concept of machine unlearning, i.e., removing the impact of a data sample to the trained model, was first introduced by Cao et al. [21]. After that, several algorithms for machine unlearning have been proposed [11–13, 22–29]. Such *centralized* methods cannot be directly applied to FL due to its distributed nature, where no single participant has access to entire data.

Unlearning in the FL setup has received relatively scant research attention, unlike the centralized setup. The federated unlearning techniques can be classified in three main categories: (i) unlearning a particular category or class [16, 17], (ii) unlearning a subset of samples [17], and (iii) unlearning a client's (party) contribution [14, 15, 18]. Liu et al. [17] consider a setup where several clients want to erase a small subset of their data or a specific category. Their guarantees hold only for convex loss functions and their techniques require each client to compute an (approximate) inverse Hessian matrix which is computationally costly. Some other works [14, 15, 18] focus on removing the contribution of a client after FL training. However, [14, 15] require the server to store the updates from each client in every round. Storing model updates at the server may not be feasible in several application scenarios, especially with strict privacy regulations. In addition, [15] requires the server to possess some extra outsourced unlabeled data, which may not be realistic in several applications. In contrast, PGDU [18] does not require the server or clients to store any client update or global update. It dynamically computes a reference model as described in Sect. 16.3.2.

Applications that are compatible with stringent privacy requirements and that require multi-party computation are quite important. For this reason, in the following we explain in detail PGDU, one of the approaches that is efficient and works with multi-party computation techniques [18].

16.3 Unlearning a Client

In this section, we present the technique we introduced in [18]. From now, we refer to it as PGDU, Projected Gradient Descent based federated Unlearning. These are its main characteristics:

1. We design an efficient federated unlearning method that erases a client by removing the influence of their entire local data from the trained global model. First, the client to be erased performs local unlearning by essentially *reversing* the learning process. Next, by using the locally unlearned model as the initialization, the server and the remaining clients can obtain the global *unlearned* model by performing very few rounds of federated learning.
2. We empirically demonstrate that PGDU achieves comparable performance as the *gold standard* of retraining from scratch, while being significantly efficient in terms of communication (and computation) costs. For instance, PGDU can reduce the communication cost compared to retraining by $5\times$ to $24\times$. We rigorously evaluate PGDU by employing three performance measures adapted from [30]: *efficacy* (which measures success in removing the influence of data to be erased), *fidelity* (which measures performance on data to be retained), and *efficiency* (which measures costs compared to retraining from scratch).
3. Our key novelty is to formulate local unlearning problem as a constrained maximization problem, wherein the client to be erased maximizes their local loss while restricting the model parameters to an ℓ_2-norm ball around a suitably chosen *reference model* obtained from the other clients' local models. Our formulation allows the client to efficiently perform local unlearning by using the Projected Gradient Descent (PGD). Starting with the *locally* unlearned model enables the server and the remaining clients to obtain the *global* unlearned model using very few FL rounds, resulting in significant efficiency gains over retraining.

16.3.1 Federated Unlearning Setup

We consider the following unlearning scenario in the FL setting. After FL training is performed with N clients for the specified T rounds, (Fig. 16.2a), a client $i \in [N]$ requests to opt out of federation and wants to remove the influence of their entire local data from the

FL model. We refer to this client as the *target client*. We focus on *approximate unlearning* with the goal of obtaining a performance *close to* retraining.

Approximate unlearning relies on the fact that randomness in training induces a probability distribution over the models in the parameter space. At a high level, approximate unlearning ensures that the distribution of the unlearned model is either *stochastically indistinguishable* from the distribution of the retrained model, where stochastic indistinguishability is typically characterized by using notions similar to differential privacy [13, 28, 30]. It is possible to formalize theoretical notions of approximate federated unlearning, similar to those in the centralized setting.

While such theoretical notions allow for designing *certified* unlearning algorithms, such algorithms are typically restricted to models with convex loss functions [13, 28, 30]. On the other hand, practical FL systems often involve deep neural networks, which have non-convex loss [20]. Therefore, we focus on the empirical evaluation of unlearning. In particular, we evaluate the unlearning algorithm by its *efficacy, fidelity,* and *efficiency* (see Sect. 16.3.3 for details).

16.3.2 Unlearning with Projected Gradient Descent

As discussed in Sect. 16.2.1, let $\mathbf{w}^T$ denote the global model after performing FL training for T rounds. We propose to perform federated unlearning in two phases: (i) the target client i performs local unlearning by essentially *reversing* the learning process (Fig. 16.2b), and (ii) the server and the retained clients start with the locally unlearned model, and perform a few rounds of federated learning to *boost* its performance (Fig. 16.2c). We now describe in detail both these phases.

Local Unlearning: We argue that a natural idea for a client to unlearn their data is to *reverse* this learning process. That is, during unlearning, instead of learning model parameters that minimize the empirical loss, the client strives to learn the model parameters to *maximize* the loss. Indeed, prior works [11, 25, 30, 31] have applied *gradient ascent* (or its variants) to find a model with *large* empirical loss. However, these works restrict their attention to the case of unlearning only a handful (even just one) samples, whereas we focus on unlearning the entire client dataset, which is typically large for the enterprise or cross-silo FL. In such cases, naïvely applying gradient ascent to maximize the loss does not work because typical loss functions in practice are unbounded (e.g., cross-entropy loss). For an unbounded loss, each gradient ascent step moves towards a model that increases the loss, and after several steps, it is likely to produce an arbitrary model similar to a random model. Thus, we formulate unlearning at the target client as a constrained optimization problem and propose to solve it using *projected gradient descent*.

To motivate our formulation, let us establish some notation for the federated training phase. In each round, the goal of a client is to learn a local model that *minimizes* the (local) empirical risk, i.e., to solve the following problem:

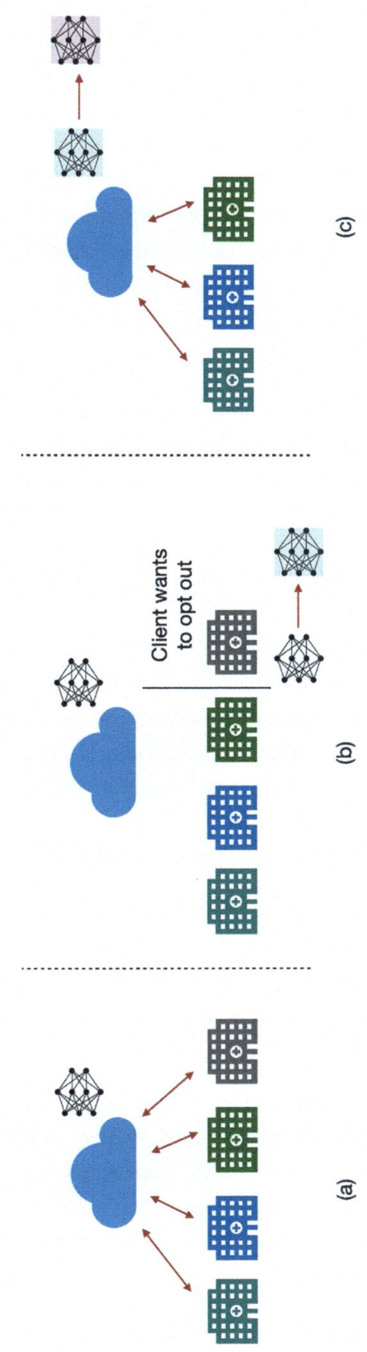

Fig. 16.2 Phases of federated unlearning: **a** First, clients and the server participate in a federated learning process to train a global model. **b** One of the clients wants to opt out of the federation, and wants to unlearn their data. The target client *i* locally runs Projected Gradient Descent (Algorithm 1) to obtain model $\mathbf{w}_i^u$. **c** The server and the remaining clients perform a few steps of federated learning with $\mathbf{w}_i^u$ as the initial point to obtain the final 'unlearned' model (Algorithm 1)

$$\text{(Train)} \quad \min_{\mathbf{w}\in\mathbb{R}^d} F_i(\mathbf{w}) := \frac{1}{n_i} \sum_{j\in\mathcal{D}_i} L(\mathbf{w}; (\mathbf{x}_j, y_j)), \tag{16.1}$$

where $L(\mathbf{w}; (\mathbf{x}_j, y_j))$ is the loss of the prediction on example $(\mathbf{x}_j, y_j)$ made with model parameters $\mathbf{w}$. Each client locally makes several passes of (mini-batch stochastic) *gradient descent* to find a model that has *low* empirical loss. (It is also possible to use other optimization algorithms, e.g., Adam.)

During unlearning, we propose to ensure that the unlearned model is *sufficiently close* to a *reference model* that has effectively learned the other clients' data distributions. In particular, we propose to use the average of the other clients' models as a reference model, i.e., $\mathbf{w}_{\text{ref}} = \frac{1}{N-1}\sum_{j\neq i} \mathbf{w}_j^{T-1}$. Note that the target client i can compute this reference model locally as $\mathbf{w}_{\text{ref}} = \frac{1}{N-1}\left(N\mathbf{w}^T - \mathbf{w}_i^{T-1}\right)$, where $\mathbf{w}^T$ is the global FL model after T rounds and $\mathbf{w}_i^{T-1}$ is the i-th client's local model update in round $T-1$. The client i then optimizes over the model parameters that lie in the ℓ_2-norm ball of radius δ around $\mathbf{w}_{\text{ref}}$. (The radius δ will be treated as a hyperparameter in our experiments.) Thus, during unlearning, the client solves the following optimization problem:

$$\text{(Unlearn)} \quad \max_{\mathbf{w}\in\{\mathbf{v}\in\mathbb{R}^d : \|\mathbf{v}-\mathbf{w}_{\text{ref}}\|_2\leq\delta\}} F_i(\mathbf{w}), \tag{16.2}$$

where $F_i(\cdot)$ is defined in Eq. (16.1).

A natural choice for solving (16.2) is to use *projected gradient descent*. More specifically, let us denote the ℓ_2-norm ball of radius δ around $\mathbf{w}_{\text{ref}}$ as $\Omega = \{\mathbf{v} \in \mathbb{R}^d : \|\mathbf{v} - \mathbf{w}_{\text{ref}}\|_2 \leq \delta\}$. Let $\mathcal{P} : \mathbb{R}^d \to \mathbb{R}^d$ denote the projection operator onto Ω. Then, for a given step-size η_u, client i uses projected gradient descent (PGD)[1] to iterate the update:

$$\mathbf{w} \leftarrow \mathcal{P}\left(\mathbf{w} + \eta_u \nabla F_i(\mathbf{w}; b)\right), \tag{16.3}$$

where $\nabla F_i(\mathbf{w}; \mathbf{b})$ is the gradient of F_i with respect to $\mathbf{w}$ computed on a batch b. To avoid learning an arbitrary model, we perform early stopping if the ℓ_2-distance of the target client $\mathbf{w}_i^{T-1}$ to the unlearned model $\mathbf{w}_i^u$ is smaller than a predetermined threshold τ (which is treated as a hyperparameter). Algorithm 1 describes the local unlearning procedure, and Fig. 16.3 provides a schematic.

FL post-training. To improve the performance of the locally unlearned model on the data of the retained clients, the server and the retained clients perform a few rounds of FL training starting with the unlearned model $\mathbf{w}_i^u$. The detailed steps are described in Algorithm 1. Interestingly, we demonstrate empirically in Sect. 16.3.3 that performing *very few rounds* of FL post-training on the unlearned model $\mathbf{w}_i^u$ gives good performance in practice.

[1] Note that Eq. (16.3) is technically projected gradient *ascent* since we are maximizing a function rather than minimizing. However, similar to adversarial machine learning literature (see, e.g., [32]), we refer to the process as projected gradient descent.

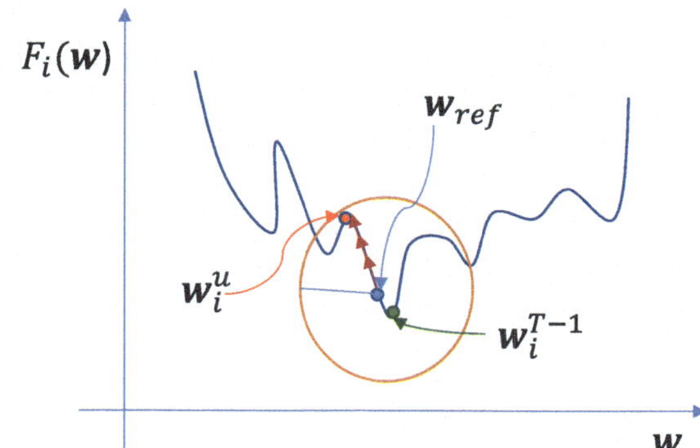

Fig. 16.3 A schematic to illustrate the main idea of the local unlearning phase in Algorithm 1

16.3.3 Evaluation

Unlearning Scenarios: We consider two scenarios to illustrate the phenomenon of unlearning: (i) removing the effect of backdoor triggers and (ii) removing the effect of flipping. At a high level, a successful federated unlearning method should produce a model that does not perform well on the target client's data distribution while keeping good performance on the other clients' data distribution. The goal in the above mentioned scenarios is to deliberately differentiate the target client's data distribution from the data distribution of the other clients.

Performance Measures: In general, an effective federated unlearning method must remove the contribution of the target client's data, maintain good performance, and be more efficient than retraining from scratch. To reflect these properties in our evaluation, we use three performance measures (similar to [30]).

Efficacy of unlearning. The efficacy of an unlearning method measures how successful it is in removing the contribution of the target client's data. We quantify the efficacy of unlearning by evaluating the performance of the unlearned model on the target client's data distribution. In particular, we use the following two metrics to measure efficacy: (i) accuracy on the target client's data distribution: depending on the scenario this will be accuracy on a hold-out test set of backdoored or flipped images; and (ii) membership inference risk with respect to the target client's dataset.

Fidelity of unlearning. The fidelity of an unlearning method measures whether it can maintain a performance close to the original model. We quantify the fidelity of unlearning by evaluating the performance of the unlearned model on the retained clients' data distribution. In particular, we use the accuracy of the unlearned model on a hold-out test set of clean images to measure fidelity.

Efficiency of unlearning. While it is straightforward to perform unlearning by retraining the FL model from scratch without the participation of the target client, such retraining incurs significant communication and computation costs. The efficiency of an unlearning method

Algorithm 1 PGDU

Local Unlearning at Client i via Projected Gradient Descent:
Inputs: learning rate η_u, batch size B_u, number of epochs E_u, clipping radius δ, and early stopping threshold τ
Set $\mathbf{w}_{\text{ref}} \leftarrow \frac{1}{N-1}\left(N\mathbf{w}^T - \mathbf{w}_i^{T-1}\right) = \frac{1}{N-1}\sum_{i \neq j} \mathbf{w}_j^{T-1}$
Define $\mathcal{P}(\mathbf{w})$ as the projection of $\mathbf{w} \in \mathbb{R}^d$ onto the ℓ_2-norm ball $\Omega = \{\mathbf{v} \in \mathbb{R}^d : \|\mathbf{v} - \mathbf{w}_{\text{ref}}\| \leq \delta\}$
Initialize unlearning model as $\mathbf{w} \leftarrow \mathbf{w}_{\text{ref}}$
$\mathcal{B}_i \leftarrow$ (split $\mathcal{D}_i$ into batches of size B_u)
for each local epoch $e = 1$ **to** E_u **do**
  for batch b in $\mathcal{B}_i$ **do**
    $\mathbf{w} \leftarrow \mathcal{P}\left(\mathbf{w} + \eta_u \nabla F_i(\mathbf{w}; b)\right)$
    if $\|\mathbf{w} - \mathbf{w}_i^{T-1}\|_2 < \tau$ **then**
      Set $\mathbf{w}_i^u \leftarrow \mathbf{w}$ and return $\mathbf{w}_i^u$ to server
    end if
  end for
end for
Set $\mathbf{w}_i^u \leftarrow \mathbf{w}$ and return $\mathbf{w}_i^u$ to server

FL post-training:
Inputs: learning rate η_p, batch size B_p, number of epochs E_p, number of rounds T_p
Server executes
Initialize $\mathbf{w}^0 \leftarrow \mathbf{w}_i^u$
for each round $t = 1$ **to** T_p **do**
  Send $\mathbf{w}^{t-1}$ to clients $[N] \setminus \{i\}$
  for each client $j \in [N] \setminus \{i\}$ **in parallel do**
    $\mathbf{w}_j^t \leftarrow$ ClientUpdate$(j, \mathbf{w}^{t-1})$
  end for
  $\mathbf{w}^t \leftarrow \sum_j \frac{n_j}{\sum_l n_l} \mathbf{w}_j^t$
end for
Set the unlearned model as $\mathbf{w}^u \leftarrow \mathbf{w}^{T_p}$

ClientUpdate$(j, \mathbf{w}^{t-1})$:
$\mathcal{B}_j \leftarrow$ (split $\mathcal{D}_j$ into batches of size B_p)
Initialize $\mathbf{w} \leftarrow \mathbf{w}^{t-1}$
for each local epoch $e = 1$ **to** E_p **do**
  for batch b in $\mathcal{B}_j$ **do**
    $\mathbf{w} \leftarrow \mathbf{w} - \eta_p \nabla F_j(\mathbf{w}; b)$
  end for
end for
Return $\mathbf{w}$ to the server

measures the reduction in communication and computation costs wrt. retraining. We evaluate the efficiency of PGDU by comparing its communication cost with that of retraining. We focus on the communication cost since it is known to be a key bottleneck in FL [20].

Datasets and Model Architecture: To evaluate the performance of PGDU, we utilize three datasets: MNIST [33], EMNIST (balanced version) [34], and CIFAR-10 [35]. For all datasets, we use a CNN from [9] with two 5×5 convolution layers, a fully connected layer with 512 units and ReLu activation, and a final softmax output layer (1,663,370 total parameters). We equally partition the training images of each dataset across N clients in the FL process, one of which is the target client.

Hyperparameters. During FL training, we use the SGD optimizer with the following hyperparameters:

- Momentum $\beta = 0.9$
- Learning rate $\eta = 0.01$
- Batch size $B = 128$
- Number of epochs $E = 1$
- Aggregation algorithm: FedAvg (described in Sect. 16.2.1).

For PGD-based unlearning, we use the SGD optimizer with the following hyperparameters:

- Momentum $\beta = 0.9$
- Learning rate $\eta_u = 0.01$
- Batch size $B_u = 1024$
- Number of epochs $E_u = 5$
- ℓ_2-norm ball radius δ is set to be one third of the average Euclidean distance between $\mathbf{w}_{\text{ref}}$ and a random model, where the average is computed over 10 random models. This value is selected because we want the model to stay closer to the reference model than a random model
- Early stopping threshold τ: Computed via a grid search over the interval [2, 6]
- Gradient ℓ_2-clipping is employed with radius 5.

For FL post-training after unlearning, we use the SGD optimizer with the following hyperparameters:

- Momentum $\beta = 0.9$
- Learning rate $\eta_p = 0.01$
- Batch size $B_p = 128$
- FL rounds T_p : see Figs. 16.8 and 16.12
- Aggregation algorithm: FedAvg.

16.3.3.1 Unlearning Scenario 1: Backdoors

We use the backdoor triggers [36] as an effective way to evaluate the performance of unlearning methods, similar to [15]. In particular, the target client uses a dataset in which a certain fraction of images has a backdoor trigger inserted in them. Because of this client, the global FL model becomes susceptible to the backdoor trigger. Then, a successful unlearning process should produce a model that reduces the accuracy of the images with the backdoor trigger, while maintaining high accuracy on regular (clean) images. For backdoors, we introduce a 'pixel pattern' trigger of size 3×3 using the Adversarial Robustness Toolbox [37], and change the label of corresponding samples to '9' for MNIST, to 't' for EMNIST, and to 'truck' for CIFAR-10. When inserting the backdoor trigger, we exclude the data sample whose label is already the target label. We consider two cases: (i) $N = 5$ clients with the target client having 66% of their images backdoored, and (ii) $N = 10$ clients with the target client having 80% of their images backdoored. We compare PGDU to retraining from scratch.

Efficacy evaluation: We analyze the efficacy of PGDU using two metrics. First, we evaluate the accuracy on a hold-out test set of backdoored images. We compute the accuracy on the backdoored data (referred as the backdoor accuracy) as the percentage of triggered data that are misclassified as the target label required by the attacker. The lower the backdoor accuracy, the better the model has unlearned the contribution of the target client's data. Figure 16.4 shows the backdoor accuracy of each model for each dataset for both cases. For both cases and all datasets, the high value of backdoor accuracy for the FedAvg model indicates that the FL model has learned the target client's data consisting of backdoor triggers. We observe that PGDU substantially reduces the backdoor accuracy, and in fact, achieves similar backdoor accuracy to retraining for all datasets. This demonstrates that the efficacy of PGDU in terms of backdoor accuracy is comparable to that of retraining.

Another metric that we use to measure the efficacy of PGDU is the membership inference risk. The goal of a membership inference attack is to determine whether a specific data sample is part of the dataset used to train the model. We leverage membership inference attacks to assess how much information from the target client's data is part of the unlearned model, similar to [14]. A successful unlearning process should produce a model that has a low membership inference risk on the data of the target client. To measure the membership inference risk, we use two well-known membership inference attacks.

Shokri's attack. Shokri et al. [38] were the first to propose a membership inference attack on machine learning (ML) models. Let M be the target model trained on dataset D. The main intuition of the attack is that ML models tend to behave differently on the training data compared to the data that they have not seen. It is assumed that the attacker knows the type and the architecture of the model M and has access to some data D_S that comes from the same underlying distribution as training data D. Thus, the attacker can train multiple shadow models M_{S_i} (one per class) that mimic the behavior of the target model. For the shadow models, the attacker has their training and test datasets and thus knows the ground

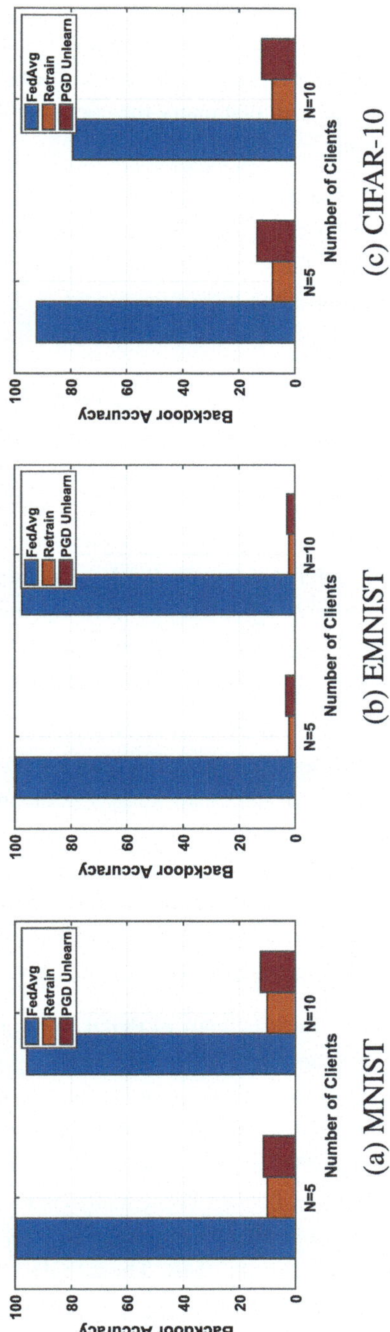

Fig. 16.4 Backdoor accuracy (efficacy) of the fully retrained and the PGD-based unlearned model in each dataset, and their comparison with the FedAvg model before unlearning. The backdoor accuracy of the PGD-based unlearned model is obtained after 1 round of FL post-training. PGDU significantly reduces the backdoor accuracy compared to FedAvg model and achieves a similar performance as retraining, which demonstrates its high unlearning efficacy

truth of the membership of the training and test data samples. Based on this, the attacker trains multiple attack models M_{A_i} (one per class) by using as input the posteriors returned by the corresponding shadow model and as labels their membership. Finally, when the attacker wants to determine the membership of a target data sample, they query the target model M, obtain its posterior probability, and with that query the corresponding attack model M_{A_i} to obtain the membership prediction.

Yeom's attack. Yeom et al. [39] assume that the attacker has white-box access to the target model M and knows its average training loss. To determine the membership of a target data sample, the attacker computes the loss of the model M on the input data sample and compares this value to the average training loss of the model M. If the loss of the target data sample is smaller than the average training loss, then it is classified as a member, otherwise as a non-member.

For the evaluation, we perform the attacks against the PGD-based unlearned model (after one round of post-training), the fully retrained model, and the FedAvg model (the global model obtained by FL before any unlearning). For simplicity, in Shokri's attack, we use the FedAvg model as the shadow model. We compute the attack accuracy as the percentage of the target client's data that are inferred as being part of the training dataset. Figure 16.5 shows the accuracy of the membership inference attacks for $N = 5$ clients. We observe that both unlearning methods achieve substantially lower membership inference attack accuracies than the one in the FedAvg model for all datasets and both attacks. In fact, for both attacks, PGDU obtains similar accuracy to retraining, which demonstrates its high efficacy.

Fidelity evaluation: We evaluate the fidelity of PGDU by computing the accuracy of the unlearned model on a hold-out test set that consists of clean images (no backdoor triggers). We refer to the accuracy computed on the clean images as clean accuracy. Note that the clean images represent the data distribution of the retained clients, and the clean accuracy indicates whether the unlearned model can maintain good performance on the retained data. In Fig. 16.6, we show the clean accuracy of the unlearned models obtained by PGDU and retraining. We observe that PGDU achieves similar clean accuracy to retraining, demonstrating the capability of PGDU for unlearning with high fidelity.

Efficiency evaluation: To evaluate the efficiency of PGDU we compare its communication cost with retraining. We compute the communication cost of a given approach as the total size of the model updates (in MB) that clients participating in the FL process communicate to the server. Figure 16.7 shows the communication cost for various clean accuracy (fidelity) values for $N = 5$ clients. We observe that PGDU is significantly more efficient than retraining while achieving similar fidelity. For instance, in the MNIST dataset, to reach a fidelity (clean accuracy) of 98.13%, PGDU requires 167 MB of communication cost, whereas retraining from scratch requires 453 MB. Thus, PGDU is 2.7× more efficient in terms of communication costs than retraining. This gap is even higher for the EMNIST and CIFAR-10 datasets. Overall, we observe that PGDU reduces communication costs by up to 24×.

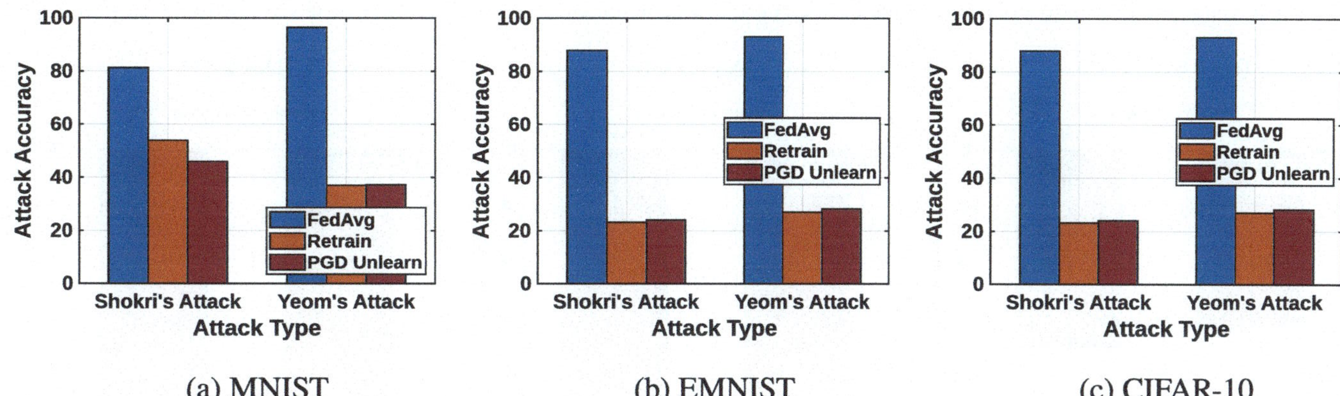

(a) MNIST

(b) EMNIST

(c) CIFAR-10

Fig. 16.5 Backdoor scenario: membership inference attacks accuracy (efficacy) for the two attacks and the three datasets for $N = 5$ clients. PGDU achieves a similar attack accuracy as retraining, which demonstrates its high efficacy

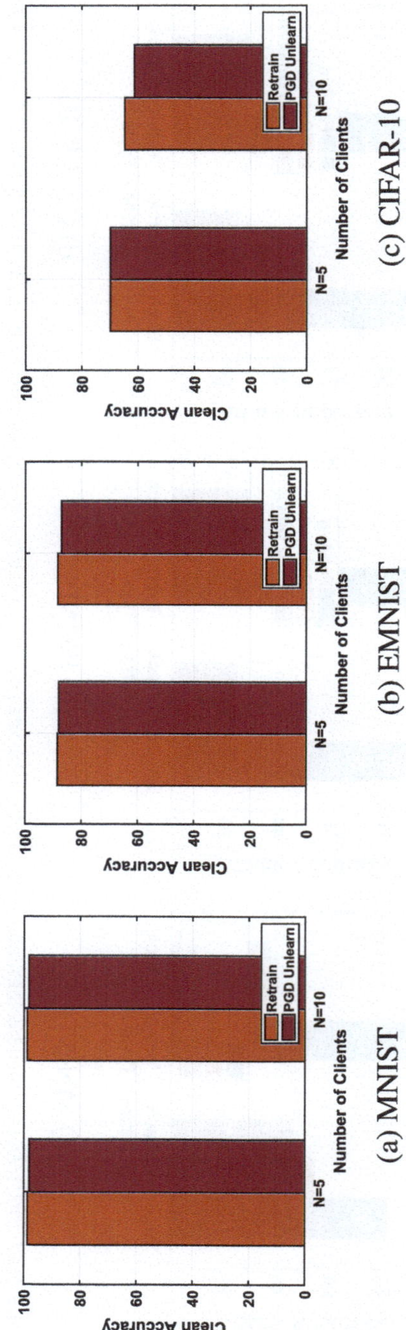

Fig. 16.6 Backdoor scenario: clean accuracy (fidelity) of the fully retrained and the PGD-based unlearned model in each dataset. The clean accuracy of the PGD-based unlearned model is obtained after 5 rounds of FL post-training. PGDU achieves similar clean accuracy to retraining, which demonstrates its high fidelity

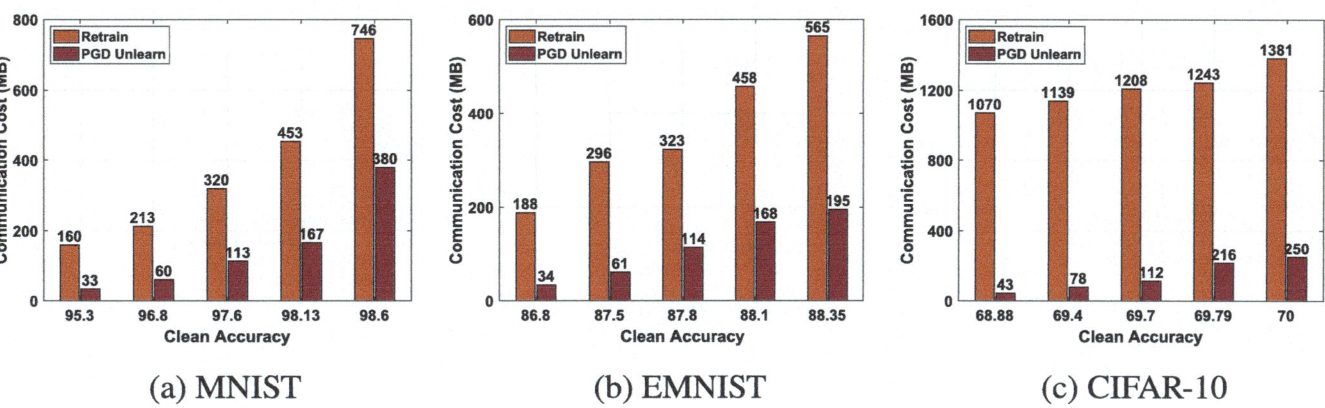

Fig. 16.7 Backdoor scenario: communication costs (efficiency) of PGDU and the baseline approach of retraining with respect to the clean accuracy (fidelity) in each dataset for $N = 5$ clients

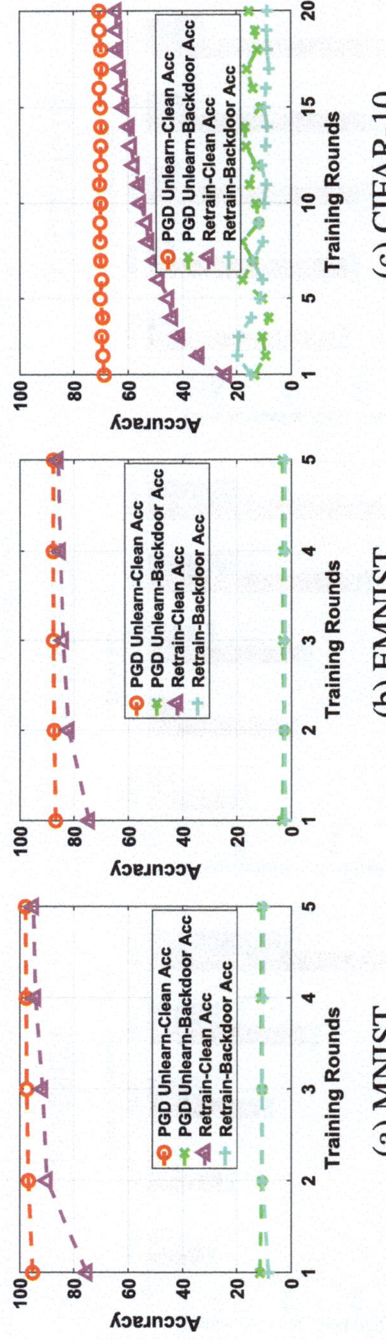

Fig. 16.8 Clean accuracy (fidelity) and backdoor accuracy (efficacy) of the PGD-based unlearned and fully retrained model with respect to the number of rounds in each dataset for $N = 5$ clients

To compare the speedup of PGDU to the baseline of retraining, we compute the clean and backdoor accuracy of both methods with respect to the number of FL training rounds. Note that, for PGDU, the FL training starts with the locally unlearned model that the target client has obtained using the projected gradient descent (as discussed in Sect. 16.3.2). On the other hand, for the baseline of retraining, the FL training starts with a randomly initialized model. Figure 16.8 shows this comparison for $N = 5$ clients. After one round of post-training, PGDU reaches a clean accuracy of 95.3% and a backdoor accuracy of 11.38% in the MNIST dataset. Retraining requires more than 5 training rounds to achieve similar performance. This shows that the PGD-based local unlearning produces an effective starting point by removing the influence of the target client's data without degrading the performance on the other clients' data.

16.3.3.2 Unlearning Scenario 2: Flipped Images

In this scenario, we consider a target client that has a dataset where a certain fraction of images are flipped. We do not apply any data augmentation on any clients' datasets. A successful unlearning method should reduce the accuracy on the flipped images while maintaining high accuracy on regular images. For evaluation, we consider two cases: (i) $N = 5$ clients with the target client having 66% of their images flipped and (ii) $N = 10$ clients with the target client having 80% of their images flipped. We flip the images horizontally and keep the label unchanged. Note that for this unlearning scenario, we do not provide the results for the CIFAR-10 dataset for the following reason. The accuracy of retraining on a hold-out test set of flipped images for $N = 5$ is 71.14%. This value is similar to the accuracy on a hold-out test set of regular images (no flipping applied), making CIFAR-10 an inappropriate dataset for this scenario.

Efficacy evaluation: To analyze the efficacy of the unlearning methods, we use the accuracy on a hold-out test set of flipped images (referred as the flipped accuracy). In Fig. 16.9, we show the flipped accuracy of the FedAvg, fully retrained, and the PGD-based unlearned models in the MNIST and EMNIST datasets for both cases. We observe that PGDU achieves a similar flipped accuracy to retraining. These results show that the PGD-based unlearning method has comparable efficacy to retraining for removing the contribution of the target client's data.

Fidelity evaluation: To evaluate the fidelity of the unlearning methods, we examine the accuracy between PGDU and the baseline approach on a hold-out test set of regular images (no flipping applied). For consistency, we refer to the accuracy on regular images as the clean accuracy. Figure 16.10 shows the clean accuracy of the fully retrained and PGD-based unlearned models in each dataset for both cases. We observe that PGDU maintains a high clean accuracy, which is similar to the baseline approach of retraining. Since the clean images represent the data distribution of the retained clients, the high clean accuracy of PGDU indicates that it can perform unlearning with high fidelity by maintaining good performance on the retained data.

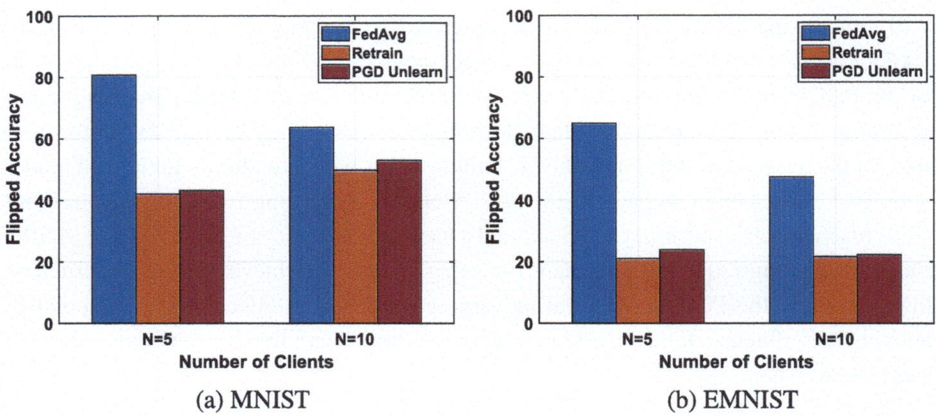

Fig. 16.9 Flipped accuracy (efficacy) of the FedAvg (before unlearning), fully retrained, and PGD-based unlearned models in the MNIST and EMNIST datasets. The flipped accuracy of the PGD-based unlearned model is obtained after 5 rounds of FL post-training. PGDU substantially reduces the flipped accuracy compared to the FedAvg model and achieves similar performance as retraining, which demonstrates its high unlearning efficacy

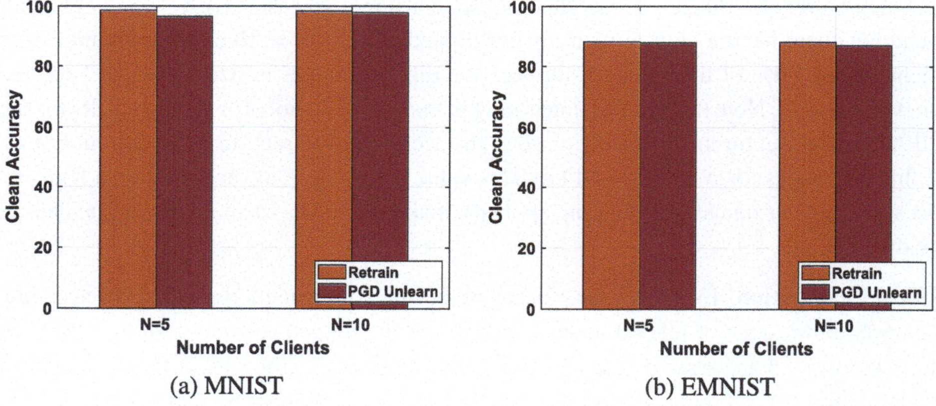

Fig. 16.10 Flipping scenario: clean accuracy (fidelity) of the fully retrained and the PGD-based unlearned models in the MNIST and EMNIST datasets. The clean accuracy of the PGD-based unlearned model is obtained after 5 rounds of FL post-training

Efficiency evaluation: We compare the communication cost of PGDU with retraining to quantify its efficiency. We compute the communication cost of a given method as the total size of the model updates (in MB) that clients send to the server. Figure 16.11 shows the communication cost for various fidelity (clean accuracy) values for $N = 5$ clients. In the MNIST dataset, to achieve a clean accuracy of 98.13%, PGDU requires 566 MB of communication costs, while retraining requires 1039 MB. Thus, PGDU is 1.8× more efficient

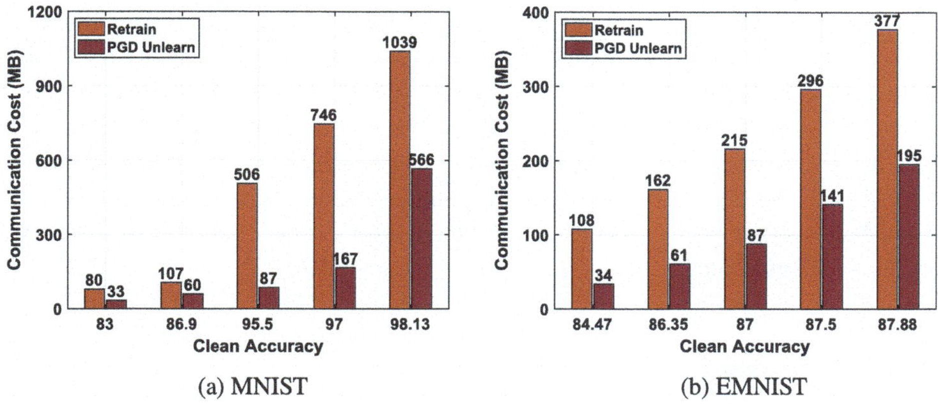

(a) MNIST (b) EMNIST

Fig. 16.11 Flipping scenario: communication costs (efficiency) of PGDU and fully retraining with respect to the clean accuracy (fidelity) in the MNIST and EMNIST dataset for $N = 5$ clients

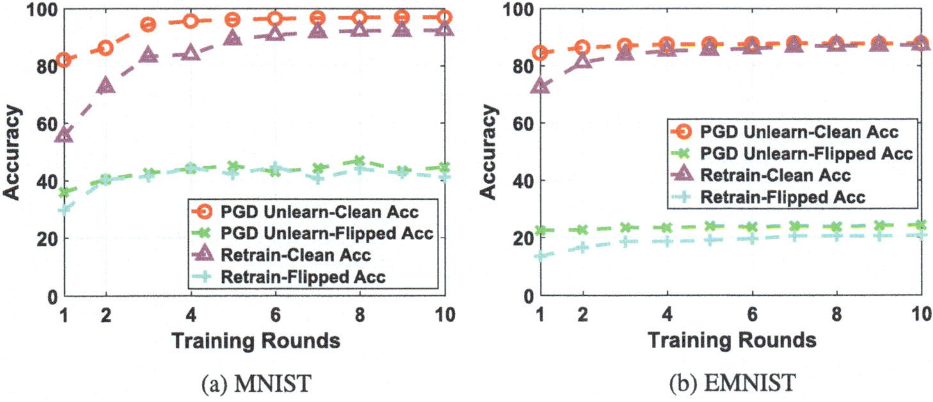

(a) MNIST (b) EMNIST

Fig. 16.12 Flipping scenario: clean accuracy (fidelity) and flipped accuracy (efficacy) of PGDU and full FL retraining from scratch with respect to the number of rounds in the MNIST and EMNIST datasets for $N = 5$ clients

than the baseline approach. We obtain similar gains for the EMNIST dataset. Overall, we observe that PGDU is up to 5.8× more efficient than retraining from scratch.

In Fig. 16.12, we show the clean accuracy (fidelity) and flipped accuracy (efficacy) of PGDU and the fully FL retraining (gold standard) with respect to the number of FL training rounds (FL post-training rounds for PGDU) for $N = 5$ clients. We observe that local unlearning at the target client bootstraps the unlearning process, and requires substantially fewer number of FL rounds than retraining from scratch. For example, for $N = 5$ in the MNIST dataset, PGDU achieves a clean accuracy of 82.13% and a flipped accuracy of 36.15% after one round of post-training, while the baseline method requires more than 3 FL rounds to achieve similar performance. These results show the efficiency of PGDU.

Overall, for both unlearning scenarios, we observe that PGDU is more efficient in terms of the communication cost on the retained clients than retraining, while achieving comparable fidelity and efficacy. It is worth noting that, even though we do not explicitly measure computation costs, PGDU also reduces the computation cost as compared to retraining since it requires much fewer rounds than retraining to achieve high performance on the retained data. We believe that lowering the communication and computation burden on retained clients is appealing in practice since these clients are not incentivized to help the target client in unlearning.

16.4 Conclusion

In this chapter, we reviewed some of the previous works in federated unlearning. We focused on the cross-silo federated unlearning scenario, where a client seeks to erase its entire contribution from the global model. We explored existing approaches and highlighted their limitations, particularly in privacy-sensitive federations. Then, we delved on PGDU, a projected gradient descent unlearning method [18], which can efficiently unlearn the contribution of any client. PGDU first performs local unlearning at the client to be erased and, starting with the locally unlearned model, performs a few rounds of FL with the server and remaining clients. Unlike prior federated unlearning works, PGDU does not require the server (or any other client) to keep track of the history of their parameter updates. We have used the backdoor triggers and flipping to effectively evaluate the performance of PGDU. We empirically demonstrated the efficacy, fidelity, and efficiency of PGDU.

References

1. Carlini, N., Tramèr, F., Wallace, E., Jagielski, M., Herbert-Voss, A., Lee, K., Roberts, A., Brown, T.B., Song, D., Erlingsson, Ú., Oprea, A., Raffel, C.: Extracting training data from large language models. In: USENIX, pp. 2633–2650 (2021)
2. Lehman, E., Jain, S., Pichotta, K., Goldberg, Y., Wallace, B.: Does BERT pretrained on clinical notes reveal sensitive data? In: Proceedings of the 2021 Conference of the North American Chapter of the Association for Computational Linguistics: Human Language Technologies, pp. 946–959. Association for Computational Linguistics (2021)
3. Carlini, N., Ippolito, D., Jagielski, M., Lee, K., Tramer, F., Zhang, C.: Quantifying memorization across neural language models. In: The Eleventh International Conference on Learning Representations (2023)
4. Voigt, P., Von dem Bussche, A.: The EU General Data Protection Regulation (GDPR). A Practical Guide, 1st edn, Springer International Publishing, Cham, 10–5555 (2017)
5. Pardau, S.L.: The California consumer privacy act: towards a European-style privacy regime in the United States. J. Tech. L. Pol'y **23**, 68 (2018)
6. Act, P.: Personal information protection and electronic documents act. Department of Justice, Canada. Full text available at http://lawsjustice.gc.ca/en/P-8.6/text.html (2000)

7. Nguyen, T.T., Huynh, T.T., Nguyen, P.L., Liew, A.W.C., Yin, H., Nguyen, Q.V.H.: A survey of machine unlearning. arXiv preprint arXiv:2209.02299 (2022)
8. Xu, H., Zhu, T., Zhang, L., Zhou, W., Yu, P.S.: Machine unlearning: a survey. ACM Comput. Surv. **56**(1), 1–36 (2023)
9. McMahan, B., Moore, E., Ramage, D., Hampson, S., y Arcas, B.A.: Communication-efficient learning of deep networks from decentralized data. In: Artificial Intelligence and Statistics, pp. 1273–1282. PMLR (2017)
10. Ludwig, H., Baracaldo, N.: Federated Learning: A Comprehensive Overview of Methods and Applications. Springer (2022)
11. Graves, L., Nagisetty, V., Ganesh, V.: Amnesiac machine learning. arXiv preprint arXiv:2010.10981 (2020)
12. Bourtoule, L., Chandrasekaran, V., Choquette-Choo, C.A., Jia, H., Travers, A., Zhang, B., Lie, D., Papernot, N.: Machine unlearning. In: 2021 IEEE Symposium on Security and Privacy (SP), pp. 141–159. IEEE (2021)
13. Guo, C., Goldstein, T., Hannun, A., Van Der Maaten, L.: Certified data removal from machine learning models. arXiv preprint arXiv:1911.03030 (2019)
14. Liu, G., Ma, X., Yang, Y., Wang, C., Liu, J.: Federaser: enabling efficient client-level data removal from federated learning models. In: 2021 IEEE/ACM 29th International Symposium on Quality of Service (IWQOS), pp. 1–10. IEEE (2021)
15. Wu, C., Zhu, S., Mitra, P.: Federated unlearning with knowledge distillation. arXiv preprint arXiv:2201.09441 (2022)
16. Wang, J., Guo, S., Xie, X., Qi, H.: Federated unlearning via class-discriminative pruning. In: Proceedings of the ACM Web Conference 2022, pp. 622–632 (2022)
17. Liu, Y., Xu, L., Yuan, X., Wang, C., Li, B.: The right to be forgotten in federated learning: an efficient realization with rapid retraining. In: IEEE INFOCOM 2022—IEEE Conference on Computer Communications, pp. 1749–1758 (2022). https://doi.org/10.1109/INFOCOM48880.2022.9796721
18. Halimi, A., Kadhe, S., Rawat, A., Baracaldo, N.: Federated unlearning: how to efficiently erase a client in FL? arXiv preprint arXiv:2207.05521 (2022)
19. Baracaldo, N., Xu, R.: Protecting against data leakage in federated learning: what approach should you choose? In: Federated Learning: A Comprehensive Overview of Methods and Applications, pp. 281–312. Springer (2022)
20. Kairouz, P., McMahan, H.B., Avent, B., Bellet, A., Bennis, M., Bhagoji, A.N., Bonawitz, K., Charles, Z., Cormode, G., Cummings, R., D'Oliveira, R.G.L., Eichner, H., Rouayheb, S.E., Evans, D., Gardner, J., Garrett, Z., Gascón, A., Ghazi, B., Gibbons, P.B., Gruteser, M., Harchaoui, Z., He, C., He, L., Huo, Z., Hutchinson, B., Hsu, J., Jaggi, M., Javidi, T., Joshi, G., Khodak, M., Konecný, J., Korolova, A., Koushanfar, F., Koyejo, S., Lepoint, T., Liu, Y., Mittal, P., Mohri, M., Nock, R., Özgür, A., Pagh, R., Qi, H., Ramage, D., Raskar, R., Raykova, M., Song, D., Song, W., Stich, S.U., Sun, Z., Suresh, A.T., Tramèr, F., Vepakomma, P., Wang, J., Xiong, L., Xu, Z., Yang, Q., Yu, F.X., Yu, H., Zhao, S.: Advances and open problems in federated learning. Found. Trends® Mach. Learn. **14**(1–2), 1–210 (2021)
21. Cao, Y., Yang, J.: Towards making systems forget with machine unlearning. In: 2015 IEEE Symposium on Security and Privacy, pp. 463–480. IEEE (2015)
22. Du, M., Chen, Z., Liu, C., Oak, R., Song, D.: Lifelong anomaly detection through unlearning. In: Proceedings of the 2019 ACM SIGSAC Conference on Computer and Communications Security, pp. 1283–1297 (2019)
23. Ginart, A., Guan, M., Valiant, G., Zou, J.Y.: Making AI forget you: data deletion in machine learning. Adv. Neural Inf. Process. Syst. **32** (2019)
24. Baumhauer, T., Schöttle, P., Zeppelzauer, M.: Machine unlearning: linear filtration for logit-based classifiers. arXiv preprint arXiv:2002.02730 (2020)

25. Golatkar, A., Achille, A., Soatto, S.: Eternal sunshine of the spotless net: selective forgetting in deep networks. In: Proceedings of the IEEE/CVF Conference on Computer Vision and Pattern Recognition, pp. 9304–9312 (2020)
26. Golatkar, A., Achille, A., Soatto, S.: Forgetting outside the box: scrubbing deep networks of information accessible from input-output observations. In: Computer Vision-ECCV 2020: 16th European Conference, Glasgow, UK, 23–28 Aug 2020, Proceedings, Part XXIX 16, pp. 383–398. Springer (2020)
27. Neel, S., Roth, A., Sharifi-Malvajerdi, S.: Descent-to-delete: gradient-based methods for machine unlearning. In: Algorithmic Learning Theory, pp. 931–962. PMLR (2021)
28. Sekhari, A., Acharya, J., Kamath, G., Suresh, A.T.: Remember what you want to forget: algorithms for machine unlearning. Adv. Neural. Inf. Process. Syst. **34**, 18075–18086 (2021)
29. Thudi, A., Deza, G., Chandrasekaran, V., Papernot, N.: Unrolling SGD: understanding factors influencing machine unlearning. arXiv preprint arXiv:2109.13398 (2021)
30. Warnecke, A., Pirch, L., Wressnegger, C., Rieck, K.: Machine unlearning of features and labels. In: 30th Annual Network and Distributed System Security Symposium, NDSS 2023, San Diego, California, USA, February 27–March 3 2023. The Internet Society (2023)
31. Jang, J., Yoon, D., Yang, S., Cha, S., Lee, M., Logeswaran, L., Seo, M.: Knowledge unlearning for mitigating privacy risks in language models. In: Proceedings of the 61st Annual Meeting of the Association for Computational Linguistics (Volume 1: Long Papers) (2023)
32. Madry, A., Makelov, A., Schmidt, L., Tsipras, D., Vladu, A.: Towards deep learning models resistant to adversarial attacks. In: International Conference on Learning Representations (2018). https://openreview.net/forum?id=rJzIBfZAb
33. Lecun, Y., Bottou, L., Bengio, Y., Haffner, P.: Gradient-based learning applied to document recognition. Proc. IEEE **86**(11), 2278–2324 (1998). https://doi.org/10.1109/5.726791
34. Cohen, G., Afshar, S., Tapson, J., Van Schaik, A.: EMNIST: extending MNIST to handwritten letters. In: 2017 International Joint Conference on Neural Networks (IJCNN), pp. 2921–2926. IEEE (2017)
35. Krizhevsky, A., Hinton, G., et al.: Learning multiple layers of features from tiny images. cs.utoronto.ca (2009)
36. Gu, T., Dolan-Gavitt, B., Garg, S.: Badnets: identifying vulnerabilities in the machine learning model supply chain. arXiv preprint arXiv:1708.06733 (2017)
37. Nicolae, M.I., Sinn, M., Tran, M.N., Buesser, B., Rawat, A., Wistuba, M., Zantedeschi, V., Baracaldo, N., Chen, B., Ludwig, H., Molloy, I., Edwards, B.: Adversarial robustness toolbox v1.2.0. CoRR 1807.01069 (2018). https://arxiv.org/pdf/1807.01069
38. Shokri, R., Stronati, M., Song, C., Shmatikov, V.: Membership inference attacks against machine learning models. In: 2017 IEEE Symposium on Security and Privacy (SP), pp. 3–18. IEEE (2017)
39. Yeom, S., Giacomelli, I., Fredrikson, M., Jha, S.: Privacy risk in machine learning: analyzing the connection to overfitting. In: 2018 IEEE 31st Computer Security Foundations Symposium (CSF), pp. 268–282. IEEE (2018)

The manufacturer's authorised representative in the EU is Springer Nature Customer Service Centre GmbH, Europaplatz 3, 69115 Heidelberg, Germany. If you have any concerns regarding our products, please contact ProductSafety@springernature.com

Printed and bound by CPI Group (UK) Ltd, Croydon, CR0 4YY
07/07/2026
02160918-0007